A Collection of Oxford Biology Readers

KV-268-452

Readings in Genetics and Evolution

Foreword by **J J Head**

Oxford University Press, Ely House, London W.1.

GLASGOW NEW YORK TORONTO MELBOURNE WELLINGTON
CAPE TOWN IBADAN NAIROBI DAR ES SALAAM LUSAKA
ADDIS ABABA BOMBAY CALCUTTA MADRAS KARACHI LAHORE DACCA
DELHI KUALA LUMPUR SINGAPORE HONG KONG TOKYO

ⓒ composite volume Oxford University Press 1973

Set in 10/11 pt Monophoto Times and printed by
BAS Printers Limited, Wallop, Hampshire

Foreword

Oxford Biology Readers are a substantial contribution to the improvement of science teaching. Each article is written by a leading authority and much of the material (including the plentiful illustrations) is otherwise available only in learned journals or expensive tertiary textbooks; but the happy co-operation of the authors with a stringent editorial process has ensured that the level of writing is not too high either for senior school students or for first year undergraduates.

The last ten years have seen the birth of many biology curriculum development projects at school, college, and university level. Natural selection has eliminated some; others survive. The broad aim of all this activity has been to make the teaching of biology 'scientific'; but it is easier to understand that the rote learning of factual material which formerly passed for biological education was not 'scientific' than it is to agree on what is. All the secondary and high school projects have been written by school teachers, often with comments and suggestions from scientists. Teachers are best equipped to appreciate the problems and difficulties of learning that their students encounter, and the limitations imposed by school laboratories and other facilities; but their writings cannot help lacking the authenticity, the command, and the authority which flow spontaneously when a leading scientist writes about his field of work; and this appears more, perhaps, in what is left out of an article than in what is included. A secondhand account lacks the human qualities, explicitly or implicitly stated, that a man involved in research communicates: his motivation for the study, his attitudes to the interpretation of data, his feelings about which lines of future investigation might prove rewarding, his selection from previous work of that which seems to him to be most significant, his awareness of the extent to which he is working within what Thomas Kuhn calls scientific paradigms. These are but a few of the qualities of a scientist, and therefore of science and of scientific method, but they are enough to make clear that an oversimplified description of a scientist as one who gathers data and then puzzles over them to formulate a hypothesis, which he then tests by experiment, is utterly unsatisfactory and cannot hope to engage in the reader any sense of wonder, any inkling that science is passionate, satisfying, and can provide areas of consummate human activity as great as those that produced *Die Fledermaus* on the one hand and *Cosi fan Tutte* on the other.

The first five Readers deal with aspects of genetics. In *Somatic cell division* and *The meiotic mechanism*, B. John and K. R. Lewis provide an authoritative and exceptionally well illustrated account of two of the basic life processes, the conservation, duplication, and exact division of the genetic material in somatic cells, and the sexual cycle which, through chromosome assortment and chiasma formation, provides by recombination much of the variation which is the raw material for evolution by natural selection. An account of mitosis and meiosis is a basic ingredient of all elementary biology courses; its repetition in accounts written by non-specialists is a prime example of de-scientizing in science teaching, and students who think that these events are fully understood and are not active centres of research will find many surprises in these two articles. One of the problems touched on in *The meiotic mechanism* is the nature and cause of chiasma formation. The fact that counting the numbers of spores of various kinds in the asci of yeasts and other ascomycetes can make a major contribution to the understanding of this, as J. R. S. Fincham reveals in *Using fungi to study genetic recombination* is both beautiful as a piece of reasoning and satisfying in showing how very simple techniques and resources can still make important contributions to biological research.

The third quarter of the twentieth century saw the unravelling of the genetic code and the understanding of much of the part played by RNAs in the transfer of genetic information from nucleus to cytoplasm. There remains the problem of differentiation: how is it that cells carrying the same genetic information develop in different ways? J. B. Gurdon is one of the leaders in this field, whose methods include the transplanting of nuclei to foreign cells and attempts to identify chemical substances which control gene expression. In *Gene expression during cell differentiation*, Gurdon concentrates on gene control since more is currently known about this than about any other aspect of the subject. An important organelle in connexion with gene expression is the nucleolus, which 'processes' the ribosomal RNA that has been made at the nucleolar organizer. The processed material then passes into the cytoplasm to be incorporated into the ribosomes. Since ribosomes control protein synthesis, the nucleolus is of fundamental importance to the life of the cell, and in *The nucleolus*, E. G. Jordan brings out how two different experimental approaches to its function, through electron microscopy and through biochemical analysis, have given remarkably coincident results.

The remaining ten Readers concentrate on various aspects of evolution. They display that evolution has taken place in various groups, and also provide evidence for its mechanism. We start with a consideration of the *The origin of life* by J. D. Bernal. Bernal was primarily a crystallographer and his interest in biological macromolecules did much to lay the foundations of molecular biology. It also led him to an interest in the origin of life, a topic which has recently passed from unsubstantiated speculation to probabilities. Miller's experiments on primitive gas mixtures indicate how life chemicals may have had their origin; molecular biology shows how information can be stored and passed on, with the vital provision for the right frequency of 'mistakes'; the electron microscope shows what the stages in the evolution of the prokaryotic and eukaryotic ultrastructure may have been. We are greatly indebted to Ann Synge for the skilful way in which she has condensed, at Bernal's request, his stimulating book on this subject.

In *Homology* by Sir Gavin de Beer we are provoked by what the author calls 'a problem that is fundamental for biology but which has never been solved, which applies to plants as well as to animals, involves at least half a dozen disciplines in biology, and should remind students that there are still fundamental discoveries to be made'. Homologous structures are not always represented in the common ancestor; homologous structures can arise from totally different positions in the fertilized egg and embryo; homologous structures can owe their origin in different organisms to different organizer-induced processes; and the development of homologous organs can be controlled by totally different genes. So what is the basis for the 'agreement' of homologous organs? Sir Gavin adopts a similarly grand perspective in writing on *Adaptation*, a phenomenon which exists at all levels, molecular, cytological, organismal, behavioural, and ecological. In this article he examines and defines the terms *teleology* and *teleonomy*, and the concept of the fitness of the environment (rather than of the organism), before leading on to a selection of examples of adaptations, the numbers of which are, in the nature of the phenomenon, legion. The existence of adaptation provides the greatest weight of evidence for evolution by natural selection.

E. B. Ford has played a major part in the development of genetics and in *Evolution studied by observation and experiment* he reveals how knowledge has been obtained about evolutionary processes in action in the world today. Before Ford's investigations it was thought (as Charles Darwin thought) that evolution took place too slowly for observation of it to be possible; but in this Reader Ford shows how it has been quantified in four different types of situation.

Turning to the evolution of various phyla, we confront in *The origin of chordates* a problem of tantalizing magnitude. The fossil record shows that the invertebrate phyla preceded the vertebrates, but from which phylum did they evolve? Quentin Bone shows how most of the evidence is derived from the comparative anatomy and embryology of living forms (since fossils of soft-bodied animals, especially ancient ones, are rare). He gives special consideration to the auricularia theory of Garstang, and the biological principle of neoteny (derivation of new species from larval ancestral forms). Some of the possible mechanisms of neoteny are dealt with by J. R. Tata in *Metamorphosis*. This term refers to the abrupt changes in growth during the course of individual development which are found in some animal groups such as insects and amphibians. These affect not only body shape and anatomy but often the function of almost all the cell types in the body. The accidental discovery in 1911 that thyroxine causes tadpoles to turn abruptly into frogs revealed that the process is under hormonal control; later investigations about the biochemistry and gene-control of the physiological processes involved have contributed a great deal to present-day notions of developmental biology and to possible mechanisms of animal evolution.

At the other end of the chordate spectrum we have the primates and, in particular, man. In *Primates and their adaptations*, J. R. Napier shows how the evolution of the group has been strongly influenced by the tree habitat. He considers the following anatomical, physiological, and behavioural adaptations of these 'professional tree livers': possession of grasping extremities; possession of nails rather than claws; possession of frontally-placed eyes; the diminished olfactory sense; and the ability to sit, stand, and walk upright. All these have had a bearing on the most characteristic feature of the primates, their behavioural individualism: they are thinkers whose responses are not predictably reflex; and the source of this adaptability is the brain. It is a natural progression to M. H. Day's *The fossil history of man*, and we cannot fail to be interested in the origin of our own ancestors. Much of the fossil material has come from the celebrated Olduvai gorge in Kenya. Three 'platforms' of human evolution have been discerned, the Australopithecine (ape-men, about 5·5m years ago), the Pithecanthropine, true men from Asia, North and East Africa who existed about 0·5m years ago, and sapient man from a range of sites some 40 000 years ago. Day illustrates each level and suggests how they might be linked to each other.

Finally we have two Readers on the evolutionary history of the major plant group, the Angiosperms. Very soon after their appearance some 135m years ago, the flowering plants underwent very rapid adaptive radiation, spreading to almost every ecological niche in the world. This spread has been one of the most important events in the history of life on earth; but there are no known fossils of 'the original flowering plants', the earliest known remains contain forty-nine modern families. In *The mysterious origin of flowering plants* K. R. Sporne reviews theories of the origin of the group from the Gnetales, the Bennetitales, the Glossopteridaceae, the Caytoniaceae, and the Cycadales, and provides a statistical review of primitiveness in an attempt to illuminate what Charles Darwin called 'an abominable mystery'. Some of the most abundant and remarkably preserved fossils are those of pollen grains and other spores, whose ornately sculptured outer walls enable accurate species identification from as far back as the Palaeozoic. A vertical borehole through a series of lake sediments gives a sequence of samples which can be analysed to reveal biological, geological, and environmental changes which took place in the region. The emerging patterns provide an understanding of the effects of changing climate on vegetation and flora and consitute the most refined and detailed knowledge of the past that has yet come to light. In *Studying the past by pollen analysis*, R. G. West, a world authority on the subject, describes the techniques in this field and gives examples of the uses made of them.

The Editor has found each of the articles in this volume exciting, and he hopes that in the complex of reactions to it, excitement will feature prominently.

J. J. Head, *Hampton, January 1973*

Contents

26

Oxford Biology Readers
Edited by J.J. Head and O.E. Lowenstein

Somatic Cell Division

B. John and K. R. Lewis

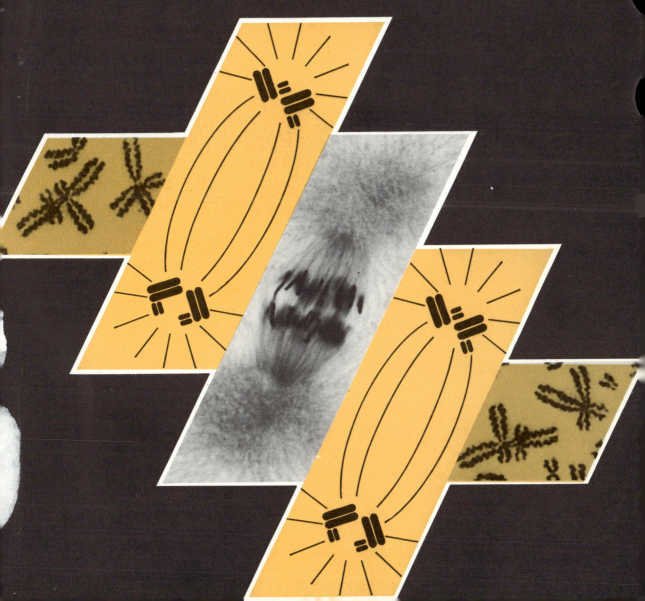

Oxford University Press, Ely House, London W.1

GLASGOW NEW YORK TORONTO MELBOURNE WELLINGTON
CAPE TOWN SALISBURY IBADAN NAIROBI DAR ES SALAAM LUSAKA ADDIS ABABA
BOMBAY CALCUTTA MADRAS KARACHI LAHORE DACCA
KUALA LUMPUR SINGAPORE HONG KONG TOKYO

ISBN 0 19 914129 0

Bernard John is Professor of Zoology in the University of Southampton. Kenneth Lewis is Lecturer in Botany and Fellow of Exeter College in the University of Oxford. They have collaborated extensively in cytogenetical research and are joint authors of numerous books and research monographs including *Chromosome marker* (Churchill 1963), *The meiotic system* (Springer-Verlag 1965), and *The organization of heredity* (Edward Arnold 1970).

PHOTOSET AND PRINTED IN GREAT BRITAIN BY
BAS PRINTERS LIMITED, WALLOP, HAMPSHIRE

Character	Cell type	
	Prokaryotic	Eukaryotic
1. Nuclear membrane	Absent	Present
2. Nucleolus	Absent	Present
3. DNA	Naked dimer (double helix)	Combined with histone
4. Linkage units	Single and circular	Multiple and linear
5. Mitosis and meiosis	Absent	Present
6. Spindle	Absent	Present
7. Enzymes of respiratory and photosynthetic electron-transport systems	Localized on cell membrane	Localized on specialized membranous organelles (mitochondria, chloroplasts)
8. Photosynthetic organelles	Simple chromophores	Complex chloroplasts
9. Ribosomes	70S(30S + 50S)	80S(60S + 40S)
10. Cell wall	Contains muramic acid and either diaminopimelic acid or lysine	Never contains muramic acid

FIG. 2. Cell organization in pro- and eukaryotes. Note that S is a Svedberg unit. It refers to sedimentation velocity during centrifugation and is therefore a reflection of molecular size.

Cell architecture and activity

The cell is the basic unit of structure and function in all organisms other than viruses. These contain only one kind of nucleic acid, DNA or RNA, which serves a genetic function. The nucleic acid is enclosed in a coat of protein which immediately determines various viral properties such as host range. Viral nucleic acid can both replicate and dictate the primary structure of virus-specific proteins. But viruses are obligate intracellular parasites and can only express their living properties when supported by the metabolic machinery of their hosts. The cell, therefore, is the lowest level of biological organization compatible with the maintenance, expression, and perpetuation of life, and many organisms consist of single cells.

Cells may assume a variety of sizes and shapes, and in complex multicellular systems a given cell may serve a very narrow range of specialized functions. However, there is accumulating evidence that even the greatest complexities of cell structure are mere elaborations of a basic pattern designed to further and favour the expression of one or some of the many universal properties of living systems. The metabolic processes that underlie these properties are of three main kinds:

1. Diffusion-control processes concerned with the selection, accumulation, and movement of raw materials and metabolic products,

2. Intermolecular interactions which do not involve sequence coding.

These may be concerned with the production of intermediate metabolites or certain end products (including some polymers), or with the acquisition, storage, and transfer of energy.

3. Polymer syntheses which require specific templates and do involve sequence coding.

These are concerned with the perpetuation and expression of molecular specificity and are of two types:

(i) Autosyntheses involving the replication of either DNA or genetic RNA. For this purpose DNA functions as its own template. Present evidence indicates that this is true also of the RNA of plant and bacterial viruses; but that of animal viruses may first specify a complementary DNA template on which the viral RNA is subsequently synthesized. If this is true the replication of animal virus RNA should be regarded as heterosynthetic.

(ii) Heterosyntheses involving polypeptide production. The genetic RNA of viruses can function directly as a template for the formation of polypeptides. But in cellular organisms this is a biphasic process because the DNA must first specify secondary mRNA templates (transcription) on which polypeptides are subsequently synthesized (translation). In either event, ribosomes and adaptor tRNA molecules are also involved.

Enzymes mediate in virtually all the reactions that occur in cells and are therefore present wherever chemical bonds are made or broken. The molecules concerned in sequence coding, like those of many other systems, are highly localized and to some degree isolated. Localization of function is reflected in structural organization which finds expression in two principal types of cell architecture. One type, the *prokaryote* (bacteria and blue-green algae) shows a minimum of organelle differentiation while the other, the *eukaryote* (plants and animals), is richly differentiated into organelles.

In prokaryotes various structural aggregates occur whose biochemical functions reflect the enzymes associated with them. Thus the DNA and the enzymes essential for its replication are concentrated in a more-or-less central nucleoid (Fig. 1) which frequently appears to be attached to a localized invagination (mesosome) of the outer cell membrane. Respiratory activity, on the other hand, is associated with membranous elements which are usually concentrated near the cell periphery; and

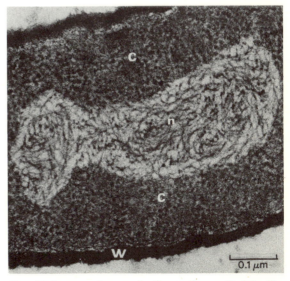

FIG. 1. Electron micrograph of part of one cell of *Bacillus subtilis* strain W 23 showing nucleoid (n). C, cytoplasm; W, cell wall. (Provided by Dr. W. van Iterson.)

3

photosynthesis, where it occurs, is localized in stacked lamellae or small membranous vesicles which, while they arise from the cell membrane, may occupy a large part of the cell interior.

The greater complexity of eukaryotic cells (Fig. 2) is in four main directions:

1. The nuclear, respiratory, and photosynthetic systems are themselves enclosed in limiting membranes.

2. They include other membrane-bound systems which do not appear to have functional counterparts in prokaryotes. These include vacuoles of various kinds, lysosomes which contain hydrolytic enzymes, and plastids other than chloroplasts.

3. The cytoplasmic matrix is frequently channelled by an extensive system of membrane-limited canals which connect the outer membrane with internal regions. This elaborate endoplasmic reticulum contrasts sharply with the limited invaginations of the bacterial plasma membrane and probably reflects the greater need for diffusion control in the much larger eukaryote cell.

4. DNA is localized predominantly in the nucleus where it is associated with basic protein (histone) to form a series of heterogeneous organelles called chromosomes.

The replication and segregation of genetic material
Prokaryotes. Bacterial cells may contain more than one nucleoid, 2–3 being usual for *Escherichia coli* under standard conditions of culture. Each nucleoid contains only a single DNA molecule. Electron microscope and autoradiographic studies of various bacteria have shown that if the DNA from gently lysed cells is allowed to disentangle slowly it forms a circle. (Lysis is the breaking open of the cell wall.) In the relatively compact nucleoid the DNA ring does not form a random tangle but is folded back and forth along itself to give the appearance of organized bundles each containing about 500 parallel threads. The nature of this rather flexible tertiary organization is not understood.

The DNA replicates continuously and retains essentially the same fibrillar structure throughout the division cycle. Meselson and Stahl investigated the replication by growing *E. coli* for a number of cell generations in the presence of heavy nitrogen (^{15}N), thereby altering the buoyant density of the DNA. An excess of normal isotope (^{14}N) was then supplied and DNA was extracted at intervals covering a number of division cycles. The position taken

by the extracted DNA in a caesium chloride density gradient showed its replication to be semi-conservative as proposed by Watson and Crick (Fig. 3).

The treatment involved in Meselson and Stahl's analysis caused breakage of the continuous DNA strand, and the fragments examined had a mean relative molecular mass of only about 7×10^6, 0·2 per cent of the genome (total gene complement). This experiment, therefore, did not provide information concerning the organization of DNA synthesis in the genome as a whole, but information on

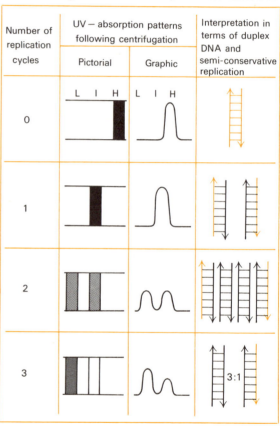

FIG. 3. The pattern of incorporation of ^{14}N (black) into uniformly ^{15}N-labelled (yellow) DNA over three replication cycles in *Escherichia coli*. The amount of DNA at various levels in the centrifuge tube is indicated by the intensity of the image in ultraviolet-photographs (pictorial) and by the tracings with the microdensitometer (graphic). The density in the caesium chloride gradient increases to the right. In terms of the hypothesis of semi-conservative replication (each polynucleotide column of the DNA dimer acting as a template to produce a new polynucleotide strand) the light (L), intermediate (I), and heavy (H) bands contain homoduplex $^{14}N-N^{14}$, heteroduplex $^{14}N-N^{15}$, and homoduplex $^{15}N-N^{15}$ molecules respectively.

this issue was provided by the autoradiographic studies of Cairns. He grew *E. coli* for 1–2 replication cycles in the presence of tritiated thymidine, which is specifically incorporated into DNA. The cells were gently lysed and after treatment to remove protein, the lysate was drained on a dialysis membrane to allow the DNA to unravel. The membrane was then covered with a photographic emulsion and kept in the dark. The pattern of thymidine incorporation was then determined from the distribution of the silver grains produced by tritium disintegration.

Cairns found that semi-conservatism characterized the whole genome and that replication proceeded sequentially in one direction from a fixed initiation point. In other words, the whole genome constitutes a single unit of replication (replicon) with a single growing point. When cell division is rapid a second replication cycle may begin before the previous one is complete. At the level of analyses allowed by autoradiography it appears that both template strands of the parental DNA duplex (double helix) are copied simultaneously. However, silver grains are about 500 times the diameter of the DNA molecule so that at the molecular level replication may be discontinuous in one or both daughter strands over distances equivalent to a few thousand nucleotides.

The mechanism that leads to the regular segregation of the replication products to daughter cells is not understood. Jacob and his colleagues have proposed that the replicon is attached to the cell membrane since growth of the membrane between the two newly replicated units would obviously provide a mechanism for their orderly separation. As yet not enough information is available to determine which part, if any, of the DNA thread is attached to the membrane or to reveal the nature of the attachment. On theoretical grounds it has been suggested that attachment occurs at the point where replication begins and that the association involves an attachment factor which is inserted into the DNA thread (Fig. 4).

In addition to the main circlet, bacteria may harbour small, independent DNA molecules some of which are concerned with conjugal mating and may be of viral origin. The replication of these supernumerary elements is usually tied to that of the main DNA component, and their segregation appears to depend on the same membrane-based mechanism.

Eukaryotes. The major part of the genetic material

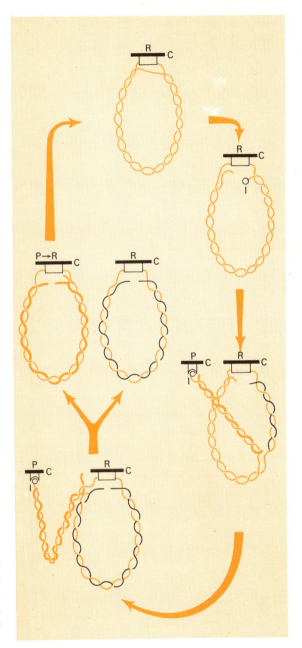

FIG. 4. Model of DNA replication in *Escherichia coli*. One strand of the dimer is presumed to be permanently attached to the cell membrane (C) by a replicator protein (R). The other is detached by an interaction with an initiator (I). This then becomes temporarily attached to the cell membrane by a third protein, the pro-replicator (P). When replication is complete the pro-replicator is converted into a replicator. (After Lark.)

of eukaryotes is distributed between two or more, often many more, complex organelles called chromosomes. These contain not only DNA but both basic (histone) and non-basic protein in the form of nucleoprotein, and small amounts of RNA. In some of the simpler plants and animals the chromosomes are always contained in a membrane-bound nucleus. In higher forms, however, the nuclear membrane breaks down during the later stages of nuclear division.

The somatic division cycle of eukaryotes can be divided into two main phases: interphase, when the nucleus is mechanically inactive, and the division phase (mitosis). Some synthetic activity takes place in the nucleus during the early stage of its division but interphase is the principal metabolic period. The reader will remember, and can see in Fig. 5, that at the onset of mitosis the chromosome is divided into two chromatids but at the end it consists of one chromatid. The syntheses that occur during interphase are concerned not only with the normal workings of the cell but also with its preparations for division, and nuclear DNA synthesis is virtually confined to a distinct part of interphase, the so-called synthetic or S-phase. The doubling of DNA content which occurs at this time is associated with the heterocatalytic doubling of histone content and the individualization of the products of replication and associated materials to give chromosomes composed of two genetically identical sister chromatids in each chromosome. Chromosomal DNA synthesis is semi-conservative and, as a rule, this property extends to the chromosome as a whole. This has been interpreted by some as evidence that the unreplicated chromosome is essentially a single DNA dimer (double helix) combined with protein, but this is debatable. There is evidence that the chromosome includes many replicons which may be organized in pairs so that synthesis proceeds coincidently in opposite physical directions from a common origin between the members of a pair.

The mechanical sequence that follows chromosome duplication may take a variety of patterns and we shall begin by considering the most common and conventional.

The standard mitotic mechanism

The basic sequence (Fig. 5). Mitosis is in essence a device to ensure the accurate and co-ordinated separation of sister chromatids. It involves two sub-cycles—a chromosome cycle and a spindle cycle. The former is one of internal coiling and condensation so that the chromosomes, which tend to be diffuse and extended in the interphase nucleus, form compact bodies (prophase) which can subsequently be moved more easily.

The spindle cycle consists of the development of a longitudinally oriented system of protein microtubules in the nuclear area at an advanced stage of the coiling cycle. When newt heart cells in culture are treated with inhibitors of protein synthesis some twenty minutes before the onset of division, spindles of reduced size are formed; and proteins with the same antigenic properties as those of the spindle apparatus can be found in pre-cleavage eggs of the

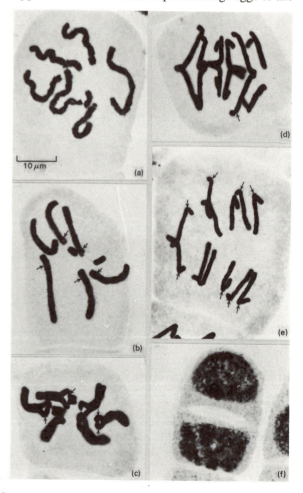

FIG. 5. Mitotic sequence taken from the root apex of *Crocus balansae*. (a) Prophase; each chromosome can be seen to consist of two chromatids; (b) pro-metaphase, positions of centromeres indicated by arrows; (c) metaphase; (d) mid anaphase; (e) late anaphase; (f) telophase–interphase reorganization. In (b), (c), (d), and (e) some of the centromeres have been arrowed. The spindle cannot be seen in this sequence but in all cases runs north–south.

sea-urchin. These observations indicate that, like DNA, spindle proteins are synthesized in more-or-less their final form during interphase, some time before the *spindle system* is organized. Spindle organization is completed when the nuclear membrane breaks down, at a time when the chromosomes have almost reached their greatest degree of compactness.

Spindles can usually be seen in living cells only by using a sensitive polarizing microscope, when they appear birefringent, or a differential interference microscope. Phase-contrast sometimes gives some indication of the spindle system (Fig. 6). Birefringence is often evident around the nuclear membrane prior to its breakdown, indicating that spindle material is formed outside the nucleus. Initially the spindle consists of *continuous fibres* which run from pole to pole. When, following the dissolution of the nuclear membrane, the chromosomes come into contact with this ordered system of microtubules they orient in relation to it (prometaphase) in such a way that the component chromatids of each chromosome are directed towards opposite spindle poles.

Orientation can occur at any point along the length and breadth of the spindle but subsequently all the chromosomes congress and move into the equatorial plane (metaphase). In most organisms these movements can be achieved only through the activity of a localized region of the chromosome, the centromere or kinetochore. This is a special genetic locus responsible in some as yet undefined way for the binding of the chromatid to the spindle microtubule protein (Fig. 7). Significantly, in this connection, centromere regions remain relatively uncoiled throughout prophase and therefore appear as *primary constrictions*. These too are double during prophase but sister chromatids are intimately associated immediately adjacent to them. Orientation involves the organization of *half-spindle fibres* which link half-centromeres to the poles towards which they are oriented.

When the stability of metaphase congression has been achieved the association of sister chromatids ends; the sister half-centromeres move apart and so the products of duplication are peeled apart to opposite poles (anaphase). This produces two groups of single stranded chromosomes which revert to an interphase state within the nuclear membranes which reform around them (telophase). Finally cytokinesis in the region of the mid spindle divides the cytoplasm into two distinct areas and

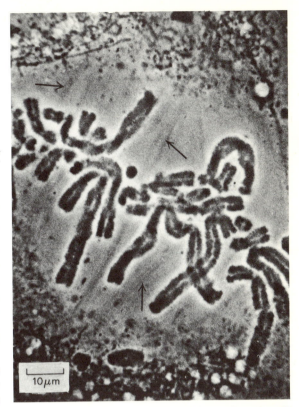

FIG. 6. Spindle fibres (arrows) in a living endosperm cell of the plant *Haemanthus katherinae* as seen at metaphase of mitosis with phase contrast. (Photo by Dr. A. Bajer.)

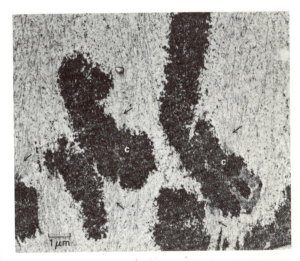

FIG. 7. Electron micrograph of a metaphase cell of *Haemanthus katherinae* similar to that shown in Fig. 6, to demonstrate centromere microtubule attachments (arrows). C, chromosome. (Photo by Dr. A. Bajer.)

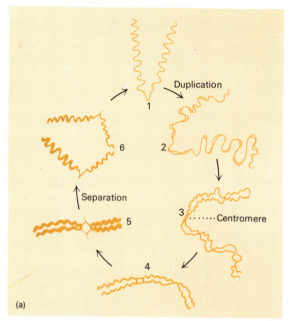

(a)

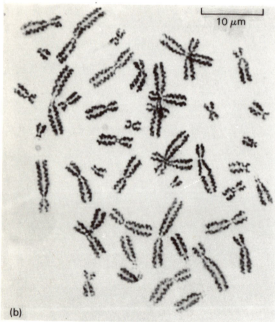

10 μm

(b)

FIG. 8. (a) The chromosome coiling cycle during mitosis. 1, telophase–interphase. 2, early prophase. 3, mid prophase. 5, metaphase. 6, anaphase.

FIG. 8 (b). Mitotic metaphase of a human leucocyte treated with a hypotonic solution involving a 4:2:0:8 mixture of an equimolar solution of KCl, NaNO$_3$, and CH$_3$COONa to reveal the spiral structure of the chromosomes. (Photo by Dr. Yasushi Ohnuki.)

so converts the one cell into two.

The nature of chromosome coiling. Coiling (Fig. 8) must involve configurational changes in the DNA molecule. Chromosome contraction is reduced when protein synthesis is selectively inhibited, so DNA–protein interactions are probably involved. Interphase nuclei often include some condensed (heteropycnotic) chromosome material. In calf thymus lymphocyte nuclei it has been shown that maintenance of heteropycnotic material depends on an association between DNA and a particular histone fraction. Acid-extracted histone from these nuclei includes four classes largely distinguished on the basis of amino-acid composition, namely lysine rich (I) arginine rich (III and IV) and intermediate (IIb). Selective removal of particular classes can be achieved by centrifuging nucleoprotein through salt solution of appropriate concentration. If the lysine-rich fraction is removed, the condensed chromatin disperses to form a diffuse network of fibrils, but the restoration of this fraction to histone-depleted nuclei restores the heteropycnotic state. These effects are not found with the arginine-rich fraction. Littau has argued that lysine-rich histones cross-link DNA fibrils. But since diffuse and condensed chromatin appear to contain the same amount (20 per cent) of this fraction it is not obvious why the proposed cross-linking should occur only under some conditions.

Spindle structure and organization. Spindle poles may be occupied by special organelles called centrioles. These are found only in organisms that have ciliated or flagellated stages somewhere in the life cycle, and even here centrioles may not be evident in all divisions. Thus the haploid myxamoebal phase of myxomycetes produce motile cells while the diploid plasmodial phase does not and centrioles occur only in the former. Centrioles appear to be involved in the formation of cilia and flagella with which they share a common ultrastructure (Fig. 9a).

Before division, centrioles are generally arranged as two adjacent pairs which occupy a cytoplasm-filled invagination of the nuclear membrane. Subsequently a radiating system of microtubules (aster) may be formed in relation to each centriolar pair (Fig. 9b). During prophase these microtubules proliferate and lengthen and the centriolar pairs move round the nuclear membrane until they occupy diametrically opposed positions. This movement is accompanied by the development of *continuous* spindle fibres between, but not connected with, the

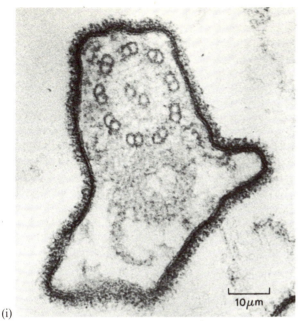

(i)

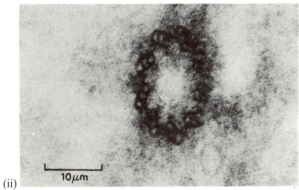

(ii)

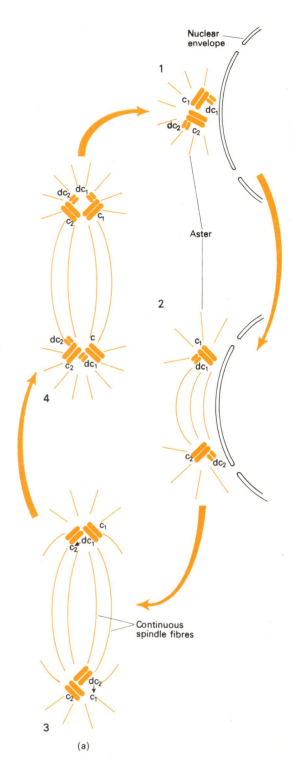

Fig. 9 (a). (i) Transverse section of flagellum of *Euglena ascus* showing 9 + 2 pattern of microtubules in the axis (9 peripheral doublets, 2 central singlets). (Photo by Dr. W. Bernhard.) (ii) Transverse section of centriole from human lymphosarcoma. Centrioles are analogous with the basal bodies of cilia and flagella, in which the 9 doublets have become skewed triplets and the central microtubules have disappeared. (Photo by Dr. G. F. Leedale.)

Fig. 9 (b). The centriolar cycle in mitosis. Centrioles always appear in pairs at interphase, 1. Each of the two bodies (c_1 and c_2) have small annular pro-centrioles (dc_1 and dc_2 respectively) lying at right angles to them. During prophase, 2, the centriolar pairs separate, through an increase in the size of the asters associated with them, and the pro-centrioles lengthen into cylinders. Elongation continues through meta-anaphase, 3. During telophase, 4, each of the two polar centrioles produce new pro-centrioles which arise from the wall of the pre-existing structures.

(a)

9

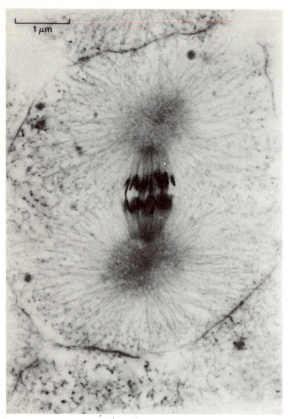

FIG. 10. Anaphase of the first cleavage mitosis of the white-fish *Coregonus clupeoides*. The asters are always large in animal eggs and are especially prominent here. (Photo by C. D. Darlington.)

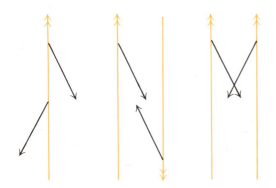

| 1 The basic motile unit – a micro–tubule possessing polarity and mechanochemical arms | 2. Antiparallel tubules interact to produce a motile force | 3. Parallel tubules do not produce a motile force |

separating centriolar pairs (Fig. 10). Astral and spindle fibres have a similar ultrastructure but the extent to which the centrioles are actually responsible for the generation of the spindle system is far from clear. It is true that when centriolar reproduction in sea-urchin eggs is disturbed by mercapto-ethanol treatment, mono- or tetra-polar spindles are formed according to the number of centriolar pairs in the cells. On the other hand Dietz has found that normal spindles can be formed even when, owing to a prior abnormality, centrioles are absent from cells that normally have them. It has been suggested, therefore, that the continuous fibres, far from being generated by the centriole, are the means of segregating them. In those cells that lack centrioles the spindle poles tend to be diffuse and the spindle may be barrel-shaped or even broader at the poles than at the equator.

Colchicine causes rapid disruption of the spindle apparatus, but when it is removed the spindle is quickly restored. The rapidity of reassembly may mean that reconstruction involves the aggregation of subunits rather than complete resynthesis; and colchine binding appears to be specific to a 6S protein unit. Inoue has therefore suggested that the microtubules of the spindle consist of subunits which may exist in an assembled (polymerized) or disassembled (monomer pool) state.

Another idea is that the spindle initially consists of both oriented and unoriented fibrillae, and at orientation in pro-metaphase the centromere is thought of as directing the linkage of unoriented units into *half-spindle* fibres. Significantly the bi-refringence of the spindle, which reflects the parallel orientation of protein fibrillae, is always greatest in the vicinity of the centromere.

Mis-oriented centromeres are quite often found at the early stages of pro-metaphase, both sister chromatids being connected to the same pole. Re-orientation to a bipolar arrangement is, however, almost invariably achieved indicating that micro-tubule–centromere connections can disrupt and reform with relative ease.

The nature of chromosome movement. A number of models of chromosome movement have been formulated. In all of them the microtubules are regarded as the basic motile units, the chromosomes being passive. We will confine attention to the most recent model since it incorporates the essential elements of many of the others. Its distinctive

FIG. 11. Properties of microtubule units assumed in the McIntosh model of mitosis (see Fig. 12).

features, which we owe to McIntosh and his colleagues, are:

1. Spindle microtubules have lateral arms which confer polarity (Fig. 11).

2. The lateral arms serve both structural and mechanochemical roles and form cross-bridges between spindle microtubules.

It is also assumed that the inter-tubular links serve to bend the normally straight microtubules into the curved paths of the spindle. Antiparallel tubules can interact so that they slide over one another. This means that the lateral arms contain enzymes capable of releasing chemical energy from ATP and converting part of it into mechanical work.

The McIntosh model, in common with all other models, fails to offer a really satisfactory explanation of the movements involved in metaphase congression; but anaphase is held to be governed by a relative sliding of microtubule units coupled with tubule shortening, which is supposed to occur not by contraction but by the disassembly of subunits from the microtubules. The half-spindle fibres are moved towards the poles while the continuous fibres slide in the direction of the opposite pole (Fig. 12). Relative sliding of antiparallel tubules then continues until a point of self-termination is reached when there are too few bridges left to move the tubules any farther.

Breakdown and reformation of the nuclear membrane. The nuclear envelope consists of two 75 Å unit membranes. Unlike the equivalent structure which delimits mitochondria and plastids it is perforated by large, diaphragm-covered pores 300–1000 Å in diameter which may occupy more than 20 per cent of its surface (Fig. 13).

Electron micrographs indicate that microtubules become attached to the outside of the nuclear membrane during prophase, distorting it. Bajer has suggested that its breakdown results from the pushing and pulling exerted by the microtubules. In animal cells, mitochondria also appear to be closely associated with the membrane at prophase.

The membranous fragments derived from the breakdown of the nuclear envelope may form rounded vesicles or flat sheets which are not easily distinguishable from the membranous elements of the endoplasmic reticulum. The reformation of the envelope at telophase involves the fusion of similar fragments which congregate near the chromosome surface at telophase. But whether these are the remnants of the original membrane, components of the

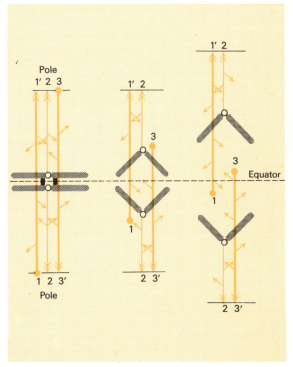

Fig. 12. The McIntosh model of mitosis. A minimum number of three microtubules is shown; 1 and 3 are polarized continuous interpolar tubules (1→1′ and 3→3′) while 2 represents half-spindle microtubules. During anaphase microtubules that lie antiparallel slide on one another. In addition the length of both the half-spindle fibres and the continuous fibres decreases by the disassembly of subunits. (After Nicklas.)

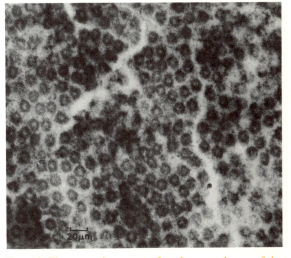

FIG. 13. Electron microscope of nuclear membrane of the sea urchin oocyte. (Courtesy of Dr. B. Afzelius.)

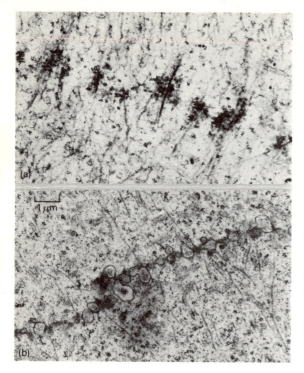

Fig. 14 (a). Electron micrograph of developing phragmo-plast from endosperm of *Haemanthus katherinae*. The phragmoplast region is characterized by a high concentration of microtubules and a large number of ribosomes some of which are grouped into polysomes. (b) Later stage in cell-plate formation than that in (a). Vesicles aligned in the equatorial plane derived from Golgi bodies, coalesce to form a cell plate. Microtubules and ribosomes are still prominent. (Photos by Dr. A. Bajer.)

original endoplasmic reticulum, or newly synthe-sized membranous material is not clear.

Cytokinesis. Nuclear division (karyokinesis) need not be followed immediately by cytoplasmic cleavage; it may not follow at all, when a multi-nucleate condition results. This occurs as a regular feature in the development of vertebrate skeletal muscle and in the formation of the liver in some mammals and the anther wall of angiosperms.

Cytokinesis takes two main forms—furrowing, which is typical of animals and the simpler plants, and cell-plate formation, which is the usual mech-anism in land plants. Cytochalasin-B is known to cause several types of animal cell to become multi-nucleate. When this drug is administered to cleav-ing marine eggs with shallow furrows the furrows disappear and cytokinesis stops. The spindle, how-ever, continues to function. Schroeder has shown that the drug causes the disappearance of a band of microfilaments which form a contractile ring loca-ted just beneath the cleavage furrow. This seems to be the contractile agent responsible for pinching the cell cytoplasm in two.

Cell plate formation in land plants involves the fusion of membrane-bound vesicles (probably derived from the Golgi system) which concentrate in the plane of cleavage guided, perhaps, by the microtubules which persist in the equatorial region even after nuclear division is complete. Under the light microscope this region appears as a dense disc of cytoplasm penetrated by fibres (phragmoplast) (Fig. 14). The cell plate generally develops centri-fugally, the phragmoplast disappearing from the centre as the plate extends peripherally. Eventually the cross plate reaches the side walls and division is complete.

Extrachromosomal elements. The cytoplasm in-cludes various organelles which are continuous through successive division cycles. These include mitochondria and plastids, both of which contain their own distinctive DNA and almost certainly show genetic continuity as well. These properties may be shared with other extrachromosomal ele-

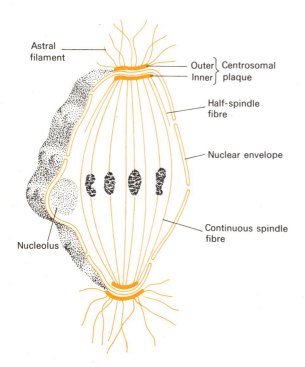

Astral filament
Outer } Centrosomal
Inner } plaque
Half-spindle fibre
Nuclear envelope
Continuous spindle fibre
Nucleolus

Fig. 15. The mitotic apparatus of the ascomycete *Ascobolus* as envisaged by Zickler.

ments, especially fibre-forming particles like the kinetoplasts of trypanosomes. Mitosis must provide for the controlled distribution of such elements and on average all the essential materials in a cell must be doubled during interphase, though where an element is represented many times a qualitatively equal segregation is compatible with quantitatively unequal distribution. However, as we have already seen, centrioles are distributed between daughter cells with an accuracy equalling that of the chromosomes. Many unicellular algae contain only one chloroplast so that its partitioning must also be well regulated; but little is known concerning the details of these aspects of the mitotic cycle.

Variations on the mitotic theme

A comparison of the somatic division sequences of diverse plants and animals reveals that the mitotic cycle has been modified in the most varied ways to adapt it for diverse purposes. There are three principal patterns of variation.

1. *Intranuclear sequences*. In ascomycete fungi the nuclear envelope remains intact throughout the entire mitosis and, as Zickler has recently shown, a centrosomal plaque, located outside the nucleus, serves as a polar focus for the *intranuclear* spindle system. This plaque consists of inner and outer zones between which the swollen nuclear envelope is sandwiched (Fig. 15). The spindle microtubules terminate at the inner zone of the plaque. The origin and mode of duplication of these plaques is, as yet, unknown. They are found only in fungi devoid of motile cells, and the phycomycetes possess differentiated centrioles.

The chromosomes of dinoflagellates (algae) also separate by an intranuclear mitosis. Here, however, the nuclear envelope appears to play an important role in the traction, orientation, and separation of the chromosomes. In addition, while there are no centrioles in the cytoplasm, specialized archoplasmic spheres are present outside the nuclear membrane. During division (Fig. 16) they give rise to a cytoplasmic invagination, containing a considerable number of microtubules, which perforates the nuclear membrane.

Dinoflagellates are the only eukaryotic group where no histone is present though definitive chromosomes are found. This, coupled with the distinctive nature of the chromosomes and the presence of such an unusual form of mitosis, has led Dodge to suggest calling the group mesokaryotes since there are sufficient differences in nuclear

organization and behaviour to merit their separation from both pro- and eukaryotes.

2. *Non-equational mitoses*. In a normal mitosis the sister chromatids of all the chromosomes separate more or less simultaneously to give two daughter nuclei both of which are genetically identical to the original nucleus. However, the normal development of some species includes specially modified mitotic sequences which lead to the non-equational distribution of chromosomes.

Sometimes the resulting daughter nuclei are identical to each other but are not exact copies of the parental nucleus. For example in nematocerous midges belonging to the family Orthocladiinae two kinds of chromosomes, S and E, can be distinguished. The S-chromosomes always behave in the normal way but at the 5th–7th cleavage divisions of the embryos all the E-chromosomes are eliminated from the *somatic* cells of males and females. They are not included in the daughter nuclei because, although their metaphase orientation appears to be normal, they stall during the early stages of anaphase and fail to complete polar movement. Comparable chromosomes in the gall midges are also eliminated but elimination in its most extreme form is found in the armoured scale insects (Hemiptera: Homoptera) where all the paternally derived chro-

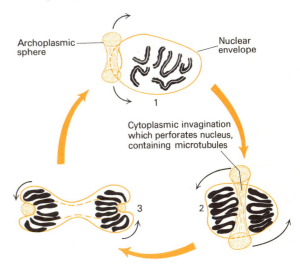

FIG. 16. Diagrammatic representation of the mitotic cycle of the dinoflagellate *Blastodinium* as conceived by Soyer. 1, prophase. 2, metaphase. 3, telophase. The small arrows indicate the pattern of movement of the archoplasmic spheres during the sequence.

mosomes are eliminated from male embryos at the blastula stage. Consequently these bugs develop and breed as though they were products of haploid parthenogenesis.

In the second type of non-equational mitosis all the chromosomes are transmitted but they are not distributed equally between the two daughters. For example XO zygotes in the creeping vole, *Microtus oregoni*, develop into females as a regular feature of the life cycle. But the single X-chromosome shows non-disjunction during the early development of the ovary, both its chromatids passing to the same daughter nucleus which thereby becomes XX. (Non-disjunction means that the chromatids do not separate at anaphase.) The fate of the cells lacking a sex-chromosome is not known but only those which have two X-chromosomes form viable oogonia, so an X-chromosome is transmitted through all the eggs.

Why particular chromosomes cohabiting a common spindle should behave differentially remains a complete enigma.

3. *Endomitotic mechanisms*. From a morphological point of view, mitosis is the principal event of the cell cycle. But from a biochemical point of view it should be regarded only as the final phase of a more complex cycle which is initiated in the interval between successive mitotic sequences. Indeed interphase is the longest event in the cell cycle and when chromosomes are cloaked in their state of interphase invisibility they are most active synthetically. Mitotic chromosomes, on the other hand, can be regarded as transient, metabolically passive elements which return to an active synthetic state with the re-establishment of their interphase expansion.

At some stage in its life a eukaryotic cell must 'decide' between one of two alternative pathways— either towards further mitosis or else towards differentiation. The path taken depends upon the state of the cell and the micro-environment operative at that time, and a common form of 'decision' involves a unique *endomitotic* cycle. The normal mitotic cycle consists in the individualization and subsequent separation into distinct nuclei of the products of chromosome duplication. Endomitosis, on the other hand, consists of one or more, often many more, cycles of chromosome duplication without nuclear division and it takes two main forms (Fig. 17).

First, chromatids may be produced, and sister chromatids may fall apart, so that the nucleus becomes endopolyploid. Second, chromosome

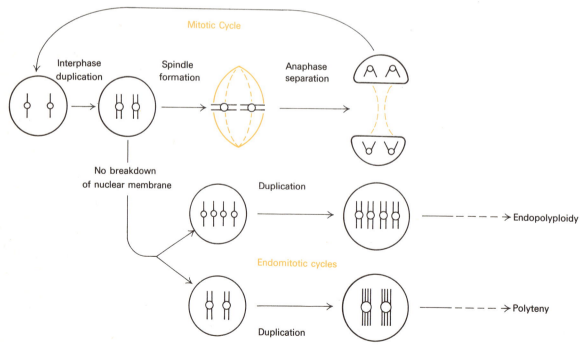

FIG. 17. The relationship between mitosis and endomitosis.

duplication may not be followed by chromatid individualization. If this happens the number of chromosomes remains unchanged but each becomes multistranded or *polytene*. A polytene chromosome may consist of hundreds or even thousands of chromatid equivalents and attain a thickness of up to ten thousand times that of the basic thread (Fig. 18).

In general, endopolyploidy and polyteny are alternative ways of securing an increase in nuclear and cytoplasmic volume without recourse to cell division. However, endopolyploid nuclei may contain polytene chromosomes and the macronucleus of certain ciliates is now known to pass through a polytene phase before becoming truly endopolyploid.

Endopolyploid somatic nuclei are widespread in many invertebrates (Fig. 19) and are known also in vertebrates and plants. Polyteny is less common but is found in various of the nuclei associated with the developing embryos of flowering plants. It is best known, however, in the larval development of dipteran flies, and notably in *Drosophila*, where it is associated with a system of transverse banding consisting of alternate chromatic zones (bands) and achromatic zones (interbands). The bands differ so specifically in appearance and sequence that they can be mapped through the greater part of the length of the polytene units. For example in *Drosophila melanogaster* some 5075 bands have been individually identified. By comparing such cytological maps with the known linkage maps of *Drosophila* it has been possible to assign particular gene loci to particular bands.

Endoreduplication of the kind found in both endopolyploidy and polyteny is a means of elevating the transcriptive capacity of a nucleus. It seems to allow nuclear material to administer a large cytoplasmic volume without recourse to cell or even nuclear division, which doubtless involve a much greater disruption of metabolic activity. This economy may be particularly important in relation to somatic cells whose products are required in large quantities over short periods of development, and endoreduplication is especially evident in secretory and nurse cells. The relative functional advantages of these two distinct methods of amplifying genetic potential are as yet obscure.

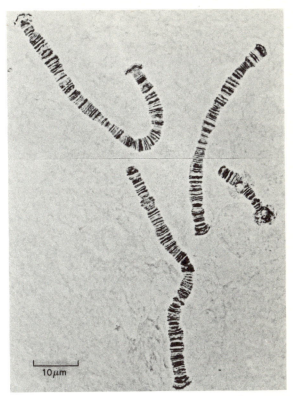

FIG. 18. The polytene chromosomes of *Chironomus obtusidens*. (Photo by Prof. H. G. Keyl.)

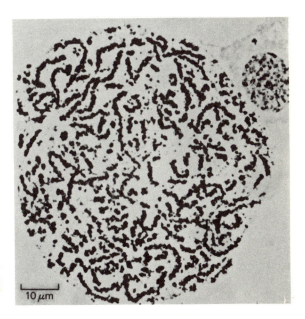

FIG. 19. Giant endopolyploid nucleus from the Australian plague locust, *Chortoicetes terminifera*. The conventional diploid interphase nucleus alongside it contains 24 chromosomes.

Types of division

Pro- and eukaryotes employ different methods of distributing genetic materials between daughter cells despite the fact that in both groups DNA replication is a necessary prelude to the separation of genetic material in cell division. The greater complexity of the mitotic machinery found in eukaryotes undoubtedly reflects the need for an efficient distribution mechanism to cope with a replicated genome which is divided into a number of distinct subunits.

Even among eukaryotes, however, there are wide variations in the mechanical details of the mitotic process. Generally somatic cell division sequences produce equal distribution of accurately replicated genetic information. But modified sequences can lead to the qualitatively unequal distribution of genetic information between daughter nuclei and even to the elimination of certain kinds of information from particular types of cell or individuals.

Thus mitosis is not a unit process and it cannot be described by any simple equation. Its essence is the doubling of all the potentialities of a cell followed by their distribution to two daughter products. There are chromosomes to be duplicated and coiled, spindle microtubules to be made, and an energy reservoir to be created large enough to sustain the entire sequence of movement. All these events are integrated spatially and temporally into a single sequence, yet the completion of some of them does not ensure the others: duplication is not always followed by effective separation, and where no effective separation occurs endomitosis ensues.

FURTHER READING

General

LUYKX, P. (1970). Cellular mechanisms of chromosome distribution. Suppl. 2. *International Review Cytology*. Academic Press, New York.

NICKLAS, R. B. (1971). Mitosis. *Adv. Cell Biol.* **2** (in press).

WENT, H. A. (1966). The behaviour of centrioles and the structure and formation of the achromatic figure. Vol. VI Gl. *Protoplasmatologia*. Springer-Verlag, Berlin.

For reference

BAJER, A. (1968). Behaviour and fine structure of spindle fibres during mitosis in endosperm. *Chromosoma* **25**, 249–81.

DIETZ, R. (1966). The dispensability of the centrioles in the spermatocyte divisions of *Pales ferruginea* (Nematocera). In *Chromosomes today* Vol. 1. Oliver and Boyd, Edinburgh.

FORER, A. (1969). Chromosome movements during cell division. In *Handbook of molecular cytology*. North-Holland Publishing Co., Amsterdam.

KUEMPEL, P. L. (1970). Bacterial chromosome replication. *Adv. Cell Biol.* **1**, 3–56.

McINTOSH, J. R., HEPLER, P. K., and VAN WIE, D. G. (1969). Model for mitosis. *Nature, Lond.* **224**, 659–63.

TUCKER, J. B. (1971). Microtubules and a contractile ring of microfibrils associated with a cleavage furrow. *J. Cell Sci.* **8**, 557–71.

ZICKLER, D. (1970). Division spindle and centrosomal plaques during mitosis and meiosis in some Ascomycetes. *Chromosoma* **30**, 287–304.

26

Oxford Biology Readers
Edited by J. J. Head

The Meiotic Mechanism

B. John and K. R. Lewis

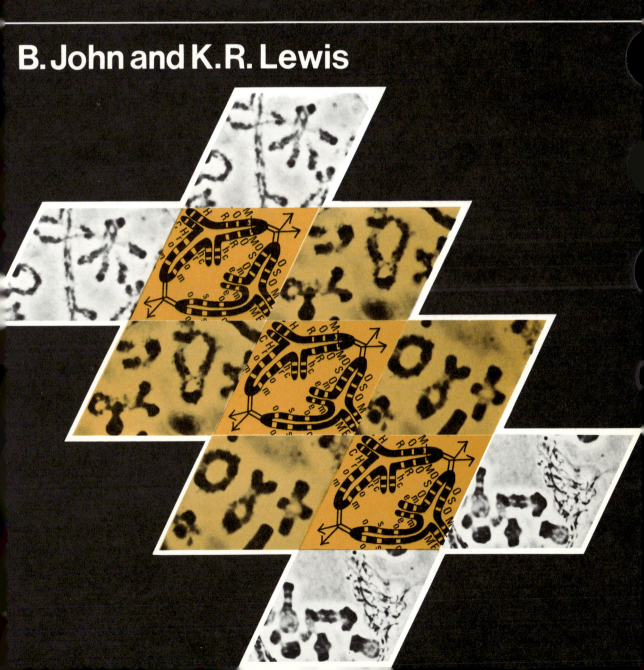

Oxford University Press, Ely House, London W.1

GLASGOW   NEW YORK   TORONTO   MELBOURNE   WELLINGTON
CAPE TOWN   IBADAN   NAIROBI   DAR ES SALAAM   LUSAKA   ADDIS ABABA
DELHI   BOMBAY   CALCUTTA   MADRAS   KARACHI   LAHORE   DACCA
KUALA LUMPUR   SINGAPORE   HONG KONG   TOKYO

ISBN 0 19 914148 7

Bernard John is Professor of Population Biology at the Australian National University, Canberra. Kenneth Lewis is Lecturer in Botany in the University of Oxford. They have collaborated extensively in cytogenetical research and are joint authors of numerous books and research monographs including *Chromosome marker* (Churchill 1963), *The meiotic system* (Springer-Verlag 1965), and *The organization of heredity* (Edward Arnold 1970).

FILMSET AND PRINTED IN GREAT BRITAIN BY
BAS PRINTERS LIMITED, WALLOP, HAMPSHIRE

The sexual system and the meiotic cycle

In many species of Protozoa, Algae and Fungi, mitosis is the only kind of nuclear division (Oxford Biology Readers 26). Their reproduction is *asexual*. But in most of these simpler organisms, and in nearly all the more complex ones, asexual reproduction is supplemented or entirely replaced by an alternative form of reproduction with a distinct and universal character of its own. This depends on two events: a distinctive type of nuclear division called *meiosis* coupled with a process of nuclear fusion called *fertilization*. The combination of these events in one life cycle is known as *sexual* reproduction.

Both meiosis and fertilization are capable of producing new nuclear compositions and sexual reproduction is a means not only of generating new but also variable individuals. In addition meiosis provides for the reduction of chromosome number to give the haploid phase of the life cycle, while fertilization combines the haploid phases, restoring the diploid state.

The position occupied by meiosis in the life cycle is variable. (Oxford Biology Readers 2.) Most animals have a gametic type of meiosis in which meiotic products are transformed into sperm cells (male) or egg cells (female) without any further division. By contrast many plants possess the sporic type of meiosis, in which reduction division takes place in the diploid sporophyte at some point between the formation of the zygote and the development of the gametes. Here the meiotic products are not gametes but spores. These grow by mitotic divisions into very much reduced gametophytes (Angiosperms) or else into the haploid phase of the life cycle (Bryophytes, Pteridophytes and many Fungi) from which gametes are subsequently produced not by meiosis but by haploid mitosis (Fig. 1).

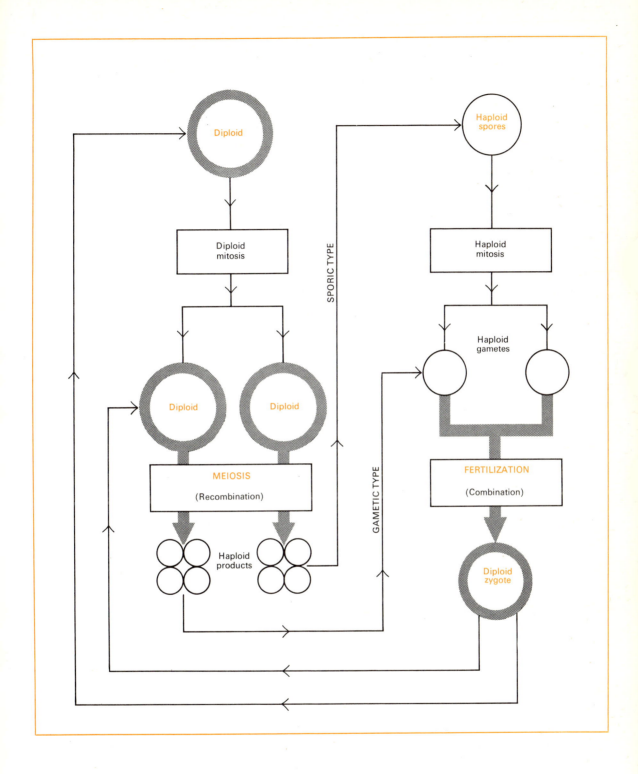

FIG. 1. Inter-relationships of meiosis and fertilization in the sexual cycles of eukaryotes.

Mitosis and meiosis compared

Although both processes are continuous, they can be described in terms of a sequence of changes in the appearance of the chromosomes. These become visible within the nucleus at the commencement of both types of division (Fig. 2). The manoeuvres of the chromosomes in both events are carried out on a transitory apparatus, a 'skeleton' of microtubules. In each case first the nucleus then the cell divides. The means of doing this is the same: duplication and spiralization of the chromosomes, the organization of the spindle, and the orientation and separation of the centromeres in relation to the spindle. But the two types of division can be distinguished in four main respects:

(1) In mitotic cycles each phase of DNA replication and chromatid formation is associated with a nuclear division. At meiosis one chromosome reproduction is followed by two successive nuclear divisions and for this reason leads to the reduction in chromosome number from which it derives its name (Greek *meioun* = to reduce).

(2) DNA-synthesis occurs about the middle of the interphase period (*the S-phase*) in mitotic cycles so that there is a more or less prolonged post-synthetic or G_2-phase before the onset of division. In the meiotic cycle, on the other hand, G_2 is either very short or non-existent. As a possible consequence the visible split into two chromatids occurs earlier in the mitotic cycle, prior to division and not after it has begun.

(3) In mitosis each chromosome represents an independent mechanical entity, so that in diploid mitosis a given chromosome behaves without reference to its homologue although, of course, the movements of all the chromosomes are co-ordinated in time. By contrast homologues are mechanically related during the first meiotic division.

(4) Whereas in a somatic cell cycle mitosis occupies but a brief period of about two hours or less, meiosis may extend for days or even weeks (Fig. 3).

FIG. 2. Meiosis in the male grasshopper ($2n = 2x = 17$; n refers to the haploid phase of the sexual cycle, x to the number of chromosome sets in the complement) as seen in lactopropionic orcein squash preparations. The uneven chromosome number results from the fact that the male has one sex (X) chromosome and the female two. In the male this X element regularly behaves as a univalent during the first meiotic division (X, a–h) and at first anaphase passes to one of the two spindle poles. Second division cells are thus of two types depending on whether (j, l and n) or not (i, k and m) an X is present. (Arrows point to centromeres.)

	Species	Chromosome number	2C-DNA Content i.e. the G_1 DNA level (picograms)	Total meiotic time (days)
Plants	Triticum aestivum	$2n = 6x = 42$	54·3	1
	Secale cereale	$2n = 2x = 14$	28·7	2
	Allium cepa	$2n = 2x = 16$	54	4
	Lilium longiflorum	$2n = 2x = 24$	106	8
	Trillium erectum	$2n = 2x = 10$	120	11
Animals	Drosophila melanogaster	$2n = 2x = 8$	0·085	4
	Locusta migratoria	$2n = 2x = 24$	13·4	8
	Triturus viridescens (Amphibian)	$2n = 2x = 24$	49	12
	Mus musculus (mouse)	$2n = 2x = 40$	5·0	13
	Homo sapiens	$2n = 2x = 46$	6·0	24

FIG. 3. The duration of meiosis in plants and animals. Equivalent times for the duration of mitosis range from 25 mins. to 4 hrs. Note that the animal species have a much longer meiotic cycle than would be expected for a diploid plant species with an equivalent nuclear DNA content (data of Bennett).

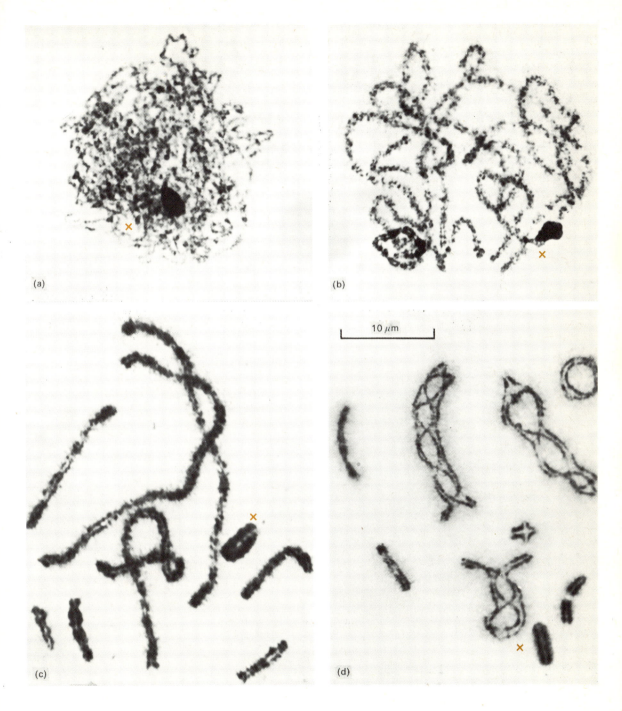

F<small>IG</small>. 2.
(a) Leptotene, prophase I—the X is condensed and very conspicuous but the sixteen remaining chromosomes (auto-somes) are long, fine, apparently single, threads which form a dense tangle; (b) zygotene, prophase I—homologous autosomes are paired and show a chromomeric organization; (c) pachytene, prophase I—paired homologues are now shorter and thicker but each still looks single (two-stranded *bivalent*); (d) diplotene, prophase I—each bivalent is now four-stranded and chiasmata are present.

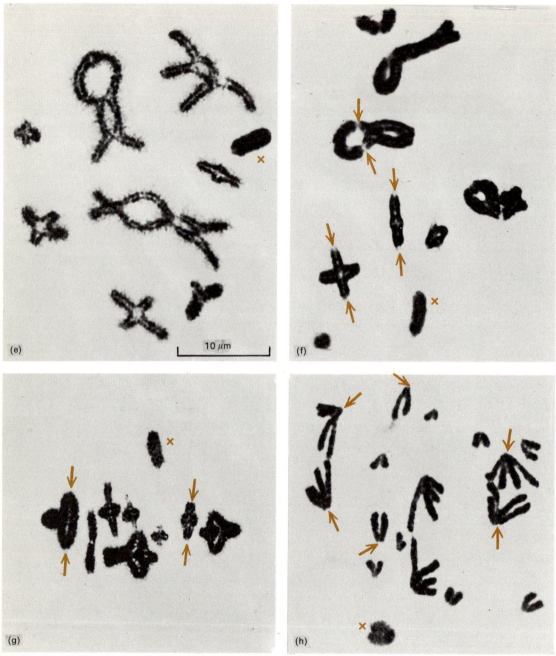

Fig. 2.
(e) diakinesis, prophase I—chromatids show increased spiralization so that the distinction between sister-chromatids is less obvious; (f) pro-metaphase I—the bivalents are oriented though not as yet congressed. The bivalent at twelve o'clock shows unipolar mal-orientation (*c.f.* Fig. 8). The spindle runs north-south; (g) metaphase I—bivalents are now congressed and co-oriented. The *X*, being univalent, lies off the main equatorial plate. The spindle runs north-south; (h) anaphase I—half bivalents separating to opposite poles, running north-south, while the *X* is oriented to the bottom pole; (i) and (j) prophase II—homologous chromatids lie splayed except in the proximity of the centromeres; (k) and (l) metaphase II—polar views of the second division spindle arrangement; (m) and (n) anaphase II—side views of the second separation sequence, spindle running north-south.

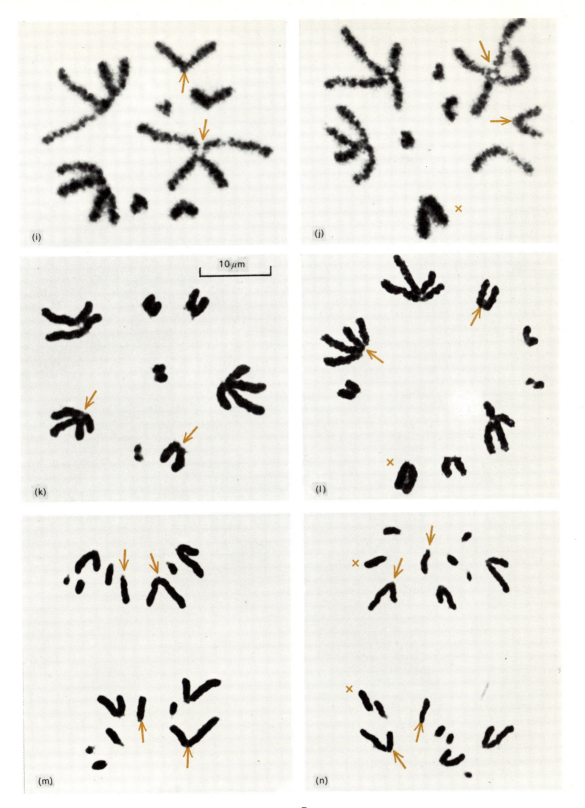

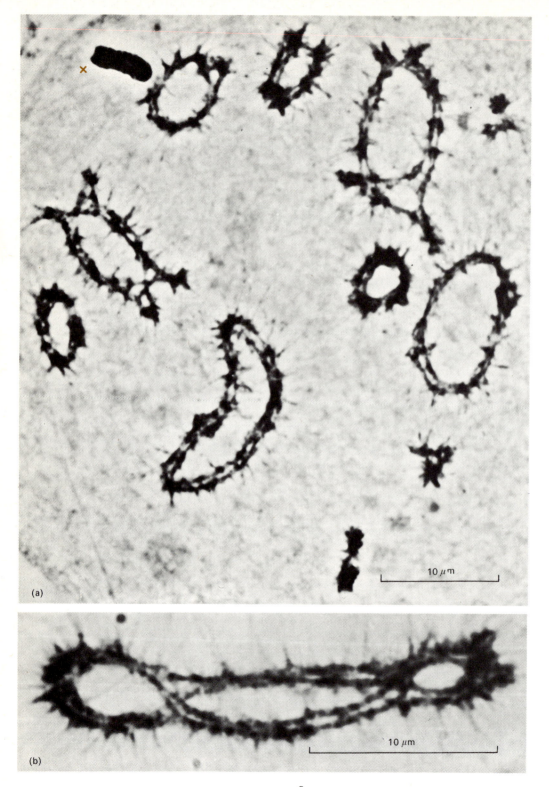

(a)

10 μm

(b)

10 μm

FIG. 4. Lampbrush diplotene chromosomes in the male meiosis of two grasshoppers, *Pezotettix giorni* (a) and *Podisma pedestris* (b).

8

The standard meiotic sequence

At the onset of meiosis, as in mitosis, homologues (homologous or partner chromosomes) are separate entities but, in contrast to mitotic chromosomes, they appear to be structurally single although DNA synthesis has been completed (*leptotene, prophase-I*). An active and specific pairing process leads, however, to the formation of the haploid number of *bivalents* (*zygotene, prophase-I*), each bivalent consisting of a pair of homologues. As a result of pairing, equivalent segments of homologous chromosomes are brought into parallel alignment. Thus each centromere is opposed to its homologous centromere and the end of each chromosome to the equivalent end in its appropriate homologue. Moreover, in many organisms at this stage each chromosome shows a visible longitudinal differentiation in the form of a sequence of darker staining blocks (*chromomeres*) separated by thinner, less dense, interchromomeric

regions. This beaded appearance reflects differential low-order coiling, and it can confer a characteristic morphology on a chromosome. These chromomeres increase in size and decrease in number as first prophase advances, and pairing frequently brings chromomeres of like size and shape into juxtaposition.

Chromosome contraction is particularly pronounced following pairing and leads to the production of much shortened and thickened bivalents (*pachytene, prophase-I*). Homologues still appear single at pachytene; it is not until the paired partners begin the separation process which initiates the opening out of each bivalent that homologues are visibly double (*diplotene, prophase-I*). On entering diplotene, or sometimes even in advance of this, the chromosomes frequently become diffuse, owing to the formation of lateral extensions from the main chromosome axis. This gives rise to the so-called *lampbrush* phase which is seen in its most

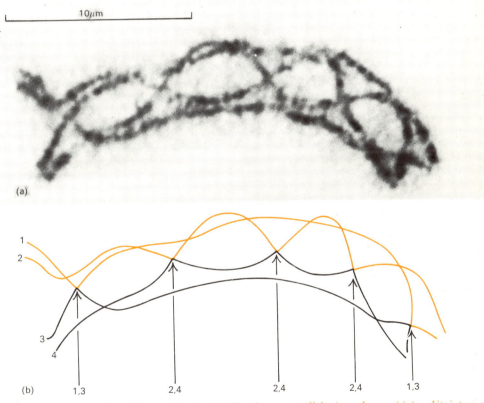

FIG. 5. The structure of a chiasma in a diplotene bivalent of *Chorthippus parallelus* (grasshopper) (a) and its interpretation in terms of chromatid organization (b). In (b), 1 and 2 represent one pair of sister chromatids while 3 and 4 represent the sister chromatids of the partner homologue. Any two non-sister chromatids can exchange at a chiasma so that in successive chiasmata the non-sister chromatids involved may be alike or dissimilar.

9

exaggerated form in animal oocytes. An equivalent phase occurs also in spermatocytes though in a much less pronounced form (Fig. 4).

At diplotene in most meiotic sequences it can be seen that, at one or more points along their length, two of the four chromatids, one from each homologue, lie in a criss-cross, X-shaped, alignment (Fig. 5). These configurations are called *chiasmata* (Greek *chiasma* = cross). The segments on opposite sides of a chiasma then rotate so that they come to lie in planes at right angles to each other (*diakinesis, prophase-I*). In this way each criss-cross configuration is transformed into an open cross. Chromatids continue their cycle of contraction during diakinesis and so acquire a smoother outline. The shape assumed by a particular bivalent depends on the position of the centromere and on the number and positions of the chiasmata (Fig. 6). Bivalents with a

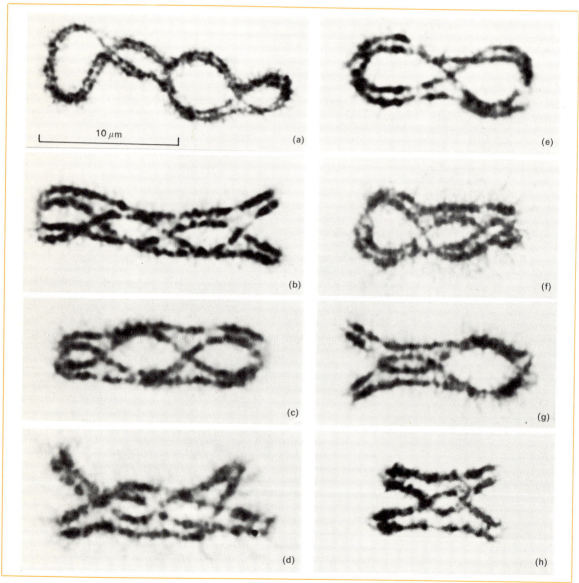

10 μm

(a) (b) (c) (d) (e) (f) (g) (h)

FIG. 6. Some common diplotene bivalent types involving multiple chiasmata. The drawings alongside illustrate their interpretation in terms of an exchange mechanism between non-sister chromatids: (a) 5 chiasmata; (b), (c) and (f) 4 chiasmata; (d), (e) and (g) 3 chiasmata; (h) 2 chiasmata. (T = terminalized chiasma.)

single chiasma appear as open crosses (Fig. 13). Where two chiasmata are present the bivalent is ring-shaped. Three, or more, chiasmata lead to the production of chain-like configurations in which successive loops lie in planes at right angles to one another. Finally, where the initial site of chiasma formation is close to the end of the bivalent, the chiasma may slip to the end of the chromosome (*terminalization*) as the bivalent opens out so that

homologues come to be associated by terminal affinity.

The disruption of the nuclear membrane and the coincident formation of the first division spindle signals the orientation of the bivalents within the spindle system (*pro-metaphase I*). Each bivalent has two pairs of sister centromeres. These orient in such a way that each bivalent establishes a bipolar arrangement. The result is that at metaphase I,

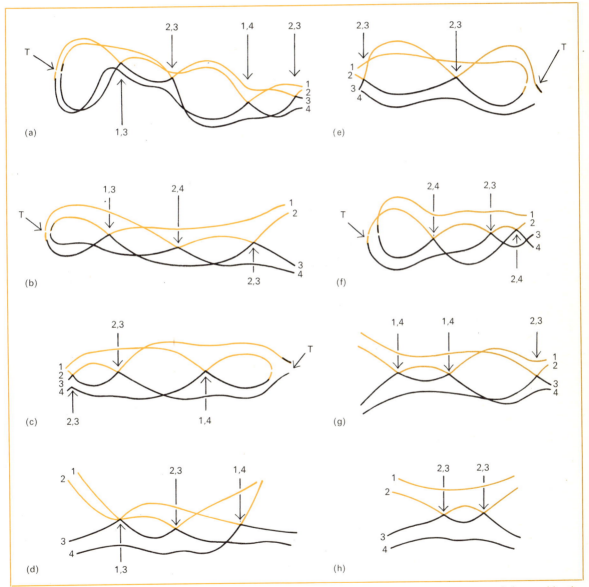

11

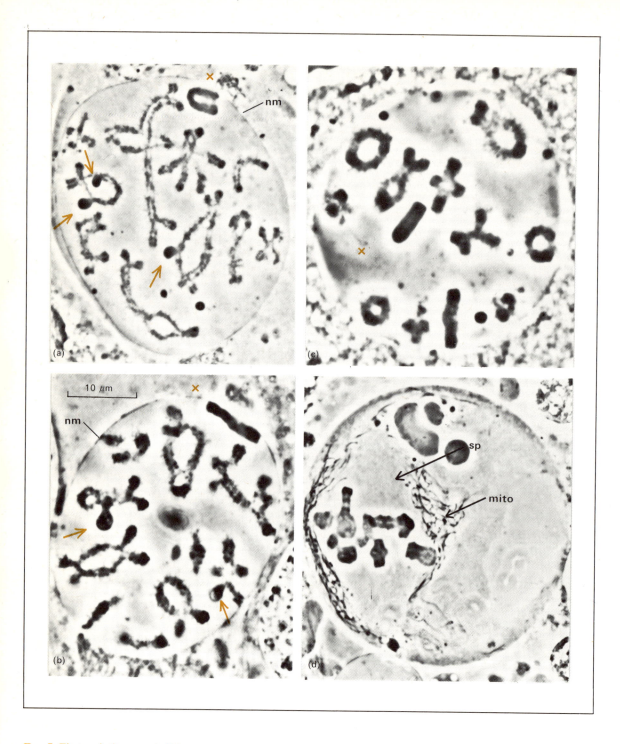

FIG. 7. First meiotic stages in living spermatocytes of the desert locust *Schistocerca gregaria* ($2n = 2x = 23$) as seen with phase contrast microscopy, nm = nuclear membrane: (a) early diplotene (b) full diplotene (c) diakinesis and (d) first metaphase; note how the mitochondria (mito) here lie on the surface of the spindle (sp) and so serve to define it. Note also the nucleoli (arrows) which are organized by and still attached to two of the autosomal bivalents.

pairs of centromeres lie equidistant above and below the spindle equator in a stable co-orientation (Fig. 7). Bivalents occasionally malorient when they first attach themselves to the spindle so that both centromere pairs of a given bivalent are directed to the same pole. This mal-orientation is unstable and usually one member re-orients to produce a stable bipolar orientation (Fig. 8). Such re-orientation must involve the loss of old spindle attachments and the formation of new ones. Nicklas has managed to induce unipolar orientation in living grasshopper spermatocytes by micromanipulation. Under these circumstances normal bipolar orientation follows within minutes of release from induced malorientation and is identical with re-orientation which occurs naturally.

Between diplotene and first metaphase, sister chromatids maintain a close parallel alignment, except, of course, at chiasma points. This association is lost after congression. The chromatids move apart and so resolve the chiasmata which, up to this point, have served to hold homologues together. Chromatid-pairs are still held together in the centromere region, however. The lapse of chromatid association allows the poleward movement of centromeres (*anaphase-I*) so that the haploid number of half-bivalents passes to each pole, each half-bivalent consisting of two diverging chromatids. It cannot be claimed that this first meiotic division reduces the number of chromosomes, because its products contain as much chromosome material as those of a mitotic division and each genetic locus is still represented twice. In fact the only quantitative difference between the products of the first meiotic division and an equivalent mitosis is that in the former homologous chromatids are held together in pairs.

Telophase, *interkinesis* and *cytokinesis* are optional stages in the first meiotic sequence. In animals a nuclear membrane is usually formed at telophase-I and the nucleus returns to the interphase state. Some plants behave in this way too, at least at microsporogenesis (*Tradescantia*, *Zea*), but in others the first anaphase chromosomes pass directly into the prophase of a second division. Even where interkinesis does occur there is no S-phase, so that the significance of this interphase period must be related to the RNA synthesis which can be demonstrated autoradiographically at this time.

Whatever procedure is followed, a second division follows quickly on the first. It is commonly claimed that this is mitotic in character. There are, however, two significant differences. First, the chromatids are widely splayed, due to the lapse of association between sister chromatids at first anaphase. This condition is never found in mitosis though it is a common consequence of colchicine treatment (c-mitosis). Second, the occurrence of chiasma formation between non-sister chromatids at first anaphase means that, while the two chromatids of any one half-bivalent are homologous, they need not be genetically identical, since it is customary to interpret each chiasma as the result of an exchange between two non-sister chromatids within the four-stranded bivalent.

At the end of anaphase-II four haploid nuclei are produced in which each chromosome is represented by a single chromatid. Thus, by two successive divisions, meiosis partitions the four chromatids of each bivalent into four separate cells.

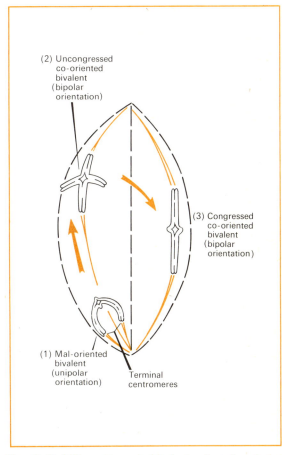

(2) Uncongressed co-oriented bivalent (bipolar orientation)

(3) Congressed co-oriented bivalent (bipolar orientation)

(1) Mal-oriented bivalent (unipolar orientation)

Terminal centromeres

Fig. 8. Stability patterns in bivalent orientation during first prometaphase of meiosis (after Nicklas).

The biochemical basis of meiosis

Mitotic and meiotic cycles differ not only in their morphological manifestations but in the time and pattern of DNA replication and histone* synthesis. As we have already seen, DNA synthesis occurs during late pre-meiotic interphase or else at early leptotene. There is also evidence, however, for a further small but quite distinct period of DNA synthesis during zygotene-pachytene. In the lily, where this prophase synthesis was first discovered by Hotta and Stern, the amount of DNA synthesized represents only about 0·3 per cent of the total and can be partitioned into two distinct phases:

(1) The DNA made during zygotene has a comparatively high GC content (Fig. 9). It has been suggested that sites of terminal DNA synthesis within individual replicon units (Oxford Biology Readers 26) are involved during this phase and that sister chromatids do not become distinguishable until their replication is complete.

(2) During pachytene there is an additional and separate synthesis of DNA amounting to about one-third to one-fifth that produced at zygotene. This DNA does not have distinctive buoyant density characteristics and is presumed to be concerned with DNA repair events at chiasma formation (see page 26).

Periodicities are also evident with regard to RNA and total protein synthesis. In fact variations in the activities of a variety of enzymes have now been recorded during the meiotic sequence and the interphase period which precedes it (Fig. 10). However, though the phases of meiosis can now be described in terms of biochemical events the mechanism, as yet, cannot.

The pattern of histone synthesis during the meiotic cycle is also characteristic. Ordinarily, DNA and histone are synthesized concurrently in the mitotic cycle. An uncoupling of DNA and histone synthesis has, however, been reported during pre-meiotic interphase in the male of the house cricket, *Acheta domestica*, where histone synthesis is not completed until pachytene-diplotene (Fig. 11).

Two additional biochemical features of meiosis are worth mentioning. First, a selective amplification of the DNA which comprises the ribosomal cistrons* (Oxford Biology Readers 16) is known to occur in the oocytes of many organisms without disturbing the normal meiotic behaviour of the chromosomes. The rDNA so formed becomes detached from the chromosomes and later serves to generate a stockpile of rRNA. This is stabilized and stored in the egg cytoplasm for use in the earliest stages of development. Second, stores of mRNA are also developed for the same purpose and this is why the first prophase chromosomes of animal oocytes show such an extensive lampbrush organization. It follows, therefore, that the early stages of animal development are largely under the control of the maternal genotype.

The genetic consequences of meiosis

The orientation behaviour of non-homologous centromere pairs at first division and of individual centromeres at second division is normally random. This means that, in an organism like *Drosophila melanogaster* with only four pairs of chromosomes, all paternal centromeres, and, consequently all maternal too, will pass to the same pole once in every eight (2^3) first anaphases. In man with twenty-three chromosome pairs it will happen only once in 4 194 304 (2^{22}) times. By the same token when four pairs of chromosomes participate in meiosis sixteen (2^4) different types of haploid cell will be produced with respect to entire chromo-

Stage	Molar base composition of ^{32}P-labelled DNA			
	C (Cytosine)	G (Guanine)	T (Thymine)	A (Adenine)
Pre-meiotic S	20	22·8	30·2	27·0
Prophase S				
(a) Zygotene S	26·1	27·0	25·0	22·7
(b) Pachytene S	25·3	25·3	25·3	24·1
Total DNA	20·2	21·4	29·6	28·9

Fig. 9. Patterns of DNA synthesis in lily microsporocytes (data of Hotta and Stern).

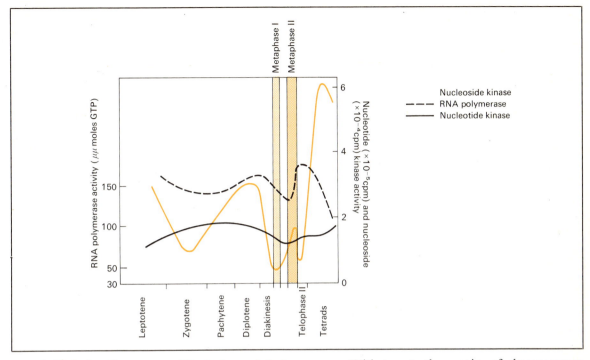

FIG. 10. Variations in enzyme activity during meiosis in the microsporocytes of *Tulipa gesneriana* (data of Hotta and Stern). (RNA polymerase is a specific enzyme which catalyses the formation of RNA from the 4′ ribonucleoside 5′ triphosphates (ATP, GTP, UTP and CTP). Kinase enzymes catalyse the synthesis of the nucleoside triphosphates from the corresponding monophosphates (nucleoside kinases) or the synthesis of the nucleotide monophosphates from the corresponding heterocyclic bases (nucleotide kinases).)

Division stage	Mean amounts in arbitrary units	
	DNA	histone
Spermatogonia		
G₁, pre-synthetic	108±5	103±4
G₂, post-synthetic	225 ±6	205±5
Leptotene-zygotene	233±2	156±3
Pachytene	260±3	196±3
Diplotene-diakinesis and metaphase-I	257±8	234±4

FIG. 11. The relationship between DNA and histone synthesis at male meiosis in the house cricket *Acheta domestica* (data of Bogdanov).

* Basic proteins found in the nucleus.
* A cistron is the series of nucleotides responsible for the amino acid sequence of a single polypeptide.

somes. With twenty-three pairs of chromosomes the number of possible gametic chromosome combinations is $2^{23} = 8\,388\,608$. Superimposed upon this is the fact that when paternal and maternal partners exchange segments by crossing-over, the chromosomes which separate at second anaphase represent new combinations of maternal and paternal segments (Fig. 12).

Thus, provided there are genetic differences between the gametes which initially fuse, the random segregation of centromeres and recombination through chiasma formation ensure that the haploid cells which result from meiosis will have different combinations of the gene forms, brought together by fertilization (Fig. 13). The sequence of meiosis therefore serves to:

(1) Reduce the chromosome complement from the diploid to the haploid number. The biological significance of this is the *segregation of allelic differences*.

(2) Segregate non-homologous centromeres randomly into different nuclei. The biological significance of this is the *recombination of non-allelic differences on non-homologous chromosomes*.

(3) Provide for the exchange of segments between homologous chromosomes. The biological significance of this is the *recombination of non-allelic differences between homologous chromosomes*.

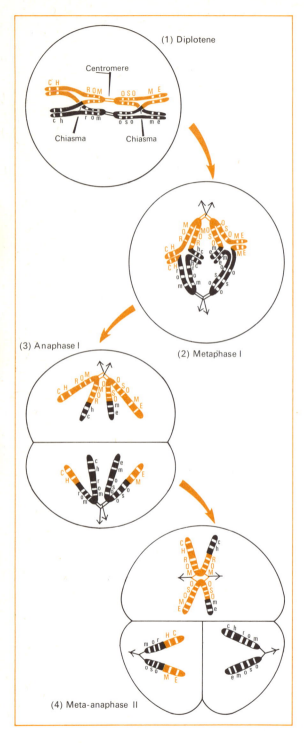

(1) Diplotene

Centromere

Chiasma Chiasma

(2) Metaphase I

(3) Anaphase I

(4) Meta-anaphase II

FIG. 12. The genetic consequences of chiasma formation in a single bivalent.

The genetic control of meiosis

Meiosis is a complex integrated sequence characterized by a spatially well-ordered and a temporally well-regulated succession of events. The behaviour of the chromosomes in this sequence must depend, in part, on their structural properties. But it depends also on the intervention and interaction of particular gene products at particular times. The role of such products can be gauged in bare outline from the fact that mutations of the controlling gene loci lead to disturbances in the meiotic mechanism. Indeed from this evidence it is clear that all the events of meiosis must be under some form of genetic control, since in both plants and animals mutant alleles affect the capacity of the chromosomes to pair, to recombine, to orient and to separate at anaphase I or II (Fig. 14).

The mutations which have major effects on meiosis and have been discovered in experimental material are largely destructive in their influence. This is partly because gross errors are conspicuous and easily detected. However, there are numerous regular inter-sexual and inter-specific variations on the standard meiotic theme which must also be genetically determined. These variants have resulted from selection among the mutant meiotic states of the past; those which proved to be adaptively superior having been favoured. Perhaps the clearest and most dramatic of these is the achiasmate meiosis. Here chiasmata are not formed. There is no inter-homologue genetic exchange and paired homologues remain in parallel alignment until the formation of the first division spindle. Such a system is found in a variety of invertebrates, including representatives of Protozoa, Annelids, Molluscs and Arthropods, though normally it is restricted to one of the two sexes. An achiasmate meiosis has also recently been described in pollen mother cells (though not embryo sac mother cells), of the endemic Japanese plants *Fritillaria amabilis* and *Fritillaria japonica*. In at least three cases (male *Drosophila* and *Phryne*, female *Bombyx*) it has been unequivocally demonstrated that the absence of visible chiasmata is correlated with an absence of genetic exchange between linked genes.

Of course, as with all other aspects of the phenotype, environmental influences, both chemical and physical, are known to modify the action of the genotype as well as to mimic genotypic systems of meiotic control. Thus temperature influences chromosome pairing and ionizing radiations alter chiasma frequency.

16

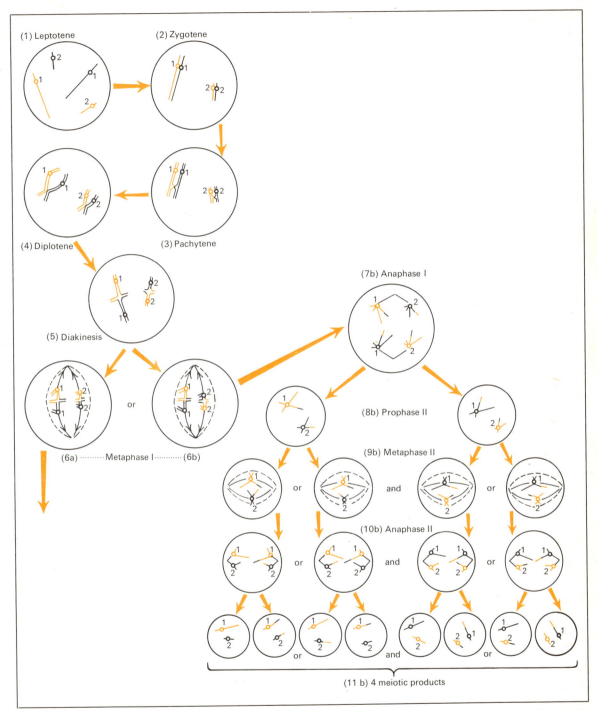

(1) Leptotene (2) Zygotene

(4) Diplotene (3) Pachytene

(5) Diakinesis

(7b) Anaphase I

(8b) Prophase II

(6a) ········ Metaphase I ········ (6b)

(9b) Metaphase II

(10b) Anaphase II

(11 b) 4 meiotic products

FIG. 13. The genetic consequences of random assortment. Two pairs of homologous chromosomes are illustrated; maternal members are coloured, paternal are black. Homologous centromeres are numbered alike. A series equivalent to that developed from stage 6b can also be devized for 6a so that, while only four meiotic products stem from any one cell which enters meiosis, a total of sixteen products can be defined in terms of all the permissible patterns of orientation at first and second metaphase.

17

The pairing process: the synaptonemal complex

At zygotene the seemingly single leptotene threads pair in a specific fashion. This association of homologues has long been one of the major puzzles of meiosis. Electron microscope studies consistently reveal the presence of a tripartite structure known as the synaptonemal complex (SC) at pachytene (Fig. 15). This is composed of one central and two axial or lateral elements. During leptotene the axial elements are unpaired and there is no central element. This central component develops, however, as homologous axial structures come into a pairing relationship. At this stage the axial elements are in close association with laterally displaced microfibrils which are assumed to loop out at right angles to them. An especially interesting feature of the complex is its frequent termination at the nuclear membrane, where dense thickenings often occur at the end of each axial component.

Indeed it is at these membrane-associated sites that central elements first develop. This suggests that the mechanism which aligns the axial elements depends on the capacity of membrane-attached ends to slide into polarized juxtapositions. In animals the basis for this polarity can be defined in relation to the centriole-associated region of the membrane; indeed in most animals this polarity leads to the development of a bouquet polarization (see Fig. 2b) in which the ends of the chromosomes are clustered towards that region of the nuclear membrane associated with the cytoplasmically-located centriolar system.

King has proposed that each chromosome contains special segments, referred to as synaptomeres, scattered along its length, which function at pairing. As the chromosomes shorten and thicken by coiling, adjacent synaptomeres are brought together. As a result the segments between successive

Phase affected	Effect of mutation	Example	
		Organism	Mutant locus
Zygotene pairing	Asynapsis, i.e. complete lack of pairing or else variable and much reduced pairing between homologues	Zea mays Secale cereale	} as
		Drosophila melanogaster	♂ } S8, O81 ♀ } c(3)G, S51, S282
Chiasma formation	Desynapsis, i.e. pachytene pairing complete but homologues subsequently fall apart as univalents	Zea mays	lo_2
Disjunction	Non-disjunction at anaphase I, i.e. homologous chromosomes fail to disjoin as half bivalents	Drosophila melanogaster	♂ } SD, RD ♀ cand
	Non-disjunction at anaphase II, i.e. homologous chromatids fail to disjoin	Drosophila melanogaster	♀ S322a
	Complete failure of second division	Drosophila simulans	♀ ca

FIG. 14. Mutant gene loci known to influence the control of meiosis.

FIG. 15. A longitudinal section through a pachytene bivalent of the ascomycete fungus *Neottiella* as seen with the electron microscope, showing the synaptonemal complex (SC); (a) photograph kindly supplied by Professors M. Westergaard and Dieter von Wettstein. (b) represents a diagrammatic representation of the EM structure (after King).

synaptomeres are forced out of the axial elements and form loops to give the bivalent its fuzzy appearance (the chromatin in Fig. 16). Rod-shaped subunits, termed zygosomes, which are synthesized outside the chromosomes, are then supposed to bind to the synaptomeres and interdigitate with one another to generate the central complex (Fig. 16).

Comings and Okada have recently shown that meiocytes* from mouse or quail testis, dispersed into a single-cell suspension and allowed to spread uniformly over distilled water, can be picked up as monolayers on electron microscope grids. Subsequent treatment of these cells with $25\mu g/cm^3$ of the enzyme DNase for 1–30 min. results in partial to complete digestion of the chromatin fibres (i.e. chromosomal nucleoprotein, see Figs. 15, 16), but the SC persists undisturbed (Fig. 17). The complex is, however, destroyed by trypsin and extensively

* A meiocyte is any cell in which the nucleus divides by meiosis.

damaged by urea and 0·2M HCl. These findings indicate that the SC is composed of protein. As far as the biogenesis of the SC is concerned, von Wettstein has produced evidence that the central region material is assembled in the nucleolus. If, as this suggests, the central element of the SC is common to all bivalents, it must constitute a relatively non-specific pairing protein, capable of moving freely in the nucleus and having access to all the chromosomes. As such it must serve to pull homologous chromosomes into approximate, but not intimate, biochemical alignment. In female *Drosophila melanogaster* it is known that the SC does not develop in oocytes homozygous for a mutant recessive cross-over suppressor in chromosome 3, symbolized as c(3)G (Fig. 14). This implicates the c(3)G⁺ locus in the synthesis of the structural subunits necessary for the formation of the central element.

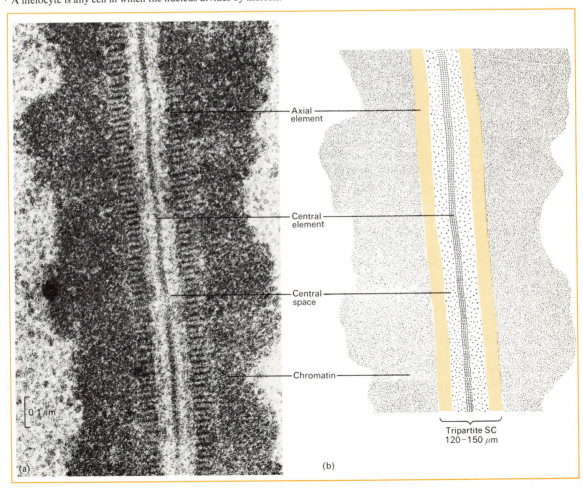

Axial element

Central element

Central space

Chromatin

Tripartite SC
120–150 μm

(a)

(b)

Pairing of partial homologues

Since pairing is normally confined to homologues it cannot be as non-specific as the formation of the central element would lead one to believe. The same argument can be levelled against the concept of synaptomeres which provide no basis for the kind of specificity required for even gross pairing. Restriction of pairing is seen under conditions of partial homology, preferential association being the rule under conditions of competitive pairing. For example, the common bread wheat, *Triticum aestivum* (Fig. 18) is a hexaploid species ($2n = 6x$

$= 42$) in which three distinct sets of 7-paired diploid species have been combined through allopolyploidy* (*AA BB DD*). The three presumed diploid progenitors show considerable genetic correspondence both in terms of chromosome morphology and activity. Consequently the allohexaploid complement can be classified into seven groups, each of three pairs, the chromosomes within each group showing marked similarities in genetic properties. Chromosomes related in this

* The multiplication of chromosome sets from different species involved in hybridization.

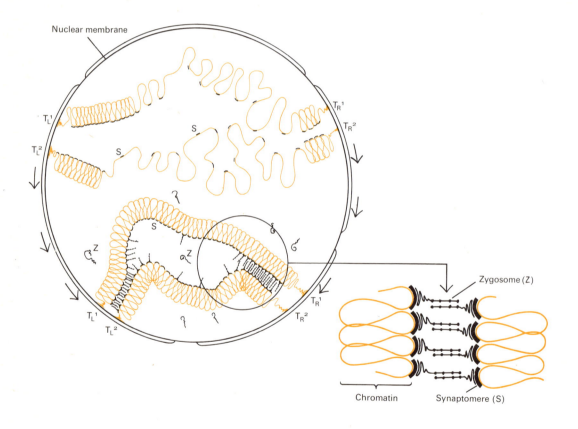

FIG. 16. King's model for the formation of the synaptonemal complex. The right (T_R) and left (T_L) telomeres of a pair of homologous chromosomes are assumed to attach to the inner membrane of the nuclear membrane. The chromosomes are assumed to shorten and thicken because of the folding that results from the association of adjacent synaptomeres (S). Zygosomes (Z) are now released into the nucleoplasm and they attach to the synaptomeres and simultaneously uncoil. Interdigitation of the free segments of the zygosomes leads to the development of the central element of the SC (after King).

FIG. 17. Electron micrographs of DNase-treated whole mounts of the synaptonemal complex obtained from the spermatocytes of the quail (kindly supplied by Dr. David Comings). The DNase treatment has removed essentially all the chromatin. L = axial elements, C = central element, F = fibres extending from the lateral elements and associated with each other to form the central element (after Comings and Okado).

20

way are described as *homoeologues*. Although homoeologous chromosomes are capable of pairing in experimentally-produced diploid hybrids between the presumed progenitor species, they do not do so in *Triticum aestivum*. This absence of homoeologous pairing in the allohexaploid can be shown to be due to the genetic activity of a single chromosome, 5B, and in all probability to a single gene locus on that chromosome. If 5B is absent from the allohexaploid then homoeologues will pair with one another.

Presumably the mutant state which suppresses homoeologous pairing arose after or at the time that chromosome 5B was incorporated into the allohexaploid. Evidently the product of this mutant gene is capable of discriminating between homologous and homoeologous chromosomes. The basis of the discrimination is not known, but Riley has suggested that the genetic restriction which maintains pairing specificity in *Triticum aestivum* can be regarded as discriminating between the sum of the correspondence or difference in the DNA component of homologous and homoeologous elements. Presumably an equivalent basis for discrimination

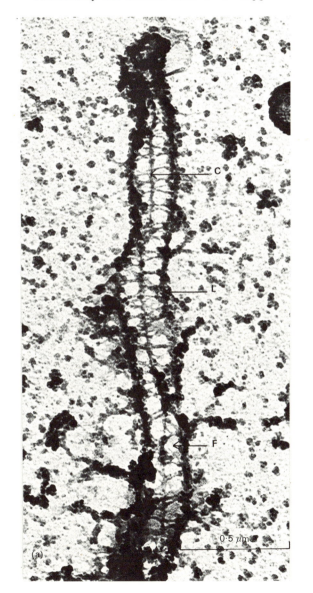

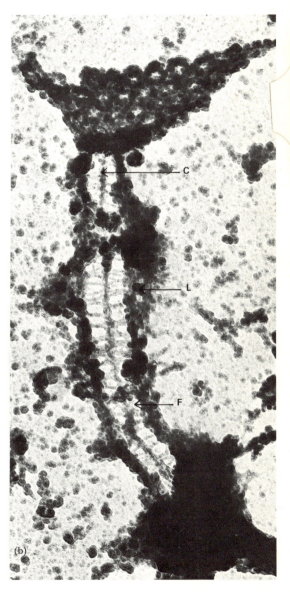

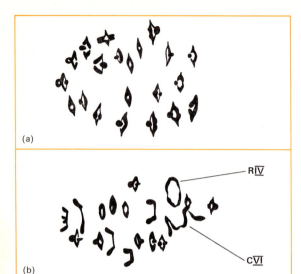

(a)

(b)

RⅣ

CⅥ

FIG. 18. Drawings of first metaphase of meiosis in *Triticum aestivum* ($2n = 6x = 42$). In the normal allohexaploid each chromosone is paired with its full homologue and there are consequently 21 bivalents (a). In a nullisomic ($2n = 6x - 2 = 40$) plant deficient for the 5B chromosome pair there are 15 bivalents and two multivalents (arrows), one of which is a ring of four chromosomes (R IV), the other a chain of six (C VI). These multivalents have resulted from the pairing of homologues and homoeologues. The normal restriction of pairing to full homologues thus results from the genetic activity of a locus in chromosomes 5B (modified after Riley).

must be involved in all pairing specificity since there must be segments with similar sequences of nucleotides in otherwise non-homologous chromosomes among many diploid species, but these do not ordinarily pair at meiosis either. Perhaps the most obvious is the recently discovered repetitious DNA which involves large scale multiplication of particular nucleotide sequences present in all the chromosomes of a given species (Fig. 19). Clearly, while homology is a necessary basis for pairing it is not sufficient to guarantee it. On the other hand a few short sections of SC have been seen in *haploid* pollen mother cells of the tomato, though whether these are indicative of some segmental homology of the type implied by a similarity of nucleotide sequences is not known.

Iso-chromosomes
Spatial considerations are also important, however, even where undoubted DNA homology occurs. Iso-chromosomes are sometimes present in a chromosome complement. These are chromosomes with equal and genetically identical arms which are mirror-images of one another. Hence the *two arms* of such an iso-chromosome can pair to form a ring univalent at meiosis (Fig. 20). Of particular importance, however, is the fact that when two such iso-chromosomes are present such internal inter-arm pairing still predominates. This results in two iso-rings in preference to a single ring bivalent.

Species	Repetitive DNA sequences		
	Percentage of total nuclear DNA	Number of copies	Number of nucleotide pairs per sequence
Calf	38 5	66 000 1 000 000	17 000 (150)
Guinea pig	5·5 2·5 2·5	2200 160 000 500 000	80 000 500 150
Mouse	25 10	(1000–10 000) 1 000 000	. . . 300
Human	3 15 10	(300) 40 000 300 000	400 000 10 000 1000
Sea urchin	20 10 3	(50) 1200 14 000	3 000 000 60 000 1600

FIG. 19. The nature and extent of some repeated DNA sequences. The values in parenthesis denote uncertainty (data of Britten and Davidson).

Temporal considerations

If meiocytes of lilies are explanted from anthers during or immediately after premeiotic S they revert to normal mitotic division after several days in culture. By explanting cells at successively later stages, following premeiotic-S but prior to leptotene, Hotta and Stern have shown a progressive commitment to meiosis. Thus, cells explanted just before leptotene enter leptotene after a few days in culture. Pairing during pachytene may then be partial or complete but no chiasmata appear at diplotene. Cells explanted at the beginning of leptotene may show a mixture of chiasmate bivalents and univalents while cells explanted later form chiasmata regularly. Deoxyadenosine (ADR) inhibits DNA synthesis preferentially, and explanted meiotic cells respond to ADR only during the zygotene-pachytene period. Explanted cells treated with ADR prior to the initiation of pairing do not develop an SC.

Explanted cells exposed to cycloheximide for 2–3 days at late zygotene and then returned to an inhibitor-free medium go through what appears to be a normal pachytene with SC formation. But no chiasmata are present when these cells reach the equivalent of diplotene. Taken together these observations suggest that one further function of DNA synthesis at zygotene is to facilitate the formation of the SC through the synthesis of a structurally associated protein. A special meiotic protein with a very high binding affinity for single stranded DNA has in fact recently been found by Hotta and Stern in meiotic cells of the lily. This protein has not been discovered in any of the somatic tissues of lily, and even at meiosis is present only during the early prophase stages. DNA-binding protein is also present in the spermatocytes of mammals. It has been suggested that such protein promotes pairing and crossing-over between homologous chromosomes.

An SC is present in the bivalents of all chiasmate forms so far studied and appears to be at least a prerequisite for chiasma formation. On the other hand the SC does not ensure chiasma formation, since an SC is also present in the achiasmate bivalents of male *Panorpa* (Mecoptera, Insecta) and male *Bolbe* (mantid). Indeed it would appear that an SC is not even necessary for securing the pairing of homologues for no SC is formed in the achiasmate male meiosis of *Drosophila*.

The pairing of homologues at zygotene is followed at the end of pachytene by their mutual

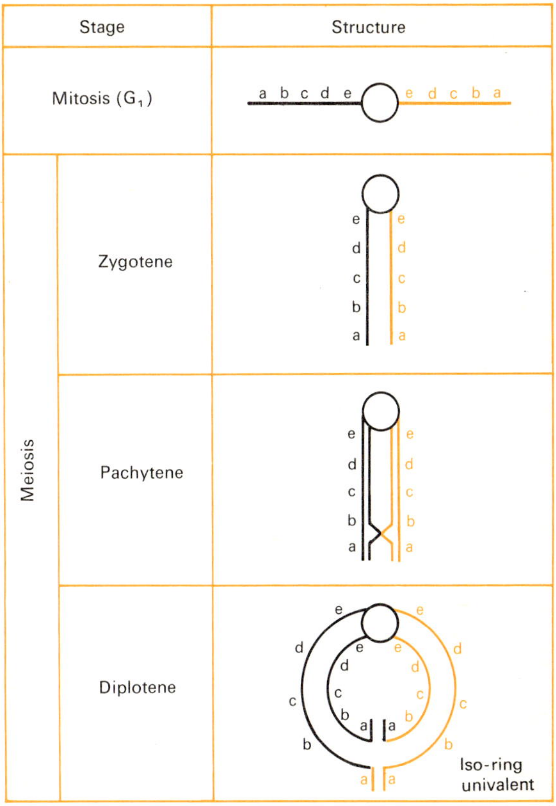

FIG. 20. The meiotic behaviour of iso-chromosome univalents.

separation. At early diplotene too it can be shown that the SC is detached and eliminated from the bivalents. From this it can be assumed that the SC has served its function by the end of pachytene. The released cores may stay intact long enough to pair laterally with one another and so form polycomplexes. No local differences in the SC that might represent sites of crossing-over have been reported, though von Wettstein claims that when the SC is shed it appears to be retained at a few points which he assumes are chiasma sites.

Chiasma formation

It is customary to interpret each chiasma as an X-type exchange between two non-sister chromatids, brought about by breakage and reunion. These events have never been seen, but genetical recombination can be shown to occur at meiosis, and first prophase is the only period during the division when homologues regularly appear to exchange partners. The most positive evidence for this interpretation comes from the recent auto-radiographic studies of Jones on the meiotic chromosomes of *Stethophyma grossa* (Grasshopper). In this species, and in *Parapleurus aliaceous* (Fig. 21), chiasmata are localized proximal to the centromeres and are restricted in number. After incorporation of tritiated thymidine at the last spermatogonial S-phase meiotic bivalents are produced in which each chromosome has one labelled and one unlabelled chromatid. If chiasma formation does involve a reciprocal exchange between homologous non-sister chromatids then proximal localization of chiasmata should be correlated with a proximal localization of label exchange when crossing-over occurs between one labelled and one unlabelled chromatid. Moreover, if chiasma formation involves any two non-sister chromatids within a bivalent at random, then one half of all the chiasmata that do form should generate a label exchange (Fig. 22). And, as predicted, one-half of the label patterns observed by Jones in this species showed the expected label exchange pattern.

The seeming unlikelihood of two breaks occurring at precisely corresponding points in two non-sister chromatids is, however, a persistent problem in the breakage/exchange concept. An examination of ordered tetrads in Fungi in recent years (Oxford Biology Readers 2) has provided evidence that recombination is not a strictly point phenomenon. Rather it appears to be but one aspect of events which extend over a segment of genetic material which, while short in terms of chromosome length, may be quite extensive in molecular terms.

The pairing achieved by the SC is at best relatively crude when viewed at the molecular level. Certainly it cannot be expected to give the intimate molecular pairing required for crossing-over. Indeed, the only process which appears to be capable of producing the necessary degree of precision is base pairing between complementary strands of DNA. Thus, on general principles, it might have been predicted that the annealing of DNA strands would have been involved in some manner in chiasma formation.

Recombination in prokaryotes

In prokaryotes ultra-violet light is known to induce the formation of dimers* between adjacent pyrimidine residues in the same strand of DNA. In these cells enzyme systems occur which are capable of recognizing and excizing these dimers, the gaps formed by excision being repaired by a cut and patch process. Mutants selected as being deficient in recombination (rec⁻) prove to be abnormally sensitive to UV which suggests that the processes involved in recombination in prokaryotes may be the same as those

* Dimers are stable compounds formed by covalent bonding between two similar (T–T, C–C) or dissimilar (C–T) pyrimidines.

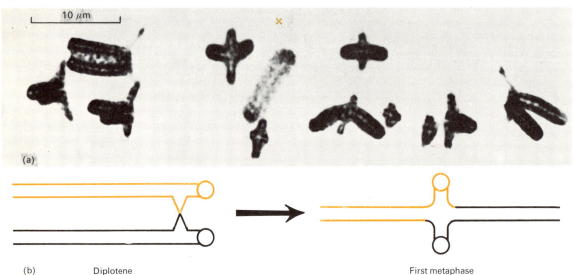

FIG. 21. First meiotic metaphase in *Parapleurus aliaceous* (a) to illustrate localization of chiasmata proximal to the centromere. In (b) the pattern of localization is shown in terms of chromatid behaviour.

24

implicated in the repair of gaps. That is, at the level of the DNA molecule, recombination appears to take place by a combination of breakage of DNA strands, homologous base pairing, repair of single strand gaps and finally the formation of covalent bonds between the recombination fragments. Indeed there is now good evidence that DNA polymerase is a repair enzyme and not a replication enzyme as was formerly assumed. Enzymes capable of breaking DNA strands, the DNases, have been known for some time, though their function in the cell has been obscure. Such DNases are broadly classified into exo- and endo-nucleases respectively according to whether they degrade DNA one nucleotide at a time from a free end or whether they break internal bonds in the molecule. Thus, as far as recombination between DNA molecules in prokaryotes is concerned the following model can be proposed (Fig. 23). Given the occurrence of single strand breaks by endonucleases, exonucleases mediate in the excision of single strand sections of the DNA in the neighbourhood of the break. Partial degradation following breakage is necessary only if the breaks in the homologous DNA molecules are not exactly aligned. This is followed by the formation of joint complexes by base pairing between non-sister polynucleotide columns and then by the repair of single strand gaps by DNA polymerase. Finally covalent bonds develop between the repaired polynucleotide columns under the influence of ligase enzymes.

Event	Behaviour at polynucleotide level
Parental DNA molecules (each a double helix)	
Breakage by endonucleases, ± partial degradation of break ends by exonucleases if break points not exactly aligned, followed by strand separation	
Formation of joint complexes by base pairing between non-sister polynucleotide columns	
Repair of single strand gaps by DNA polymerase (repair synthesis)	
Formation of new covalent bonds by ligase enzymes giving rise to recombined DNA molecules	

FIG. 23. A model to explain recombination between DNA molecules in prokaryotes.

FIG. 22. The correlation between localization of chiasma formation and localization of label exchange as determined by autoradiography in the grasshopper *Stethophyma grossa* (based on the observations of Dr. G. H. Jones). Yellow strands represent tritium labelled chromatids, black strands unlabelled chromatids.

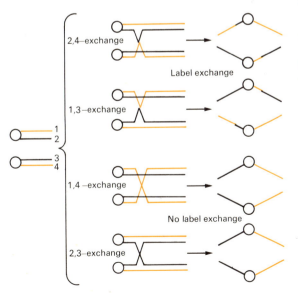

Recombination in eukaryotes
In eukaryotes the lateral unit of recombination is the chromatid so that this simple model can be applied to chiasma formation only if the chromatid can be equated with the DNA duplex. Nevertheless, some facts point to similarities with the proposed prokaryote mechanism:

(1) Tritiated thymidine is incorporated into the chromosomes of human cells in tissue culture following UV-irradiation. In controls there is no such uptake except, of course, by cells in the S-

25

phase. This suggests that treated cells have a mechanism for the repair of UV-induced DNA damage which involves the incorporation of thymine. There is an inherited deficiency of this DNA repair system in the human disorder known as *Xeroderma pigmentosum* where the skin is extremely sensitive to UV-light and numerous skin cancers develop after exposure to sunlight. Tissue cultures of fibroblasts from patients with this disorder do not incorporate tritiated thymidine after UV treatment as efficiently as those from unaffected individuals. This deficiency, which can also be observed *in vivo*, and appears to affect all cells in the skin, is the cause of the skin cancers and supports the existence of a DNA repair mechanism equivalent to that known in prokaryotes.

(2) An infertile human male has recently been described in whom chromosome behaviour at meiosis appears to be normal up to pachytene but few chiasmata and many univalents are present at diakinesis. Fibroblasts from the skin of this individual showed a reduced facility for repairing radiation-induced chromosome breaks. Such a reduction in relation to the repair of DNA breaks during chiasma formation could explain the observed failure in this process in the abnormal male.

Major gene mutants affecting meiotic recombination fall into two categories. In one the synapsis (pairing) of chromosomes is faulty and for this reason the mutants are called asynaptic. In the other, termed desynaptic, chiasma formation does not always follow apparently normal pairing so that a proportion of univalents is present at first metaphase. By comparing the frequency of radiation-induced aberrations in X-irradiated and control seeds of normal (*Ds Ds*) and mutant desynaptic (*ds ds*) genotypes of barley, Riley and Miller found that an increased sensitivity to radiation is associated with a reduced chiasma frequency in the desynaptic mutant. However, Searle, Berry and Beachley using the frequency of X-ray induced rearrangements as markers of radiation sensitivity, have shown that in mice having different chiasma frequencies reduced radiation sensitivity is associated with a lower chiasma frequency. Even so, both results can be interpreted theoretically in terms of enzyme systems which regulate the repair and/or excision process.

(3) As we have already seen (page 14), there is evidence in lily anthers for a small burst of DNA synthesis during pachytene, and Hotta and Stern have suggested that this forms part of a DNA-breakage repair mechanism involved in chiasma formation. In these anthers they have also found an endonuclease capable of producing scissions in the DNA chain, together with a ligase enzyme capable of restoring breaks. The endonuclease appears to reach maximum activity between zygotene and pachytene while the ligase, which is active throughout interphase, falls in activity during leptotene and then returns to a maximum during zygotene-pachytene (Fig. 24).

Models of chiasma formation in eukaryotes

Despite the lack of direct evidence on the biochemical basis of chiasma formation, models have been proposed which involve the removal and replacement of portions of DNA in a hybrid region together with repair synthesis. Two such models with many features in common are currently in vogue (Fig. 25). Both postulate that:

(1) A meiotic chromatid consists of a single DNA-duplex (double helix).

(2) Regions of hybrid DNA are produced which consist of one polynucleotide chain from one chromatid and another from an homologous, non-sister, chromatid. This involves chain separation following breakage and the development of a new pattern of strand association. It also involves a repair system which can recognize and correct

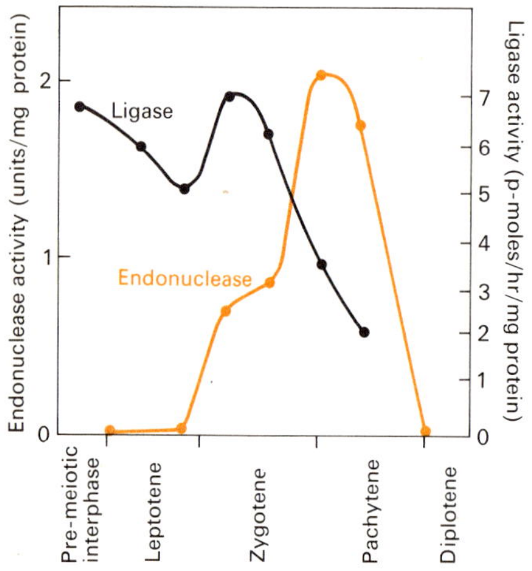

FIG. 24. Endonuclease and ligase activity in microsporocyte extracts of the lily (data of Howell and Stern).

departures from the regular double helix.

(3) A reconsitution of DNA molecules occurs in new ways following breakage, as a result of the lateral association of complementary segments from homologous regions.

The two models differ, however, in their proposals for the origin of the regions of hybrid DNA. In the model of Holliday the initial breakage occurs in DNA strands of like polarity while in that of Whitehouse it occurs in strands of opposite polarity. (The sugar-phosphate backbone of a polynucleotide chain has a polarity defined by the direction of its 3–5 linkages.) They also differ in that the Holliday model does not require DNA synthesis other than as a repair involvement. In the Whitehouse model, on the other hand, DNA synthesis is obligatory because of the need to produce newly synthesized chains.

Of course, it may be that the initial step in chiasma formation is not the breakage of nucleotide chains but rather a failure of such chains to join up at replicon junctions after pre-meiotic DNA synthesis. Thus if the GC-rich DNA synthesized in early zygotene does represent late

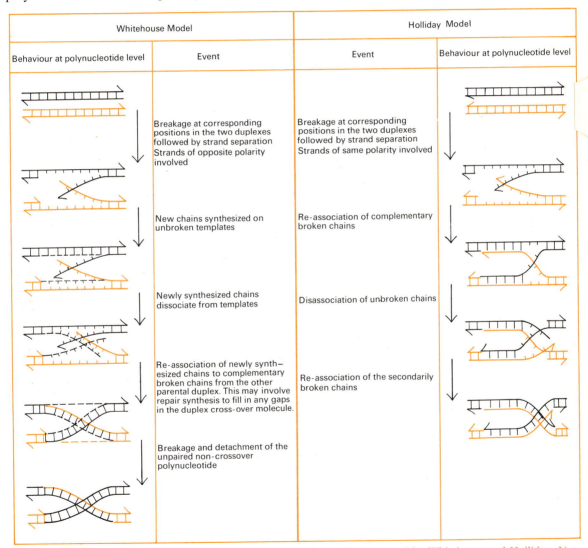

Whitehouse Model		Holliday Model	
Behaviour at polynucleotide level	Event	Event	Behaviour at polynucleotide level
	Breakage at corresponding positions in the two duplexes followed by strand separation Strands of opposite polarity involved	Breakage at corresponding positions in the two duplexes followed by strand separation Strands of same polarity involved	
	New chains synthesized on unbroken templates	Re-association of complementary broken chains	
	Newly synthesized chains dissociate from templates	Disassociation of unbroken chains	
	Re-association of newly synthesized chains to complementary broken chains from the other parental duplex. This may involve repair synthesis to fill in any gaps in the duplex cross-over molecule.	Re-association of the secondarily broken chains	
	Breakage and detachment of the unpaired non-crossover polynucleotide		

FIG. 25. A comparison of the molecular models for chiasma formation proposed by Whitehouse and Holliday. Note that in both models only two of the four chromatids of the meiotic bivalent are shown and each is assumed to consist of a single DNA duplex.

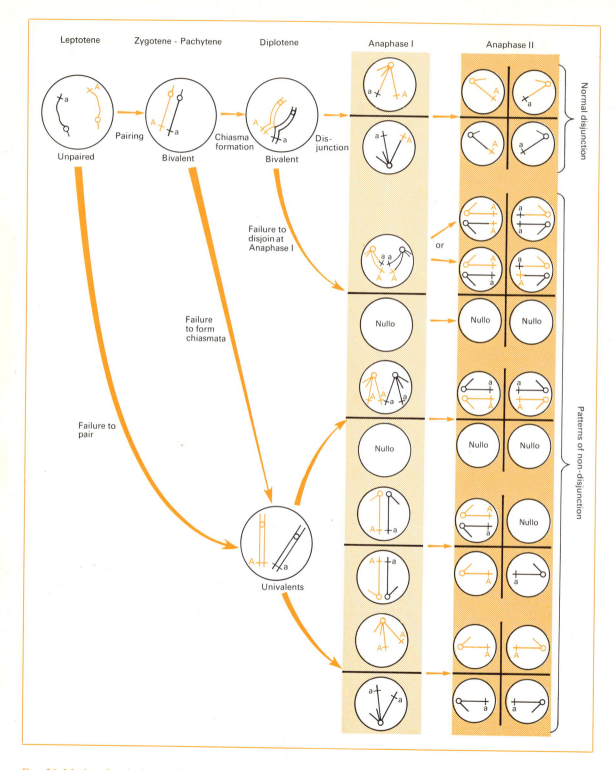

Fig. 26. Modes of meiotic non-disjunction (after Bartales and Baramki).

28

synthesis at chain ends then a failure of such synthesis rather than an actual breakage of polynucleotide columns might be the critical event which triggers the exchange mechanism.

While the elementary events of genetic recombination envisaged in these meiotic models are much the same as those proposed for prokaryotes there must be substantial differences in the overall process. Thus a comparison of the frequency of recombination per nucleotide pair in pro- and eukaryotes makes it clear that recombination has been considerably curtailed in the latter. In fact the recombination frequency per nucleotide pair is approximately a thousand times greater for phage and approximately a hundred times greater for bacteria than for a eukaryote like *Drosophila*. To take a specific example, in the lily there is 1.1×10^{-10}g DNA per meiotic nucleus and about 0.1 per cent (1.1×10^{-13}g) of this is synthesized at repair replication. From this Howell and Stern have estimated that the minimum number of repair-replication sites per meiotic nucleus is in the order of 6.6×10^5. On average some thirty-six chiasmata are present per nucleus in the lily, which in turn implies that the number of single strand breaks involved in successful chiasma formation is 1.4×10^2 which is equivalent to one break per 1.6×10^9 phosphodiester bonds. Whether this low frequency of breakage and rejoining events in eukaryotes depends on undetermined specificities of the breakage-repair enzymes or whether it is a consequence of the elaborate pairing structure, the synaptonemal complex, which is such a distinctive feature of pre-recombination organization in eukaryotes, remains to be clarified.

Segregation

Segregation at meiosis depends, as we have seen, on two events. First, the pairing of homologous chromosomes, or chromosome regions, at zygotene. Second, the formation of chiasmata between paired homologous, though non-sister, chromatids. This ensures that the pairing initiated at zygotene is maintained until co-orientation is complete. Apart from any genetic function they subserve, chiasmata play a vital mechanical role in regulating segregation. Thus, individual chromosomes generally behave independently unless physically connected in some way, and chiasmata normally provide such a physical connection. When pairing and/or chiasma formation fail, homologous chromosomes generally assort at random (Fig. 26). It is true that, in special cases, modes of association other than chiasmata may serve to secure the segregation of homologues. For example in many beetles the sex chromosomes are associated by means of a common and persistent RNA-body reminiscent of a nucleolus. This first appears at pachytene and does not disappear until first anaphase. The mutual association of the sex chromosomes with this organelle provides the basis for their co-orientation and subsequent segregation (Fig. 27).

Cases are also known where particular meiotic chromosomes show non-random orientation, and hence segregation, in the absence of any physical contact between them. Such cases fall into two groups:

(1) *Interactions between unpaired chromosomes in a normal spindle system.* Males of the mole cricket, *Gryllotalpa hexadactyla*, have twenty-two autosomes and a single X-chromosome. One pair of autosomes is regularly heteromorphic consisting of a large (*L*) and a small (*S*) element. At first metaphase of meiosis the *L* and the X invariably orient to the same pole and they move to it independently at first anaphase. No physical connection can be demonstrated between them in living cells when either is moved with a microneedle. If the X is detached from the spindle at pro-metaphase I and moved to the *S*-pole by micromanipulation it re-orients and returns to the *L*-pole. On the other hand when the heteromorphic *L*–*S* bivalent is experimentally detached and inverted it remains in stable orientation. The X, however, re-orients and moves to that pole to which the *L* is oriented (Fig. 28). It would appear, therefore, that it is the capacity of the univalent X-chromosome for re-orientation which determines the non-random segregation of the *L* and *S* members of the heteromorphic pair in this instance.

In many male tipulid flies the two sex chromosomes form a pair of hereditary sex univalents. These tend to move poleward at pro-metaphase I but eventually re-orient and move more or less independently to the equator. By late first metaphase they have taken up adjacent positions on the spindle equator where they remain until the autosomal half-bivalents move to the poles. Then, at late first anaphase, they move from these adjacent positions towards opposite poles, each one, however, maintaining half-spindle fibre connections with both poles (Fig. 29). Dietz has found that if, as a result of a sex-autosome translocation, only one sex univalent is present at first meiosis, it usually

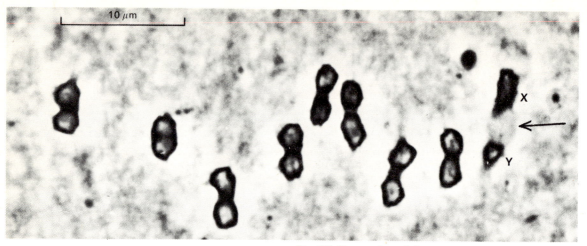

FIG. 27. Segregation of the sex (X and Y) chromosomes by means of a persistent RNA body (arrow) as seen with phase contrast at first metaphase of meiosis in the beetle *Dermestes lardarius*.

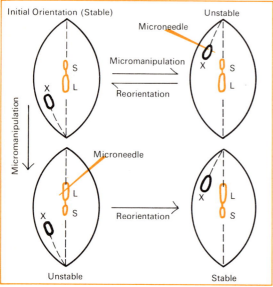

FIG. 28. Orientation stability of the X univalent in the mole cricket *Gryllotalpa* (based on the observations of Camenzind and Nicklas).

remains permanently at the equator. This implies that the segregation of the X and Y chromosomes is determined by an interaction between their spindle fibre systems.

That spindle organization is likely to play an important role in the segregation of non-conjoined elements is suggested also by the behaviour of the giant sex chromosomes found in fleabeetles. Here, as Virrki has shown, the X and Y do not exchange

chiasma; nevertheless they segregate at first anaphase by regularly orientating to opposite poles and so forming a distance-pseudobivalent at first metaphase. This is achieved through the organization of a specialized sex-spindle, separated from the main spindle system by a well defined zone of cytoplasm (Fig. 30). The physical basis of the interdependency of the unconnected sex chromosomes within this sex-spindle is not understood as yet.

(2) *Interactions between unpaired chromosomes in a spindle system with differentiated poles.* In *Sciara* (Diptera) the spermatocytes are disposed radially within a common cyst. The first division spindles are asymmetrical, consisting of a discrete pole located away from the centre of the cyst, and a diffuse pole directed centrally. None of the chromosomes pairs at first prophase and at first metaphase all are arranged with their centromeres apparently oriented to the discrete pole. Despite this, maternally and paternally-derived homologues respond differentially to that pole. Maternal chromosomes remain oriented towards it while paternal chromosomes back away in an asynchronous fashion and are finally pinched off in a small cytoplasmic bud at the centre of the cyst.

At second division a bipolar spindle forms, but now the maternal X responds preferentially to the peripherally sited pole and undergoes non-disjunction, both chromatids passing to the only pole in the second division at which a definitive gamete is produced.

30

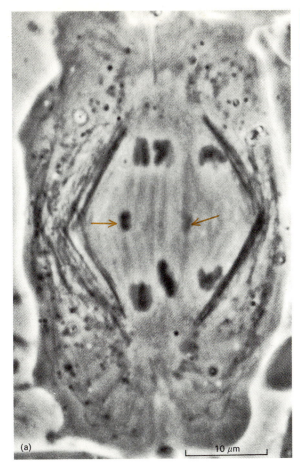

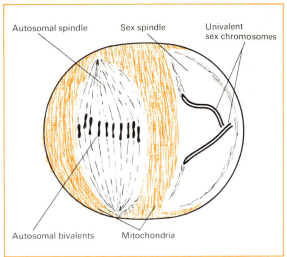

FIG. 30. First metaphase in the fleabeetle *Oedionychu bicolor* to show the dual nature of the spindle system (base on the observations of Dr. N. Virrki).

Labels in figure: Autosomal spindle, Sex spindle, Univalent sex chromosomes, Autosomal bivalents, Mitochondria

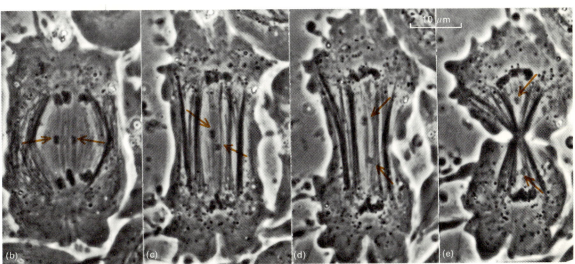

FIG. 29. Anaphase-I behaviour of the hereditary sex univalents (arrows) as seen in living cells of the crane fly *Pales ferruginea* by phase contrast microscopy. There are only three autosomal bivalents in this species. Note how the spindle surface is delimited by the bundles of thread shaped mitochondria (photographs kindly supplied by Dr. Roland Dietz).

31

Variations on the meiotic theme

Minor variations on the standard pattern of meiosis are known not only between species but also within them, between strains, or even between sexes. These, as we have seen earlier, may involve differences in the modes of pairing, of chiasma formation or of segregation adopted by particular chromosomes. More drastic modifications occur in cases where entire groups of chromosomes show an unconventional behaviour pattern. This occurs most commonly where, as a result of evolution, the chromosome complement has been radically reconstructed in terms of structure (complex chromosome-heterozygotes, where multiple rings or chains are regularly encountered at meiosis because terminal segments of non-homologous chromosomes have been reciprocally translocated), or number (polyploids, especially those of hybrid origin). The most dramatic variants are found in species which have dispensed with the conventional sexual cycle and in its place adopted a parthenogenetic pattern of propagation where the organism develops from an unfertilized egg. Under such circumstances one of the principal consequences of meiosis, reduction, must be avoided for clearly the system can only work given the formation of unreduced eggs. Two possibilities present themselves. Reduction must either be suppressed or it must be compensated for in some way.

The simplest situation is that where haploid males arise from unfertilized eggs but fertilization is still required for the production of diploid females. In such a system, which is common in the Hymenoptera, there is no meiosis in the males; rather male gametes arise by an equational division of the haploid set. More complex situations arise where unfertilized eggs give rise to females only. Here there are many ways of avoiding reduction while retaining a semblance of meiosis. Since normal meiosis involves two divisions of the nucleus but only one phase of chromosome duplication, reduction can be avoided by a premeiotic doubling of the entire chromosome set, by suppressing one of the meiotic divisions, by a postmeiotic fusion between two of the haploid meiotic products or even by a fusion of two of the haploid cleavage nuclei at an early stage of the parthenogenetic development of the reduced zygote. Examples of all these types are known. Some of them are, to all intents and purposes, equivalent to asexual systems and provide an alternative means of reproduction where the sexual process is prevented by

failure of fertilization or failure of meiosis. Others conserve those favourable gene combinations which the sexual process has created but which by the very same mechanisms of meiosis and fertilization it tends also to destroy.

FURTHER READING

General

COMINGS, D. E. and OKADA, T. A. (1972). The architecture of meiotic cells and mechanisms of chromosome pairing. *Advances in Cell and Molecular Biology* **2** (in press).

JOHN, B. and LEWIS, K. R. (1965). *The meiotic system, Vol. VIFI, Protoplasmatologia*. Springer-Verlag, Vienna.

KING, R. C. (1970). The meiotic behaviour of the *Drosophila* oocyte. *Inter. Rev. Cytol.* **28**, 125–168. Academic Press, New York.

RILEY, R. (1966). Genetics and the regulation of meiotic chromosome behaviour. *Scient. Prog.* **54**, 193–207.

For reference

CAMENZIND, R. and NICKLAS, R. B. (1968). The non-random segregation in spermatocytes of *Gryllotalpa hexadactyla*. A micromanipulation analysis. *Chromosoma* **24**, 324–335.

DIETZ, R. (1969). Bau und Funktion des Spindelapparats, *Naturwissenschaften* **56**, 237–248.

GALL, J. G. (1969). The genes for ribosomal RNA during oogenesis. *Genetics suppl.* **61**, 121–132.

HOTTA, Y. and STERN, H. (1971). Analysis of DNA synthesis during meiotic prophase in *Lilium. J. Mol. Biol.* **55**, 337–355.

HOWELL, S. H. and STERN, H. (1971). The appearance of DNA breakage and repair activities in the synchronous meiotic cycle of *Lilium. J. Mol. Biol.* **55**, 357–378.

LUYKX, P. (1970). Cellular mechanisms of chromosome distribution. *Suppl. 2. Inter. Rev. Cytol.* Academic Press, New York.

PEARSON, P. L., ELLIS, J. D., and EVANS, H. J. (1970). A gross reduction in chiasma formation during meiotic prophase and a defective DNA repair mechanism associated with a case of human male infertility. *Cytogenetics* **9**, 360–467.

65

2

Oxford Biology Readers
Edited by J. J. Head and O. E. Lowenstein

Using Fungi to Study Genetic Recombination

J. R. S. Fincham

3

Oxford University Press, Ely House, London W.1

GLASGOW   NEW YORK   TORONTO   MELBOURNE   WELLINGTON
CAPE TOWN   SALISBURY   IBADAN   NAIROBI   DAR ES SALAAM   LUSAKA   ADDIS ABABA
BOMBAY   CALCUTTA   MADRAS   KARACHI   LAHORE   DACCA
KUALA LUMPUR   SINGAPORE   HONG KONG   TOKYO

J. R. S. Fincham, F.R.S., is Professor of Genetics in the University of Leeds. He is author of a number of books on genetics including *Fungal genetics* (with P. R. Day) (Blackwell 1963), *Microbial and molecular genetics* (E. U. P. 1965), and *Genetic complementation* (Benjamin, N.Y. 1966).

PHOTOSET AND PRINTED IN GREAT BRITAIN BY
BAS PRINTERS LIMITED, WALLOP, HAMPSHIRE

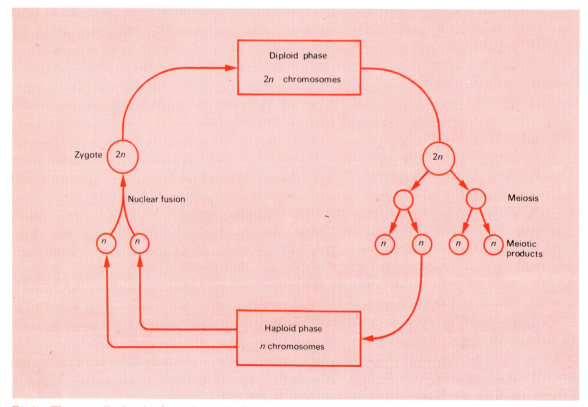

FIG 1.   The generalized cycle of sexual reproduction.

2

Introduction

The essential feature of sexual reproduction is the alternation of haploid and diploid phases. In the haploid phase, represented in higher plants and animals solely by the germ cells and their immediate precursors, the cell nuclei each have a single set of chromosomes, whereas the diploid phase, represented in higher organisms by all the other cells of the body, has nuclei each with two chromosomes of each kind. The transition from haploid to diploid is accomplished by the fusion of sexual cells and their nuclei, while the regeneration of the haploid phase is brought about by two successive nuclear divisions accompanied by only one division of the chromosome set. These two special divisions together constitute the process of meiosis, and the products of a single meiosis are always a *tetrad* of four nuclei, though it may be, as in egg formation in animals and higher plants, that only one member of each tetrad is used for further development. Figs. 1 and 2 summarize the sexual cycle and variations on it found in some important groups of organisms.

The biological importance of meiosis lies in the genetic recombination which it brings about. Not only are the members of different chromosome pairs segregated independently of one another into haploid products, so that free reassortment of unlinked gene differences takes place, but exchanges occur, during the first meiotic division, between corresponding (homologous) parts of the same chromosome pair. Thus, as a result of these exchanges, recombination can occur even between genes in the same chromosome, the recombination frequency depending on the probability of an exchange between the two genes which, in turn, depends on the distance separating them. An important goal of genetics is to understand the process of meiosis and, in particular, the mechanism of exchange between homologous chromosomes.

It is enormously helpful for the study of meiosis if *all four* products of a single meiosis can be isolated and analysed genetically. This is not possible in higher organisms, but it *is* possible in many fungi.

In the Ascomycetes, which are the most convenient group for experimentation, meiosis occurs in a diploid cell called the ascus. The life history of a typical Ascomycete (the yeasts include the sole important exceptions) is predominantly haploid, nuclear fusion occurring only at the time of ascus formation. Without going into the details of fertilization and fruiting, it is also necessary to point out that the haploid nuclei which fuse in the young ascus either can or must (depending on the species) come from different strains and so may constitute a genetic cross. The four cell nuclei resulting from meiosis are either enclosed at once in four ascospores or (much more commonly) they undergo one more division after which the eight resulting nuclei are enclosed in eight ascospores (Fig. 3). In either case the ascospores are held for a while within the ascus wall, before being released.

In the higher Ascomycetes, such as the genus *Sordaria* with which this article is much concerned, numerous asci are produced in a single fruiting structure, and the spores are usually violently discharged in a coherent group of eight ascospores

Type of organism	Diploid phase	Products of meiosis	Haploid phase	Nuclear fusion
Animals	The animal body	Eggs and sperm	Eggs and sperm only	Following fertilization of egg by sperm
Flowering plants	The plant body	Embryo sac and pollen grain	Developing embryo sac and germinating pollen grain	Following fertilization of egg by pollen tube nucleus
Yeast	Free budding cells eventually forming asci	Four ascospores in ascus	Free budding cells	Following fusion of haploid cells
Other Ascomycete fungi	The young ascus only	Four nuclei in ascus dividing to give 8 ascospores	Free living mycelium	In ascus formation

FIG. 2. Comparison of the sexual life histories of different organisms.

3

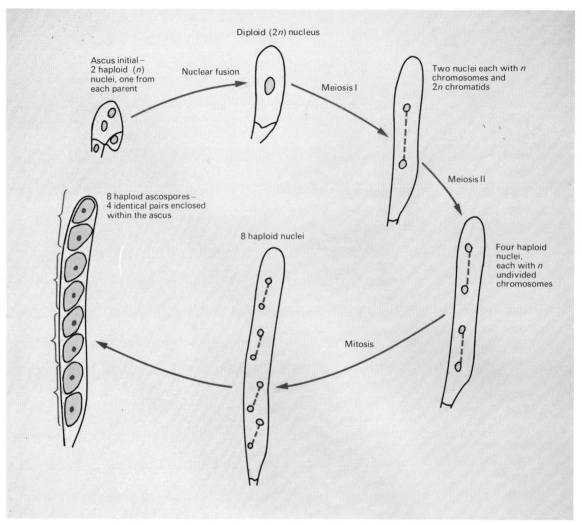

Fig. 3. Development of the ascus in an Ascomycete fungus, such as *Sordaria* or *Neurospora*.

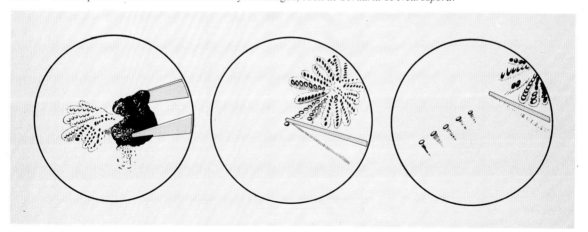

as each ascus becomes thoroughly ripe. By dissecting out the asci from the fruit body shortly before they are due to discharge their spores it is possible to isolate all the spores from a single ascus and to germinate and grow them as eight separate haploid cultures (Fig. 4). It is sometimes convenient, not so much in *Sordaria* as in other useful genera such as *Ascobolus*, to allow the spores to be discharged on to a clean surface and to isolate the groups of eight after discharge.

The special virtue of *Sordaria*, *Ascobolus*, and a few other Ascomycete genera, is that they have pigmented ascospores the colour of which is autonomously determined by their own genes and which can be altered in various ways by specific gene mutations. Thus one can get a good deal of information about the constitutions of the spores with respect to colour-determining genes simply by looking at the colour of the ascospores in the intact asci without having to do any laborious dissection or spore isolation. This convenient feature makes it possible to detect and estimate the frequency of rare as well as common types of recombination, since thousands of asci can be examined under the microscope in the course of a day's work. In the remainder of this article I will describe how this convenient experimental system can be used, first to confirm some of the basic postulates of classical chromosome and recombination theory and, second, to demonstrate very important departures from it.

General confirmation of Mendel's first law

Mendel's first law, the law of segregation, states, in effect, that when the parents of a cross contribute two different forms (*alleles*) of a gene to a diploid zygote, then these alleles will be segregated from one another at the following meiosis, each of the products of meiosis having one allele or the other with equal probability. In terms of chromosome theory this is completely understandable, since each haploid product of meiosis can carry one or other member of a chromosome pair but not both. When, in *Sordaria* or *Ascobolus*, we cross a wild-type strain, producing normal black (*Sordaria*) or dark green (*Ascobolus*) ascospores, with a mutant strain producing pale or transparent ascospores, almost every ascus formed by the cross shows four dark and four light spores.

Fig. 5 shows this result from a cross in *Sordaria brevicollis*. Note that the two nuclei from each of the third (post-meiotic) divisions in the ascus go into adjacent ascospores, so we see four pairs of identical twin ascospores representing the four products of meiosis, collectively known as a *tetrad* (cf. Fig. 3). This kind of demonstration shows more than did the classical Mendelian experiments on peas since the 1:1 ratio of alleles is seen not merely as a statistical regularity but as a result of a strict 2:2 segregation at every, or nearly every, tetrad of spore pairs arising from meiosis.

FIG. 4. Method of ascus dissection in *Neurospora crassa*. The asci are squeezed from the fruiting body (perithecium) with sterile forceps on to the surface of a block of stiff agar gel. The ascospores from a single ascus are dissected out in order with a fine glass needle operated free-hand. The spores are moved to well separated positions on the edge of the block. Each spore is cut out in a cube of agar with a sterile flattened wire and transferred to a separate culture tube for germination.

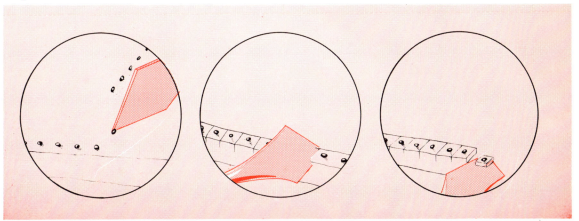

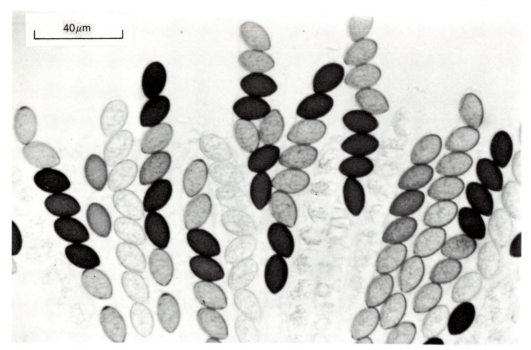

FIG. 5. Asci from a cross between *Sordaria brevicollis* wild-type and a pale-spored mutant (*yellow*). Note the 4 dark (wild-type) and 4 mutant spores in each ascus (apart from immature asci in which all spores are pale). Some asci show first division segregation and some second division segregation (compare Fig. 8). Exceptional asci, showing 6 : 2, 2 : 6, 5 : 3 and 3 : 5 ratios, occur with a low frequency, and cannot be seen here.

Demonstration of the reciprocal nature of crossing-over

According to classical theory, recombination between linked genes occurs by reciprocal and equal exchanges (*cross-overs*) between homologous chromosomes during the early part of the first division of meiosis. A corollary of this hypothesis is that each recombinant product should be accompanied in the same tetrad by another recombinant product of reciprocal constitution. Thus in a cross $a\ b^+ \times a^+\ b$, a^+ and b^+ being two linked genes of the normal, or wild-type, and a and b their respective mutant derivatives, the same reciprocal cross-over as generates a wild-type $a^+\ b^+$ recombinant should form, in the same tetrad, the double mutant $a\ b$ as the reciprocal product.

Fig. 6 shows a rather elegant demonstration of this principle using two mutants of *Sordaria brevicollis* affecting ascospore colour. One mutant, when crossed to itself, produces yellow (*y*) while the other produces buff (*b*) ascospores. The *b* and *y* genes are linked on the same chromosome. A cross was made between the two strains, and the con-

stitution of the resulting asci can be represented as $\dfrac{b^+\ y}{b\ y^+}$, since the chromosome carrying y has the wild-type (b^+) allele of the buff gene, and conversely, the chromosome from the b parent carries y^+. Examination of the figure will show that many asci have two pairs of y and two pairs of b spores—in other words these asci show segregation of chromosomes of the two parental types without recombination. The predominance of this type shows that the genes are fairly close together and do not recombine very freely. In the minority of asci in which other types of ascospores appear there is always one pair of black spores, which are of the wild-type constitution $b^+\ y^+$ and, in the same ascus, a spore pair of a type paler than either yellow or buff and which can, in fact, be shown to be of the double mutant constitution $b\ y$. Evidently the two mutations have a cumulative effect in diluting spore colour when present together. The other two spore pairs are always respectively buff and yellow.

6

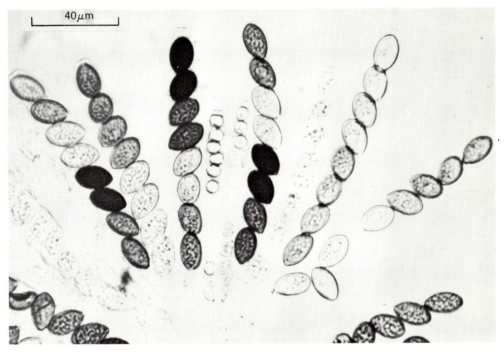

FIG. 6. Asci from a cross of two linked mutants of *Sordaria brevicollis*, *buff* and *yellow*, both giving relatively pale spores. The double mutant shows a cumulative effect of the two mutations and has completely colourless spores. Note three non-recombinant asci with all pale or medium-dark spores (actually 4 buff and 4 yellow) and three others with two reciprocal recombinant spore pairs (wild-type black and double-mutant colourless) as well as *buff* and *yellow* pairs of the parental types. The proportion of asci showing recombination is about 20 per cent of all asci in this cross.

This result, which remains true even when thousands of recombinant asci are looked at, shows not only that crossing-over is reciprocal but also that it affects only two of the four meiotic products. This, in turn, is in harmony with the view, which is actually well substantiated on other grounds, that crossing-over occurs at a time when the chromosomes have already divided into half-chromosomes, or *chromatids*, and involves only one chromatid of each divided chromosome. The four chromatids are afterwards distributed to the four members of the meiotic tetrad as a result of the two successive nuclear divisions. The stages are summarized in Fig. 7. Fungal tetrads certainly afford the most direct and convincing proof of this general principle of crossing-over at the four chromatid stage.

First and second division segregation
Before dealing with exceptions to classical cross-over theory it is worth pointing out that reciprocal crossing-over at the four-strand stage will also explain another feature of genetic segregation in asci which is obvious in Fig. 5. In many asci the four spores in one half of the ascus are the same colour (dark or light) while the four in the other half are of the other type. Thus, counting spore pairs from the tip to the base of the ascus we see dark-dark-light-light or light-light-dark-dark. Looking at Fig. 3 you will see that this means that one of the two nuclei produced by the first division carries and transmits only the wild-type allele while the other carries only the mutant allele. This is expressed by saying that the gene difference is *segregated at the first division*. Some asci, however, show alternating patterns of dark and light spores with the two spore pairs in one half of the ascus differing from each other. There are four possible patterns of this type. Counting spore pairs from the tip of the ascus to the base we can have light-dark-light-dark, light-dark-dark-light, dark-light-dark-light and dark-light-light-dark. All these types of asci occur with about equal frequency, and they indicate that the dark and light

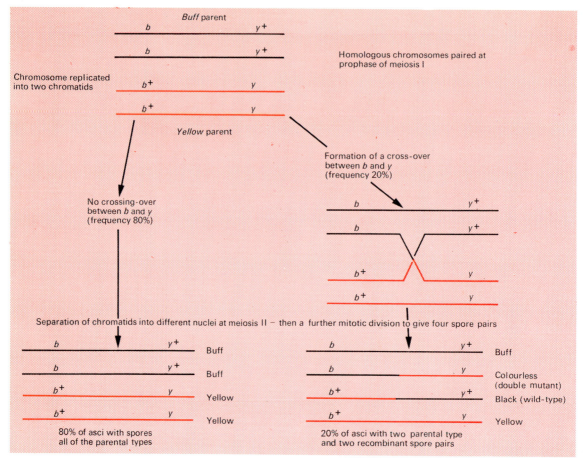

determining alleles segregated only at the *second division of meiosis*.

These two types of behaviour are simply explained by the fact that, although the chromosomes are divided into chromatids before the first division of meiosis, the chromatids of each chromosome are held together in a region called the *centromere*, which does not divide until the second division. Fig. 8 shows that crossing-over between the centromere and the gene whose segregation is being followed will lead to second division segregation; absence of such crossing-over will give first division segregation. Thus the frequency of second division segregation is a direct reflection of the frequency of crossing-over in the gene–centromere interval, and is a measure of the gene–centromere distance.

Exceptions to simple Mendelism: gene conversion
Tetrad analysis, however, continues to support classical genetic rules only so long as fairly small numbers, of the order of a few hundred, of tetrads are looked at. In larger samples it becomes evident that neither of the principles considered above – 2 : 2 segregation of allelic differences and the reciprocal nature of recombination – are absolutely true.

In both *Sordaria* and *Ascobolus* it is found in practically all crosses showing segregation with respect to ascospore colour that, while 2 : 2 segregations (that is 4 : 4 patterns of ascospores) are the *general* rule, 6 : 2 and 2 : 6 patterns also occur with significant frequencies which vary according to the fungal strain and the gene mutation concerned. Fig. 9 shows an example. The frequency of the

8

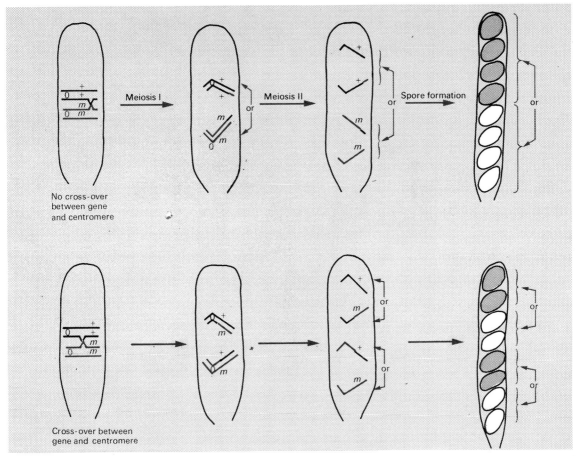

FIG. 8. Explanation of first and second division segregation patterns in asci (compare Fig. 5). The symbol m stands for a mutation giving pale ascospores. Arrows indicate equally probable alternative arrangements.

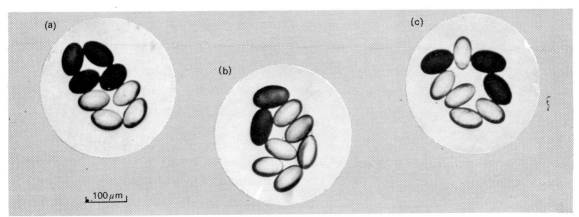

FIG. 9. Examples of aberrant segregations due to gene conversion in a cross of wild-type × pale spore mutant in *Ascobolus immersus*. The ascospores from each ascus have been discharged as a coherent group of eight. (a) The normal segregation of 4 wild : 4 mutant. (b) Aberrant group with 2 wild : 6 mutant. (c) Aberrant group with 3 wild : 5 mutant.

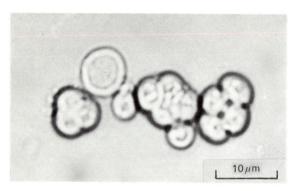

FIG. 10. A yeast ascus. Owing to its small size this ascus cannot be dissected free-hand as in Fig. 4, but requires the use of a micromanipulator.

exceptional asci in *Sordaria* and *Ascobolus* is usually measured in terms of a few per thousand or per ten thousand, but it may be much higher. In yeast, where tetrad analysis can also be performed, though without the help of spore colour mutants, frequencies of 3:1 and 1:3 tetrads (there are only four spores per ascus in yeast, see Fig. 10) commonly amount to several per cent. The *overall* 1:1 segregation among all meiotic products usually holds fairly precisely, since the 3:1s tend to balance the 1:3s, but at the tetrad level the 'exceptions' sometimes become so common as to cease to seem exceptional, and one of the supposedly solid foundations of genetics comes to seem precarious indeed.

This phenomenon of 3:1 or 1:3 segregation where 2:2 is expected is known as *gene conversion*. As the term implies, it is as if an allele present on one chromosome strand at the beginning of meiosis can be *converted* unilaterally, and at frequencies much higher than those of ordinary spontaneous mutations, to the type of allele present on the homologous chromosome. We know that it is a matter of specific conversion rather than of induced mutation of a random kind because, when the change is from mutant to wild-type, it is always a perfectly normal strain of the parental wild-type that results, while, when it is from wild-type to mutant, the resulting 'extra' mutant in the tetrad is always indistinguishable by all available criteria, both genetic and biochemical, from the mutant type that went into the cross.

Extensive studies of gene conversion have established several basic facts. Firstly, as has already been stated, conversion can occur in either direction. The frequency of conversion from mutant to

wild is often (and, in yeast, nearly always) about equal to the frequency in the opposite direction. In some cases in *Sordaria* and *Ascobolus* (for example see Rossignol 1969) the two frequencies are clearly unequal but, overall, there is no general bias for or against wild-type alleles as against mutants. Where a bias exists either direction of conversion may be favoured, wild to mutant or mutant to wild, depending on the specific mutant concerned.

Before any further discussion of the nature of gene conversion it will be as well to remind the reader of the structure of DNA, the material of which genes are made, and to indicate the modern view of the way in which DNA structure relates to gene structure. Fig. 11 shows a simplified picture of the well known double-stranded structure of a DNA molecule, with the complementary pairing of nucleotide bases – adenine paired with thymine and guanine with cytosine – holding the strands together. A gene is a stretch of DNA, probably usually of the order of a thousand nucleotide pairs in length, concerned in some *single* biochemical function. Most genes constitute codes for specific polypeptide chains of proteins, each amino acid being coded for by a sequence of three bases. The commonest type of gene mutation consists of the replacement of one base-pair by another – say adenine-thymine (A-T) for guanine-cytosine (G-C). Every base pair is theoretically able to be changed in this way and so many different mutational changes can be identified at different sites in one gene, all resulting in loss of or change in the normal gene function. Conversion from mutant to wild or from wild to mutant during meiosis can be viewed biochemically as the replacement of a DNA base pair with a differing base pair of a kind 'copied' from the homologous chromosome.

A second fact of great importance, however, is that the replacement of material involved in conversion is not usually confined to a single base pair. Rather it involves, at least frequently and probably always, the removal and replacement of a fairly extensive length of DNA which may include two or more identified mutant sites. This can be readily shown by a cross in which the parents differ at two different sites in the same gene. One of the best examples was provided by Kitani and Olive, using *Sordaria fimicola*. They had two kinds of mutation in one gene affecting ascospore colour. Several mutations at different sites gave colourless (*hyaline* or *h*) ascospores, while the other kind, which was only found once, gave *grey* (*g*) spores, the wild-type being black. It was clearly shown by genetic analysis that the site of the *g* mutation was some little distance away from each of the sites of the various *h* mutations, though both *g* and *h* sites are within the same gene and very close together by ordinary standards. Thus we can represent a cross between *grey* and *hyaline* as:

$$\frac{g \qquad h^+}{grey} \quad \times \quad \frac{g^+ \qquad h}{hyaline}$$

with h^+ and g^+ representing the wild-type situation

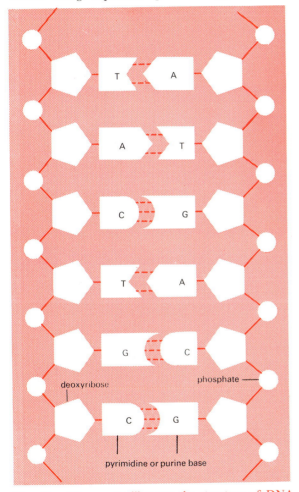

FIG. 11. Diagram to illustrate the structure of DNA and the role played in its structure by the specific fit of pairs of complementary bases: guanine (G) with cytosine (C), and adenine (A) with thymine (T). The DNA strands should in reality be helices wound round one another but are drawn straightened out for the sake of clarity. The commonest kind of mutation cinsists of the replacement of just one base pair (say A-T) by another (T-A, C-G, or G-C). Broken lines represent hydrogen bonds.

at the two sites concerned. If conversion occurs at one site and not at the other we get one of the tetrad types I–IV shown in Fig. 12, depending on which site is involved and which way the conversion goes (the converted allele in each case is printed in green). Note that type II is not distinguishable from absence of conversion since $g^+ \rightarrow g$ makes no further difference to colour when the spore already carries *h*. Tetrad types I, III, and IV are indeed found. In addition, however, it rather frequently happened that the conversion of one site was accompanied by the simultaneous conversion of the other, giving type V or type VI shown in Fig. 12.

Ascus type VI is indistinguishable, without genetical analysis, from type IV, but type V is distinctive and only easily explained on the basis of double-site conversion. This simultaneous conversion of closely-placed sites is a very general feature of gene conversion. Several other investigations, especially on yeast and *Ascobolus*, have shown beyond doubt that the chance of simultaneous conversion at two mutant sites is greater the closer they are together, and that for the most closely linked sites conversion of one without the other is exceptional. This result means that conversion usually involves a stretch of DNA of appreciable length, and it has been calculated by Fogel and Mortimer that this length may be of the order of several hundred nucleotide pairs in yeast.

One final important feature of gene conversion, which I have neglected up to now, but which is immediately obvious in some of the *Sordaria* and *Ascobolus* crosses involving ascospore colour mutants, is that conversion does not always affect *whole* meiotic products, but sometimes only *half* meiotic products. Thus in a mutant × wild (*m* × +) cross (the mutant site being represented as *m*), where the four chromatids at the beginning of meiosis are of constitutions *m*, *m*, +, +, one sometimes gets a result as if only half a chromatid had undergone conversion, the other half retaining its original constitution. The internally hybrid (*m*/+) chromatid goes into one of the four products of meiosis and the *m* and + components are segregated from each other into different ascospores at the post-meiotic nuclear division (Fig. 13). Thus half-chromatid conversion contravenes yet another basic rule of genetics – that segregation is completed by the end of meiosis and that meiotic products are genetically pure. The result, in an 8-spored Ascomycete, is an ascus showing a 5 : 3 or

11

3 : 5 ratio of spores, with one pair of spores, which ought to be identical, showing the segregating difference (Fig. 9c). Asci of these kinds may be as frequent as those of the 6 : 2 and 2 : 6 classes, or even more frequent, especially in *Sordaria*. The other features of whole-chromatid conversion I have mentioned – occurrence in either direction and involvement of an extended length of DNA – are equally characteristic of half-chromatid conversion.

The concept of a chromatid being internally divided into halves, with the possibility that the two halves may be genetically different, is rather a new one in genetics, the conventional view being that the chromatid is always genetically pure. Certainly the easiest way of visualizing the situation is to equate the two halves of the chromatid, revealed by post-meiotic segregation, with the two strands of the DNA. It is widely believed that the genetic material of a chromatid consists essentially of a single DNA molecule or a series of

DNA molecules end to end. This view, though still open to some doubt, is certainly the simplest one. If it is true then the two halves of the chromatid, revealed by post-meiotic segregation, can only be the two strands of the double-strand DNA molecule, and half-chromatid conversion must involve the replacement of a section of just one of the two DNA strands, leaving the other unchanged. We will explore the consequences of this view further at the end of this article.

Connection between conversion and crossing-over
Gene conversion, a non-reciprocal transfer of genetic information, seems at first sight altogether different from ordinary crossing-over which always appears strictly reciprocal. Yet both modes of genetic recombination must depend on a close association of interaction between homologous chromatids and so it would not be surprising if they had some common basis. One question which may be asked, and which has been to a large extent

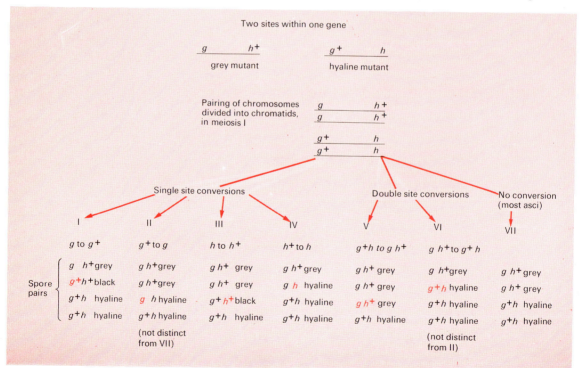

FIG. 12. The types of asci in *Sordaria fimicola* generated by whole-chromatid gene conversion in a cross of the mutants *grey* and *hyaline* which carry mutations at different sites *g* and *h* within the same gene. The constitutions of the four spore pairs of the relevant ascus types are shown and the genetic sites which appear in extra copies as the result of conversion are printed in green. For explanation see text. Ascus types resulting from half-chromatid conversion (leading to post-meiotic segregation and non-matching spore pairs) are not considered, and neither are the rare asci showing apparent reciprocal crossing-over between *g* and *h*.

12

answered by tetrad analysis, is whether the occurrence of conversion within a particular gene is correlated with crossing-over in the same region of the chromosome. The usual way in which this problem has been approached is to make a cross in which the parents differ with respect both to two different mutations within a gene *and* to genetic markers outside the gene but fairly close to it on each side.

Such experiments were carried out by Kitani and Olive (1967, 1969) with the *hyaline* and *grey* mutants of the *g* gene of *Sordaria fimicola*, mentioned above. They used two mutations outside the *g* gene which affected the appearance of the mycelium of the fungus in different ways. One of these, *mat*, lies close to *g* to the left and the other, *corona* (*co*) is close and to the right. Different mutations in *g* were combined with either *mat* or *co* by appropriate breeding operations and crosses were made between strains differing in respect of these mutants as well as in the *g* gene. Asci showing conversion in *g* (i.e. 6 : 2, 5 : 3, 3 : 5 or 2 : 6 ratios for

g versus wild) were dissected, the ascospores were grown separately, and the constitutions of the resulting cultures with respect to *mat* and *co* were determined. The striking result was that conversion at either the *g* or the *h* sites was often, but not always, accompanied by crossing-over between *mat* and *co*, to give meiotic products *mat⁺ co*, *mat⁺ co⁺*, *mat co* and *mat co⁺*. In unselected asci such crossing-over was no more than a few per cent in frequency, *mat* and *co* being quite close together, but among the asci showing conversion in *g* this frequency was nearly 50 per cent. Thus where a conversion occurs there is a greatly enhanced probability of crossing-over in the same region.

The same conclusion has been reached in numerous similar experiments involving different genes and different fungal species, and it applies both to whole-chromatid and to half-chromatid conversion. Where conversion and crossing-over in the same region occur together the chromatid showing the conversion is nearly always one of those involved in the crossover.

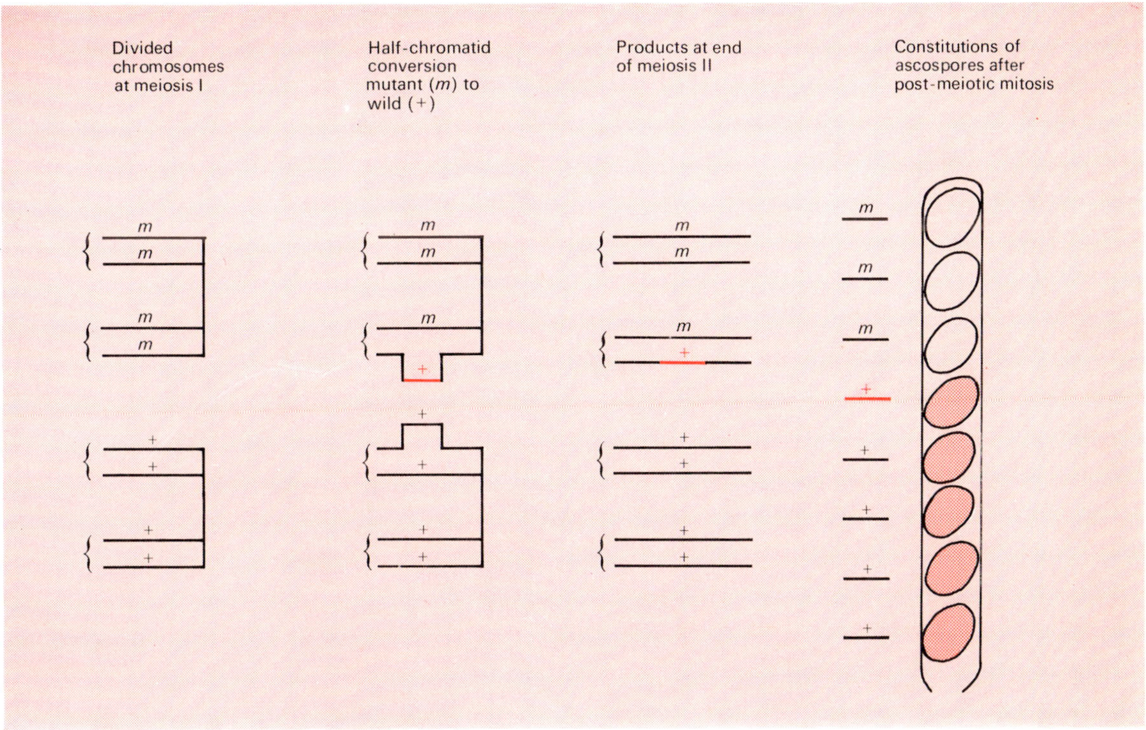

FIG. 13.   To illustrate half-chromatid conversion and post-meiotic segregation. Each chromatid, or unsplit chromosome, is shown as a pair of genetic strands (equivalent to single strands of a double-stranded DNA molecule) bracketed together.

13

An approach to a general theory of recombination

I have concentrated on a few of the most important facts to emerge from tetrad studies on recombination in general and gene conversion in particular, and there are many other complexities. Recombination still remains mysterious in many of its details, and there is as yet no universal agreement even on its basic mechanism. Concentrating, however, on the main conclusions outlined in this article one can interpret them plausibly in terms of a general theory of the kind first proposed by Whitehouse and Hastings and independently by Holliday. One can abstract the common features of the various forms of the theory and state them in general terms as follows. During the early part of the first division of meiosis pairs of homologous chromosomes become associated section-for-section along their lengths, but their molecular structures only interact effectively in a limited number of rather short segments, the number and position of which vary from one meiosis to another. In these short segments, pieces of single-stranded DNA (equivalent to pieces of half-chromatid) become unwound from their parent duplexes, exchanged between homologous chromatids, and rewound so that a certain amount of *hybrid* double-stranded DNA is produced with strands of different parental origins. If the hybrid region includes a site of mutation which results in different base pairs being present in the two parents, there being a normal base pair in one and a different, mutant, base pair at the corresponding position in the other, then there will, in the hybrid, be a pair of mismatching or non-complementary bases which do not fit well together in the DNA structure. This is thought to be the basis of gene conversion, the idea being that there is a mechanism in the cell which 'corrects' such a mismatched base pair by removing a sequence of nucleotides including one member of the pair and replacing the excised sequence with a new one properly complementary to the other strand. This will bring about conversion in either direction depending on whether it is the wild-type or the mutant member of the mismatched pair which is attacked and removed. The absence of correction in a hybrid chromatid would lead to post-meiotic segregation in the corresponding meiotic product and to half-chromatid, instead of whole-chromatid, conversion.

The regions of close interaction in which hybrid DNA is being formed must also be regions of potential crossing-over. Very probably (as supposed by Whitehouse and by Holliday) the exchange of lengths of single-stranded DNA is a first step in the formation of complete cross-overs. If so, the exchange process seems, about half the time, to be abortive so far as crossing-over is concerned but even where crossing-over fails gene conversion can still occur. Fig. 14 sketches one possible scheme. Alternatively, it may be that hybrid DNA formation is not in itself necessary for crossing-over but merely depends on the same sort of close association between chromatids as is necessary for crossing-over. Regardless of the number of steps which conversion and crossing-over may or may not have in common it is still interesting to ask whether *all* cross-overs are likely to have, closely associated with them, the kind of interaction between chromatids (hybrid DNA formation by hypothesis) which may also lead to gene conversion. In fungi rough calculations, which can best be made for yeast, indicate that conversion is quite frequent enough to make it likely that *all* cross-overs are liable to be associated with gene conversion if the genetic markers necessary for revealing it are present in their near vicinity. To put it another way, recombination of a non-reciprocal nature probably occurs in the immediate vicinity of all cross-overs. We will, of course, only detect these non-reciprocal events if we have a gene difference close by the cross-over whose subsequent segregation we can follow. The reason why cross-overs usually appear to be exactly reciprocal is that we are usually looking at genetic markers which are, in molecular terms, far away from the cross-overs. It is evident from the genetic analysis of recombination of close markers that the short-range effects associated with crossing-over are more complex and, in a way, messier than one would suppose from the exactly reciprocal recombination of more widely spaced markers.

To elucidate the various mechanisms of recombination is not within the power of genetics alone. But the purely genetic analysis of Ascomyete tetrads of the type I have tried to outline has succeeded in identifying many of the key questions and even in suggesting possible biochemical mechanisms whose validity can eventually, one hopes, be tested by biochemical means.

14

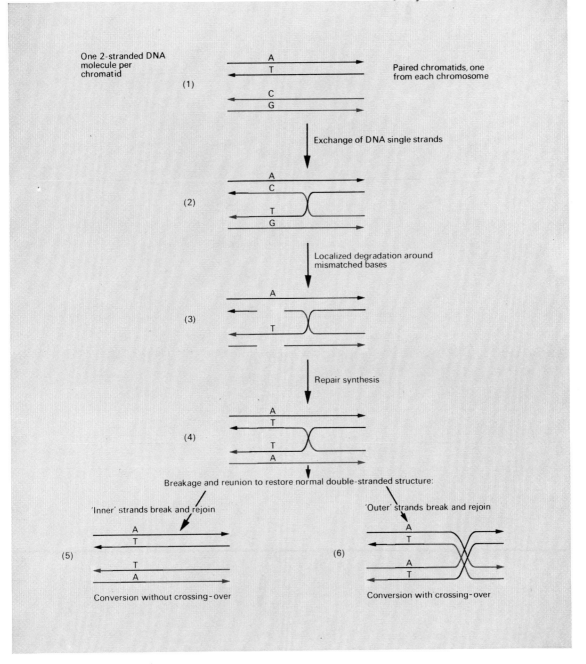

FIG. 14. One possible scheme for the origin of gene conversion and the association of crossing-over with it. Hybrid DNA formation, by unravelling and exchange of single DNA strands (as in (2)), leads to correction of mismatched base pairs, with consequent genetic conversion (3 and 4) and may (5) or may not (6) also lead to crossing-over. The participating chromatids, one from each chromosome, are shown as DNA duplex molecules in which the strands are drawn for the sake of simplicity as parallel lines rather than as coiled in double helices. A genetic difference at one site is shown as an A-T base pair on one chromatid and a G-C base pair on the other. Correction is shown as occurring on both chromatids so as to lead to the same base pair on both, but this is only one possibility. The different colours indicate material derived from, or copied from, chromatids of different parental origin. (After Holliday 1964.)

15

FURTHER READING

References to original papers

FOGEL, S., and MORTIMER, R. K. (1969). Informational transfer in meiotic gene conversion. *Proc. natn. Acad. Sci. U.S.A.* **62**, 96–103.

HOLLIDAY, R. (1964). A mechanism for gene conversion in fungi. *Genet. Res.* **5**, 282–304.

KITANI, Y., and OLIVE, L. S. (1967). Genetics of *Sordaria fimicola*. VI. Gene conversion at the *g* locus in mutant × wild-type crosses. *Genetics, Princeton* **57**, 767–82.

—— —— (1969). Genetics of *Sordaria fimicola*. VII. Gene conversion at the *g* locus in interallelic crosses. *Genetics, Princeton* **62**, 23–66.

ROSSIGNOL, J.-L. (1969). Existence of homogeneous categories of mutants exhibiting various conversion patterns in gene 75 of *Ascobolus immersus*. *Genetics, Princeton* **63**, 795–805.

WHITEHOUSE, H. L. K., and HASTINGS, P. J. (1965). The analysis of genetic recombination on the polaron hybrid DNA model. *Genet. Res.* **6**, 27–92.

More extensive reviews for further reading

EMERSON, S. (1954). Methods in the biochemical genetics of micro-organisms. In *Handbuch der Physiologisch-Chemischen Analyse,* Vol. 2. Springer-Verlag, Heidelberg.

—— (1969). Linkage and recombination at the chromosome level. In *Genetic organization* (eds. E. W. Caspari and A. W. Ravin). Academic Press, New York.

FINCHAM, J. R. S., and DAY, P. R. (1971). *Fungal genetics*, 3rd ed., chapter 8. Blackwell Scientific Publications, Oxford.

Acknowledgements

I thank Dr. Sterling Emerson for permission to use his drawing for Fig. 4, Dr. S. Fogel for the photograph for Fig. 10, and Professor G. Rizet for the *Ascobolus* cross from which the photographs for Fig. 9 were obtained. The *Sordaria brevicollis* mutants used in the cross shown in Figs. 5 and 6 were originally supplied by Dr. H. L. K. Whitehouse and were isolated in his laboratory. Mrs. Barbara Matthews took the photographs for Figs. 5, 6, and 9.

2

25 Oxford Biology Readers
Edited by J. J. Head and O. E. Lowenstein

Gene Expression During Cell Differentiation

J. B. Gurdon

Oxford University Press, Ely House, London W.1

GLASGOW   NEW YORK   TORONTO   MELBOURNE   WELLINGTON
CAPE TOWN   SALISBURY   IBADAN   NAIROBI   DAR ES SALAAM   LUSAKA   ADDIS ABABA
BOMBAY   CALCUTTA   MADRAS   KARACHI   LAHORE   DACCA
KUALA LUMPUR   SINGAPORE   HONG KONG   TOKYO

J. B. Gurdon, F.R.S., is Lecturer in Zoology at the University of Oxford and is a Research Student at Christ Church, Oxford.

Figs 14 and 16 drawn by Derek Whiteley.

FILMSET AND PRINTED IN GREAT BRITAIN BY
BAS PRINTERS LIMITED, WALLOP, HAMPSHIRE

The characteristics of cell differentiation

Cell differentiation is the process by which stable differences arise between the cells of an individual. An essential, and perhaps even diagnostic, feature is the persistence of these differences, over several cell divisions, in the absence of the condition which first brought them about. This distinguishes cell differentiation from comparable phenomena such as enzyme induction and sporulation in bacteria. In these cases, new conditions outside a cell (such as the provision of metabolic substrates or a hormone, or exposure to poor growth conditions) lead to a change in cell composition by the synthesis of enzymes or of materials needed for bacterial spore formation; however, a few hours after the removal of the inducing conditions, the synthesis of enzymes and of spore-forming material ceases. In contrast, once bone marrow or epidermal cells have become committed to blood cell formation and haemoglobin synthesis, or to skin cell formation and keratin synthesis, they and all their mitotic progeny are irreversibly committed to the same kind of differentiation. There is no known way of making a cell committed to one kind of differentiation change into a cell of another kind; cell differentiation seems to involve something more than the induction and repression of enzyme activity, and this is one reason why it is a subject of considerable current interest.

In fact, the recognition of stable cell differentiation is a little less straightforward than has been implied. What is stabilized is the *capacity* to differentiate in a particular way (the 'determined' state), and not the overtly differentiated state itself. The difference between determination and overt differentiation is best exemplified by the celebrated work of Hadorn on *Drosophila* imaginal disc cells (Fig. 1). Larvae contain several separate discs, each consisting of apparently unspecialized cells which are indistinguishable between one disc and another. When discs are passed through the pupal stage to the adult, each disc shows individual differentiation into leg, wing, antenna, etc. In the larva, the disc cells are already determined for a particular type of specialization, and the determined state is stable enough to be propagated by serial grafts in the adult haemocoel for several hundred cell generations (Fig. 1). At any time the determined cells can be made to undergo overt differentiation, though this is not necessary for the propagation of the determined state.

Control of gene expression

There are good reasons for believing that one of the most important aspects of cell differentiation is the control of gene expression. All major differences between cells either involve, or result from, differences in protein content, and proteins are the eventual products of genes. Although some significant differences between cells may involve carbohydrates and lipids, these are usually formed in cells as a result of the action of enzymes, which are themselves proteins and therefore gene products.

Three aspects of control of gene expression are discussed in this article: (1) the level of control—DNA synthesis, RNA synthesis, or protein synthesis; (2) the nature of cell components which regulate gene expression, and the way these come to be distributed unequally in the cytoplasm of a cell so as to lead to differences between its daughter cells; (3) the way these regulatory cell components interact with genes to produce the characteristic stability of cell differentiation.

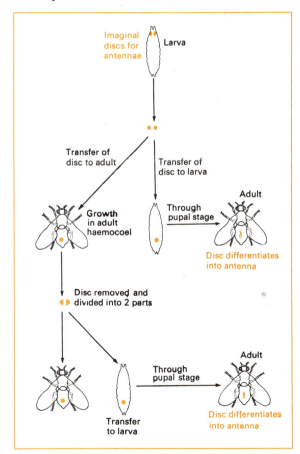

Levels of gene control

Differential replication. One kind of mechanism which could lead to the synthesis of different proteins in different cells is an alteration in the number of genes in their nuclei. There is no reason in principle why red blood cells should not have many times more haemoglobin genes than any other cell-type, or conversely why all other cells should not lose their genes for haemoglobin. Either condition would be sufficient to account for the large amount of haemoglobin synthesized in red blood cells, and for the lack, or undetectably low level, of haemoglobin synthesis in other cell types.

Quantitative changes. The genes in a nucleus can undergo two kinds of change, namely quantitative and qualitative. (Those quantitative changes which involve an equal increase in the number of all genes in a nucleus, as in polyploid cells, are not of much interest from the point of view of cell differentiation, because the relative gene *proportions* for each cell are not altered.) An increase in one gene independently of others is called *gene amplification*, and has been established in two instances. The clearest case is in amphibian oocytes, i.e. growing egg cells (Fig. 2). In young oocytes, multiple copies of the genes which make ribosomal RNA are formed, and these extra genes, together with associated materials, appear in the nucleus as free nucleoli. The proof that these nucleoli really contain extra ribosomal RNA genes lies in the cytological demonstration that they contain DNA, and in the biochemical identification of the extra DNA as ribosomal because it forms specific molecular hybrids with added ribosomal RNA.

The second demonstrated instance of gene amplification is in the nuclei of the salivary gland

FIG. 1. Determination and differentiation. Insect larvae contain paired groups of cells (each group is called an imaginal disc), which differentiate in the adult into various external appendages, such as antenna, leg, wing, etc. In Hadorn's experiments (illustrated), an imaginal disc was removed from a larva and cut into two parts. One part was injected directly into the haemocoel of an adult fly, and these cells continued to grow and divide, but not differentiate; the other part of the disc was injected into another larva which was allowed to pupate and metamorphose into a fly, when the disc cells then differentiated into antennae (now located in the fly's abdomen). This grafting operation can be continued indefinitely and at any time the determined cells can be tested for their differentiative capacity by passage through the pupal stage to an adult. In this way it was shown that the determined state can be propagated for several hundred cell generations. (After Hadorn, 1968.)

3

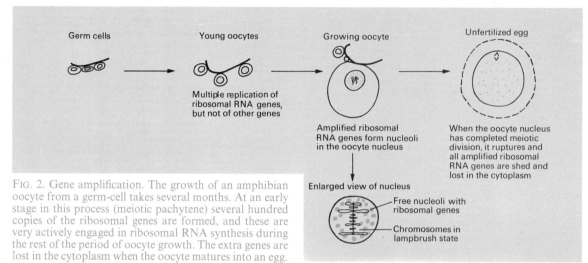

Germ cells

Young oocytes

Growing oocyte

Unfertilized egg

Multiple replication of
ribosomal RNA genes,
but not of other genes

Amplified ribosomal
RNA genes form nucleoli
in the oocyte nucleus

When the oocyte nucleus
has completed meiotic
division, it ruptures and
all amplified ribosomal
RNA genes are shed and
lost in the cytoplasm

Enlarged view of nucleus

Free nucleoli with
ribosomal genes

Chromosomes in
lampbrush state

FIG. 2. Gene amplification. The growth of an amphibian oocyte from a germ-cell takes several months. At an early stage in this process (meiotic pachytene) several hundred copies of the ribosomal genes are formed, and these are very actively engaged in ribosomal RNA synthesis during the rest of the period of oocyte growth. The extra genes are lost in the cytoplasm when the oocyte matures into an egg.

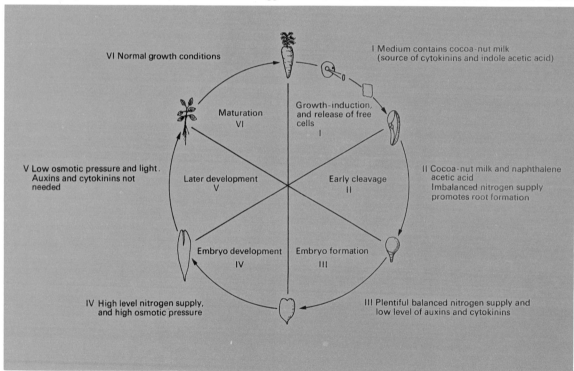

VI Normal growth conditions

I Medium contains cocoa-nut milk
(source of cytokinins and indole acetic acid)

Maturation
VI

Growth-induction,
and release of free
cells
I

V Low osmotic pressure and light.
Auxins and cytokinins not
needed

Later development
V

Early cleavage
II

II Cocoa-nut milk and naphthalene
acetic acid
Imbalanced nitrogen supply
promotes root formation

Embryo development
IV

Embryo formation
III

IV High level nitrogen supply,
and high osmotic pressure

III Plentiful balanced nitrogen supply and
low level of auxins and cytokinins

FIG. 3. The growth of a whole carrot plant from a single phloem cell. Pieces of root tissue are incubated in a basal nutrient medium supplemented with a complex of factors such as those contained in cocoa-nut milk. During incubation, quiescent cells of the storage root are stimulated to grow and proliferate; some free cells are released from the growing explants, and are cultured further in hormone-supplemented media, until eventually a complete flowering plant is formed. At each stage, normal growth and development can be promoted by the appropriate nutrients (e.g. nitrogen supply), hormones (e.g. indole and naphthalene acetic acids), and physical conditions (e.g. light, osmotic pressure). It is important to appreciate that the same conditions can promote different responses when administered to embryos at different stages of development. The correct *relative* concentrations of components must be provided; for example auxins, when supplied in the presence of low nitrogen, can produce abnormal or deficient cotyledonary primordia during organogenesis (stage III). (After Steward, 1970.)

cells of some dipteran larvae. These contain multistranded (polytene) chromosomes, having numerous transverse bands, each of which is believed to mark the location of one or a few genes. These bands may expand, or 'puff' (Fig. 6), and some puffs are associated with DNA synthesis on a larger scale than occurs over most of the chromosome set. It is not known what proteins, if any, are coded for by the genes in the DNA puffs.

It is still uncertain how general gene amplification is. This is because normal chromosomes cannot be studied in such detail as polytene chromosomes, and because no other animal genes can be recognized as easily as those for ribosomal RNA; however, it is known that in the oocytes of some insect species, and in some adult tissues which are particularly active in ribosomal RNA synthesis, amplification of ribosomal RNA genes does *not* take place. Therefore the two definite conclusions which can be drawn at present are: (1) a mechanism exists in animal cells by which some nuclear genes *can* be selectively replicated independently of others, and (2) gene amplification does not account for all cases of differential gene expression during cell differentiation.

Qualitative changes. This category includes changes in the sequence of nucleotide pairs as well as the loss or permanent inactivation of genes. It seems that such changes do not take place in plants during cell differentiation, because the appropriate use of hormones and culture media has made it possible to grow complete plants from carrot phloem cells and from single callous cells of the tobacco stem pith. This eliminates the possibility that any essential genes are lost or permanently inactivated during phloem or stem pith cell differentiation (Fig. 3). In animals it is not possible to grow a complete organism from a single somatic cell (somatic cells are all cells other than sperm, eggs, and germ cells), and qualitative gene changes in these cells have to be looked for by means of *nuclear transplantation.*

This technique (Fig. 4) involves the insertion of the nucleus of an embryonic or specialized cell into an unfertilized egg whose nucleus has been removed or inactivated. If the transplanted nucleus has a normal set of genes (i.e. the same range of genes as a fertilized egg), and if the nuclear transfer operation is carried out successfully, an entirely normal adult animal can result from the combination of egg cytoplasm and the transplanted nucleus. In practice the chance of achieving normal develop-

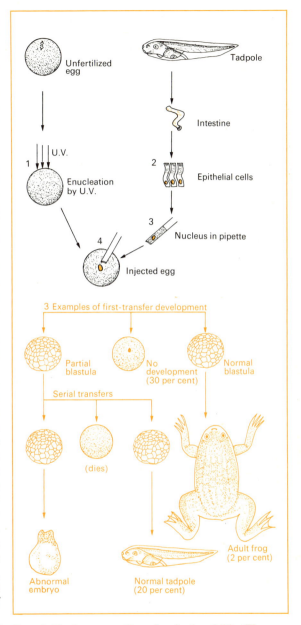

FIG. 4. Nuclear transplantation in Amphibia. The procedure involves (1) the destruction of the egg nucleus by u.v.-irradiation, (2) the separation of donor cells, (3) the rupture of a donor cell by sucking it into a micropipette just too small for it, and (4) the injection of the broken cell and intact nucleus into the recipient cell. Serial transfers are made by separating cells from a first-transfer blastula (10 000 cell stage), and injecting single nuclei into a further set of recipient eggs. This diagram shows the results obtained when the procedure was used with the nuclei of the specialized intestinal epithelial cells of a feeding tadpole. (From Gurdon, 1968).

ment is much increased by using an embryo (blastula), itself the result of nuclear transplantation, for a second set of nuclear transfers; this is called serial nuclear transplantation. If the results of nuclear transfer experiments are to be interpreted with complete confidence, it is essential to use donor nuclei which are genetically distinguishable from nuclei of the recipient egg strain. This provides proof that the nuclei of the transplant embryo are derived from the donor nucleus and not the egg nucleus. In the experiment illustrated by Fig. 4, the donor nucleus came from a mutant strain which possessed one nucleolus per nucleus, whereas the nucleus of the recipient cell was of the 2-nucleolate wild-type strain.

Serial nuclear transfer experiments on the frog *Xenopus laevis* have yielded entirely normal and fertile adult males and females (Fig. 5) after transplantation of nuclei from the intestinal epithelium cells of tadpoles. These cells are differentiated, as is evident from their possession of a striated border, a structure specialized for absorption. In other experiments normal feeding tadpoles have been prepared from nuclei transplanted from adult frog skin cells. Cultured cells were grown out from a small piece of adult skin; just before they filled up completely with keratin, their nuclei were used for single, and later serial, nuclear transfers. The tadpoles and frogs obtained from both types of transplantation contained normal muscle, blood, nerve, lens, and other cell types. The nuclei of these various specialized cells are all the direct mitotic products of nuclei that had previously promoted intestinal epithelium or skin cell differentiation. The general conclusions to be drawn are that the differentiation of a cell does not normally entail the loss or permanent inactivation of genes no longer needed for that type of differentiation, and that differential gene expression is not usually achieved by altering the numbers or kinds of genes in specializing cells.

Differential transcription. If we assume that most cells of an individual have the same numbers and kinds of genes, the next question to ask is whether the genes within the nucleus are transcribed (read into RNA) at different rates. Direct evidence for differential transcription is of three kinds. The first is provided by the polytene chromosomes of the insect salivary glands already mentioned. These have been studied in great detail in various species of Diptera by Beermann, Clever, and others, and the following facts have been established. Several bands are enormously expanded at various times

FIG. 5. An adult nuclear-transplant frog. This is an adult female of *Xenopus laevis*, which resulted from the serial transplantation of a larval intestine nucleus. It was fertile, and after mating with a normal male, its offspring were reared to maturity. (From Gurdon, 1968.)

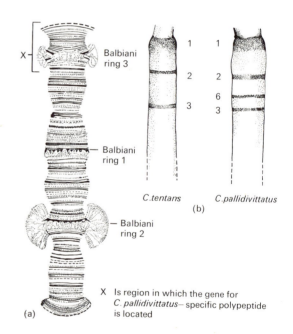

FIG. 6. Cytological evidence for differential gene activity. The left hand figure (a) is a diagram of chromosome IV isolated from the salivary gland cells of the midge *Chironomus pallidivittatus*; transcriptionally active regions of the chromosome are called Balbiani rings. This diagram is based on a detailed cytological study of Beermann. The right hand figure (b) shows an acrylamide gel electrophoresis pattern of protein sub-units extracted from the salivary gland secretion; band 6 is a polypeptide whose synthesis is correlated with the presence of genes at the distal end of chromosome IV where a special Balbiani ring occurs in chromosomes of the salivary gland, but not in chromosomes of other cell types. (After Grossbach 1969).

into puffs or 'Balbiani rings' (Fig. 6), and these are seen by autoradiography to be associated with intense RNA synthesis compared with other bands on the same chromosome. Characteristic changes in the pattern of puffed bands are observed at different developmental stages and in different tissues, and also after treatment with hormones such as ecdysone, as would be expected of differential gene activity. An important point recently established by Grossbach is that one of the puffed bands on chromosome IV of *Chironomus pallidivittatus* is the site of the gene(s) for a salivary protein, and this formally justifies the view, for which there is much other evidence, that puffs provide a uniquely favourable opportunity to observe genes in action.

The second source of direct evidence for differential transcription comes from the biochemical analysis of early development in several species. There are susbstantial variations in the relative rates of synthesis of the major classes of RNA in several species. In amphibian development, for example, RNA synthesis can not be detected for the first few hours after fertilization; then during the next twelve hours, the predominant class of RNA to be synthesized changes from DNA-like RNA (including messenger RNA), to transfer RNA, to ribosomal RNA. Clearly these major classes of genes for transfer, ribosomal, or messenger RNA, can be transcribed at different rates, and this happens during normal development.

The most important question concerning transcription is whether the genes which code for cell-type-specific proteins are differentially transcribed. At present it is very difficult to answer this question conclusively because it is not yet possible to identify the mRNAs for more than a very small number of the many different proteins synthesized in an organism. The best that can be done is to compare the ability of the RNA molecules synthesized in two different kinds of cells to form hybrids with the same sample of DNA. The ability of an RNA molecule to form a longitudinally base-paired structure with a strand of DNA enables different gene messages to be compared. Hybrid formation is a very complex reaction when it is not known how many copies of each gene are present in the sample of DNA, nor how many molecules of each kind are present in the RNA samples, nor the extent to which different mRNA molecules have part of their base sequence in common. The reaction is carried out in such a way that stable hybrids are formed only when the RNA and DNA molecules are closely matched in base sequence, and it is possible to make some allowance for the fact that a significant part of the DNA sample consists, in many species, of repeated sequences that form hybrids very much faster than other rarer genes. Until it is possible to carry out these reactions with purified genes or purified mRNA samples, the results must be treated with caution. However, all of the hybridization experiments which have been carried out so far clearly indicate substantial differences between the RNA populations of different cell-types. It seems almost certain that some of these differences will turn out to include mRNAs. This is the third kind of evidence for differential gene transcription.

In conclusion, differential transcription of genes certainly takes place in some instances; there is also some evidence for, and none against, the view that it takes place in all differentiating cells.

Differential translation. If all cells in an organism had the same genes from which the same number of mRNA molecules were transcribed, it would still be possible for them to synthesize quite different populations of proteins; this would happen if the various mRNAs in each cell were to make protein (be translated) with different efficiencies.

We know that mRNAs in the same cell *can* be translated at different rates. For example, the administration of cortisone to a certain line of cultured hepatoma (cancerous liver) cells can cause a doubling in the amount of synthesis of one enzyme, with little effect on other proteins being made, while all transcription is suppressed. Evidently the hormone stimulates preferentially the translation of one kind of mRNA. The hormone does not induce the synthesis of a protein which the cell was not making before; rather, it increases the rate of translation of a message which is already being translated in that cell. The same situation seems to apply to the molecules released from ribosomes by weak KCl treatment. When tested in cell-free protein-synthesizing systems (consisting of ribosomes, mRNA, etc.), these molecules are message-specific; that is, haemoglobin mRNA is translated much more efficiently with those from blood cell ribosomes than with ones from muscle cell ribosomes, and vice versa for myosin mRNA (Fig. 7). These experiments have led to the view that all cells synthesize messages for many different proteins and that the question of which message will be translated is largely determined by the distribution of message-specific components, such as

Source of ribosomes	Source of ribosome factors	Myosin synthesis (relative percentage)
Muscle (complete ribosomes)	—	100
Muscle (salt-washed)	—	14
Muscle (salt-washed)	Reticulocyte	13
Muscle (salt-washed)	Muscle	62
Reticulocyte (complete ribosomes)	—	0
Reticulocyte (salt-washed)	—	0
Reticulocyte (salt-washed)	Reticulocyte	0
Reticulocyte (salt-washed)	Muscle	28

FIG. 7. Translational specificity of ribosome factors. A cell-free protein synthesizing system was composed of chick muscle mRNA, ribosomes, and ribosome factors (materials washed off ribosomes by 0·5M KCl), as well as other basal components. Myosin mRNA is translated successfully only when ribosome factors from muscle are provided; the source of salt-washed ribosomes (i.e. ribosomes from which the factors have been removed) is unimportant. (After Heywood, 1969.)

factor-bearing ribosomes.

It has now become possible to test this view in living cells, by injecting purified mRNA molecules for rabbit haemoglobin into unfertilized frog eggs, which never normally synthesize haemoglobin, but which do so when injected with this RNA (Fig. 8). The fact that the injected frog eggs synthesize rabbit haemoglobin demonstrates conclusively that there is no mechanism in cells to prevent the translation of messages which they do not normally possess. The clear implication is that any message formed in a cell can be translated, and therefore that translational control cannot itself account for the widely different populations of proteins found in different cells. Instead, translational control seems to be used to modulate to a limited extent the exact amount of protein synthesized by cells already committed to producing certain kinds of proteins.

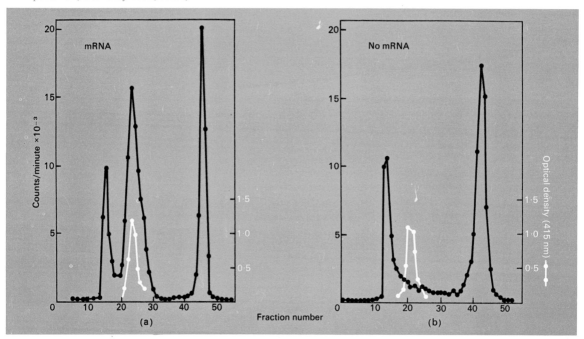

FIG. 8. The synthesis of rabbit haemoglobin in frog oocytes. The mRNA for haemoglobin is extracted from rabbit blood cells and injected into frog oocytes. The ribosomes, transfer RNA, etc., of the frog cell cytoplasm use the rabbit mRNA to synthesize haemoglobin. The figures show the results of passing molecules extracted from oocytes through a Sephadex column, which separates molecules according to their size. Most of the proteins synthesized by oocytes not injected with mRNA are large, in the black peak on the left of each figure; the black peak on the right contains unincorporated ^{3}H-histidine. The coloured peak in the middle of each figure marks the position of rabbit haemoglobin which was added to the extracted oocytes in sufficient quantity to be recognized by optical density at 415 nm. (a) shows that oocytes injected with mRNA have synthesized molecules of the same size (central black peak) and, judging from other experiments, of the same amino acid sequence as the rabbit haemoglobin. (From Lane, Marbaix, and Gurdon, 1971.)

Brief mention should be made of another type of translational control often thought to exist in sea urchin development. A fifteen-fold increase in the overall rate of protein synthesis takes place at fertilization, in the absence of any new RNA synthesis, and it has been suggested that previously formed mRNA molecules are unmasked. However, it has not yet been proved that this activation of protein synthesis affects some kinds of proteins more than others, and unless it does, the mechanism involved would not be of major interest for cell differentiation.

The general conclusion reached so far is that it is mainly at the level of transcription, rather than of replication or translation, that responsibility lies for the synthesis of different kinds of proteins in differentiating cells.

Donor cell nucleus	Recipient cell	Activity of transplanted nuclei
Adult brain $\left(\begin{array}{c}RNA^+\\DNA^-\end{array}\right)$	→ Growing oocyte (RNA$^+$ DNA$^-$)	RNA$^+$ DNA$^-$
	→ Oocyte in meiotic division (condensed chromosomes)	condensed chromosomes
	→ Activated egg (RNA$^-$ DNA$^+$)	RNA$^-$ DNA$^+$
Neurula $\left(\begin{array}{c}DNA\text{-like RNA}^+\\transfer\ RNA^+\\ribosomal\ RNA^+\end{array}\right)$	Activated egg (RNA$^-$) ↓	RNA$^-$ (all kinds)
	Mid-blastula (DNA-like RNA$^+$) ↓	DNA-like RNA$^+$
	Late blastula (DNA-like and transfer RNAs$^+$) ↓	DNA-like and transfer RNAs$^+$
	Neurula (all 3 classes of RNAs$^+$)	All 3 classes of RNAs$^+$

FIG. 9. The cytoplasmic control of nuclear activity. Nuclei from one type of cell are injected into the cytoplasm of another cell-type. After incubation for a few hours, the general nature of nuclear activity in the injected cells is determined by histological, autoradiographic and biochemical methods. Each cell injected with brain nuclei received many nuclei, and for this reason did not develop normally; neurula nuclei were injected singly into eggs which did develop normally. + or − indicates the level of activity, e.g. RNA$^+$ DNA$^-$ refers to a cell whose nucleus is actively engaged in RNA synthesis, but rarely or never synthesizes DNA. (From Gurdon, 1968.)

Identity of substances which control gene activity

Having outlined the reasons for devoting our main attention to gene transcription, we may now ask how this is controlled. There are many possible mechanisms, including one in which changes in the ionic composition of the nucleus, mediated across the nuclear membrane, could affect the likelihood of different genes being transcribed. However, most workers currently favour the view that gene transcription is regulated by large molecules which associate directly with genes on the chromosomes, partly because this type of mechanism has been demonstrated in bacteria, and partly because it suggests more experimental approaches than other ideas.

Kinds of molecules likely to affect gene transcription. Relevant experiments are orientated towards two problems. First there is the problem of isolating the kinds of molecules which seem likely to affect gene transcription, and secondly the problem of testing whether such molecules do in fact control transcription. One approach to the first problem is to isolate substances found in chromosomes or known to be involved in RNA transcription, and this has led to the separation of various kinds of RNA polymerases (the kind of enzyme which transcribes RNA from DNA). Another obvious approach is to investigate conditions known consistently to alter gene expression. There are very few ways of inducing major reproducible changes in gene expression of a kind resembling those known to take place in normal cells; one is to alter the cytoplasmic environment of a nucleus, as happens after nuclear transplantation and cell fusion. In artificially or spontaneously fused cells, two nuclei share the cytoplasmic mixture derived from the parent cells. A rather clear result that has emerged from nuclear transfer and cell fusion experiments of this kind is that gene activity is governed by the kind of cytoplasm in which a nucleus resides. This is evident from experiments in which the nucleus of a cell characterized by one kind of synthetic activity is combined with the cytoplasm of a cell characterized by quite a different kind of activity. For example, nuclei can be switched from DNA synthesis to RNA synthesis and vice versa by injecting them into the cytoplasm of eggs or oocytes (Fig. 9). Changes in transcription can be observed by transplanting an embryonic nucleus which is synthesizing each major class of RNA to an enucleated egg; the transplanted nucleus and its mitotic products at once cease to synthesize RNA, and then go through the sequential changes in the pattern of RNA synthesis characteristic of early development (Fig. 9). Somewhat comparable experiments can be carried out with the cytoplasm of specialized adult cells by fusing cells together with the aid of certain inactivated viruses. When a hen's red blood cell is fused with a cultured mouse cell, a negligible amount of blood cell cytoplasm is incorporated into the hybrid cell, and the blood cell nucleus is in effect transferred to cultured cell cytoplasm. As a result, some of the blood cell genes which have become inactive during blood cell differentiation are reactivated by the growing cell cytoplasm (Fig. 10). The conclusion from all experiments of this type is the same: the cytoplasm tells the nucleus what to do, and therefore presumably contains components capable of directly or indirectly regulating gene transcription.

It is of great importance to remember that nucleo-cytoplasmic combinations of the kinds outlined, especially nuclear transfers to eggs, demonstrate a type of cytoplasmic control that is characteristic of entirely normal cell function. For this reason there is much interest in the mechanism by which cytoplasmic components induce changes in gene activity. Among those that have been described, one that may well be of special importance is the entry of cytoplasmic proteins into nuclei transplanted to eggs, and into red blood cell nuclei fused to other cells (Fig. 11). This passage of proteins may well be closely connected with the induced changes in nuclear activity since proteins do not accumulate in damaged nuclei. It is not known what kinds of proteins enter nuclei, nor whether molecules other than proteins do so, but the phenomenon clearly suggests one way of trying to identify cytoplasmic molecules which influence gene activity. *Tests of molecules likely to control transcription.* The second problem concerning molecules which regulate transcription is that of finding a satisfactory assay system; that is, a way of testing which of the many kinds of molecules sometimes associated with genes do in fact regulate their activity. Two types of assay system have been used with some degree of success. The first is a cell-free system which synthesizes RNA; the one which is most representative of conditions in living cells consists mainly of chromatin—chromosome-like material extracted from isolated cell nuclei—to or from which various molecules are added or removed. Some very interesting results have come out of experiments in which changes in the kinds and

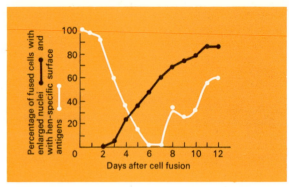

FIG. 10. The cytoplasmic control of gene expression. A cultured mouse cell is fused with a hen erythrocyte under conditions which eliminate the erythrocyte cytoplasm, so that the hen nucleus is transferred to the cultured mouse cell. The erythrocyte genes for surface antigens have become inactive in the normal course of cell differentiation, and surface antigens have disappeared by the 6th day after fusion. Between 8 and 12 days after fusion, new hen cell-surface antigens appear and demonstrate that the cultured cell cytoplasm has reactivated the hen erythrocyte genes. Reactivation is accompanied by nuclear enlargement and uptake of proteins. (From Harris, *et al.*, 1969)

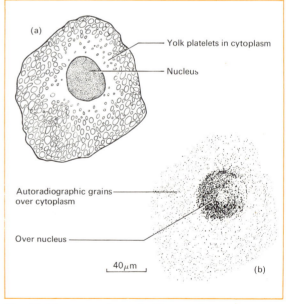

FIG. 11. Movement of proteins from nucleus to cytoplasm. (a).Diagram of a cell in a blastula which was derived from an enucleated egg and transplanted nucleus. The egg was provided with ³H-amino acids before the subsequent injection of a nucleus and puromycin to stop further protein synthesis. Labelled proteins move from the cytoplasm into the nucleus where they become concentrated, as shown by an auto-radiograph (b).

amount of RNA synthesized are related to the removal or subsequent addition of chromosomal proteins. For example, the removal and subsequent addition of histones (basic proteins found in the nucleus) is associated with a pronounced increase, and subsequent decrease, in the amount of RNA synthesized, a result which would be expected if histones are gene repressors (Fig. 12). It is unlikely, however, that histones can recognize different genes, and more recent experiments in which the kinds of RNA synthesized by chromatin were tested by RNA:DNA hybridization have suggested that it may be the non-histone proteins that determine which genes are transcribed (non-histone proteins include the RNA polymerases and their cofactors). Some of the difficulties in this work are indicated by the fact that another research group using a very similar experimental approach has reached the conclusion that a class of small RNA molecules, and not proteins at all, determines the kinds of RNA transcribed. The resolution of these anomalies, and the more refined use of chro-

matin in cell-free systems, is impeded by the difficulty of estimating the amounts of each kind of RNA synthesized. Hopefully, it may soon be possible to combine a cell-free system making RNA from chromatin with one that makes protein from mRNA. Since it is fairly easy to identify different kinds of proteins very accurately by peptide analysis and by other methods, this would greatly facilitate efforts to identify molecules which regulate gene transcription.

The second type of useful assay system involves the injection of macromolecules directly into cells, and the assessment of their affect by observing changes in RNA synthesis in the injected cells. With very few exceptions, large molecules enter cultured cells very slowly, and it is only by injection that sufficient can be got inside cells to provide a worthwhile test of their function. Amphibian oocytes can be used for this purpose; they withstand the injection of quite large volumes (50 nl) of solution and can be grown in culture for several days, during which time RNA can be conveniently labelled via the medium. Such experiments have shown that a protein fraction extracted from amphibian cells is able specifically to repress ribosomal but not other kinds of RNA synthesis. When histones are injected into oocytes they rapidly accumulate in the nucleus, but certainly have no immediate effect on the amount or kinds of RNA synthesized. It is much too early to conclude that any of the molecules so far tested by injection into oocytes are, or are not, concerned with the normal regulation of gene expression. The more interesting aspect of these experiments is the indication that oocytes may provide meaningful test systems for analysing the control of transcription in living cells.

The present position regarding molecules which regulate gene transcription can be summarized as follows. Progress in this area has been severely limited because there is no entirely satisfactory assay system for testing molecules of known potential regulatory capacity, though there is hope that this may soon be possible both *in vitro* and *in vivo*. Such experiments as have been carried out indicate that molecules which regulate gene transcription commonly exist in the cell cytoplasm as well as on chromosomes, and are likely to include non-histone proteins, a category which includes RNA polymerases and their cofactors.

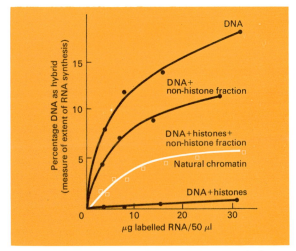

FIG. 12. Chromosomal proteins and the control of gene transcription *in vitro*. Chromatin (a soluble preparation of chromosome material extracted from isolated nuclei) is treated in various ways so as to remove some or all of the chromosomal proteins. It is then incubated with added bacterial RNA polymerases and nucleoside triphosphates (one of which is labelled) for 1½ hours. The kinds and amounts of RNA synthesized by the chromatin are estimated by extracting the labelled RNA molecules from the incubation mixture and hybridizing them to single-stranded DNA. This figure shows that the removal of proteins from chromatin increases the range of RNA molecules which it synthesizes *in vitro*, as judged by the higher percentage of test DNA to which it can be hybridized. (From Paul, 1970.)

The intracellular distribution of regulatory molecules
Even when molecules which control gene expression
have been identified, this is by no means the end of
the problem of differentiation. We must then explain how genes are regulated in one way in one
region of an embryo or organism, and in another
way in another region. This means asking what
determines the spatial distribution of regulatory
molecules. It is convenient to think of cell differentiation as consisting of a series of steps, at each of
which two daughter cells differ from each other or
from the parent cell. We might therefore expect to
find a parent cell with one or more distinguishable
regions of cytoplasm (containing regulatory molecules) which are distributed unequally between its
two daughter cells; in fact situations of this kind
have been described in the eggs of several different
animal species.

One of the earliest and most impressive examples
is that described by Conklin for the ascidian *Styela*.
The fertilized egg has regions of cytoplasm which
are naturally coloured yellow and various shades of
grey. The coloured regions can be followed during
development and are seen to become associated
with particular cell-types. In a classical series of
experiments on sea urchin embryos, Hörstadius
found that cells which contained cytoplasm from
the vegetal half of the egg will always form endodermal (stomach and gut) cells, even when some
animal pole cells are removed, or when isolated
from the other cells of the embryo (Fig. 13). A
number of experiments have been carried out on
the eggs of other animals which also show consistently a relationship between the arrangement of
materials in the egg and subsequent morphogenesis.
In amphibia, for example, the redistribution of
yolk by inversion of an egg, and local damage to
the region of the egg known as the grey crescent
(which forms opposite the point of sperm entry),
cause consistent developmental abnormalities.
However, much the best example of a region of
cytoplasm which has a specific morphogenetic
effect is that of the pole-plasm of insects or the
germ-plasm of amphibia (Fig. 14). In both these
cases a histologically recognizable region of cytoplasm at the vegetal pole of the egg becomes associated with one or two early cleavage nuclei and
is specifically related to germ-cell (eventually egg
and sperm) formation. By a variety of ingenious
experiments, which include localized u.v.-irradia-
tion, low-speed centrifugation, and constriction
techniques. it has been possible to prevent the

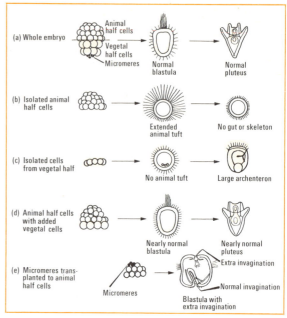

FIG. 13. The relationship between regions of egg cytoplasm
and cell differentiation. Cleaving sea urchin embryos have
layers of cells which can be separated and experimentally
rearranged. Isolated cell layers develop more or less
according to their expected fate, but micromeres can
correct the animal-type differentiation of animal cells
alone, (a)–(d). The micromeres promote invagination (and
gastrulation) even when they have been grafted to the
animal half of the embryo (e). (After Hörstadius, 1939).

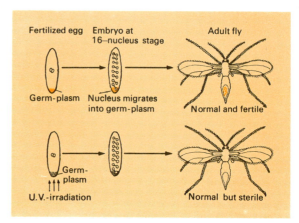

FIG. 14. Germ-plasm and germ-cell differentiation in
insect development. The unfertilized eggs of many insects
contain a cytologically specialized region of cytoplasm
called the pole-plasm. After a few (nuclear) divisions a
nucleus at the vegetal end of the egg enters the pole-plasm,
and gives rise to a line of cells from which the germ-cells
(eventually eggs and sperm) are formed. The pole cells are
prevented from forming germ cells by the u.v.-irradiation
of the pole-plasm. (After Bantock, 1961.)

nuclei which normally associate with the germ-plasm from doing so and thus stop germ cell formation; this has given the clearest demonstration of the relationship that exists between a particular region of cytoplasm and a particular kind of cell differentiation.

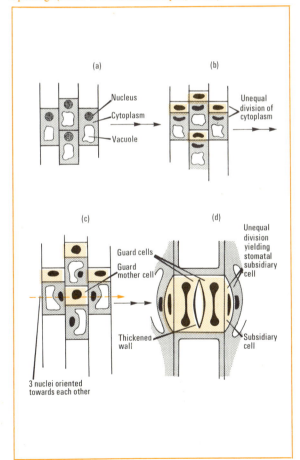

Fig. 15. Unequal cell division and cell differentiation in plants. The diagram shows dividing epidermal cells in a grass leaf. In (a) and (b), four cells undergo unequal cytoplasmic division, one of each pair of daughter cells becoming vacuolated. In (c), the nuclei of the vacuolated daughter cells become orientated towards the nuclei of non-vacuolated daughter cells. In (d), the vacuolated daughter cells undergo another unequal cytoplasmic division, after which the non-vacuolated daughter cell becomes a stomatal subsidiary cell. At the same time, the guard mother cell undergoes an equal division into the two stomatal guard cells, whose nuclei become dumb-bell shaped. The guard cells acquire thickened walls on each side of the stomatal opening. (After Clowes and Juniper, 1968.)

It is hard to prove the same relationship with cells other than eggs because of their small size. However, this has been done with grasshopper neuroblasts. These are cells of the neural tube which undergo a series of divisions, after each of which one daughter cell becomes another neuroblast and the other cell a non-dividing nerve cell. Just before each division, a region of clear cytoplasm forms on one side of the parent neuroblast, and the daughter cell which receives this cytoplasm is always the one which becomes the nerve cell. Carlson was able to rotate the mitotic spindle of a dividing neuroblast (so altering the relative disposition of chromosomes and cytoplasm), and to show that it really is the cytoplasm, and not a particular set of chromosomes, which determines nerve as opposed to neuroblast differentiation. This pattern of differentiation, in which one daughter cell differentiates into a terminally specialized cell, and the other continues as a dividing 'stem cell', is typical of many adult animal organs; it is therefore quite likely that in all these cases the unequal distribution of parent cell cytoplasm may be connected with cell differentiation. In differentiating plant cells, it commonly happens that a parent cell gives rise to two quite different daughter cells, as in the epidermal cells of monocotyledon leaves. Unequal divisions of this kind are related, possibly casually, to an unequal distribution of cytoplasm in the parent cell (Fig. 15).

It is extremely difficult to analyse the nature of cytoplasmic substances which exert their effect on cell differentiation. Almost nothing is known about this except that, in the case of pole-plasm and germ-plasm, the effective components seem to include RNA. Unfortunately it has not been shown that regionally distributed cytoplasmic components exert their effects by acting on the nucleus. It is hardly conceivable that the pole- and germ-plasms do not act eventually on the nuclear genes, but analysis is difficult because these cytoplasmic regions seem to contain many substances with different effects, all needed for germ-cell differentiation.

The main purpose of this section has been to summarize the reasons for believing that cell differentiation depends to a significant extent on the unequal distribution of parent cell materials to daughter cells. Only when it is known what the effective cytoplasmic components are can we attempt to answer the key question of how they come to be distributed in the ways observed.

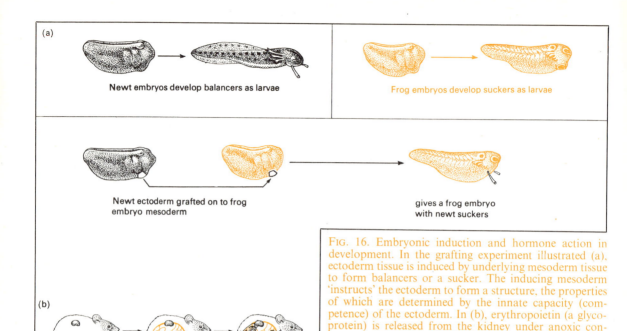

Newt embryos develop balancers as larvae

Frog embryos develop suckers as larvae

Newt ectoderm grafted on to frog embryo mesoderm

gives a frog embryo with newt suckers

(b)

Erythropoietin released by kidney under anoxic conditions

Erythropoietin dispersed in blood circulation reaching all tissues

Bone marrow cells only respond to erythropoietin by specialization to red blood cells

Fig. 16. Embryonic induction and hormone action in development. In the grafting experiment illustrated (a), ectoderm tissue is induced by underlying mesoderm tissue to form balancers or a sucker. The inducing mesoderm 'instructs' the ectoderm to form a structure, the properties of which are determined by the innate capacity (competence) of the ectoderm. In (b), erythropoietin (a glycoprotein) is released from the kidney under anoxic conditions, and stimulates bone marrow cells to undergo erythropoiesis (a series of cell divisions in the course of which a bone marrow cell passes from a proerythroblast to a mature erythrocyte). Although erythropoietin reaches all cells in the body, only the bone marrow cells are competent to respond to the hormone by undergoing erythropoiesis.

Hormones and inducers

In addition to the intracellular rearrangement of materials, some extracellular agents, such as hormones and inducer substances, may play a part in promoting regional differentiation.

A good example of a hormone of major developmental importance is erythropoietin, which stimulates the early stages of red blood cell formation. As is true with other hormones, however, erythropoietin would not be of developmental importance if it induced the same changes in all cells. The cell-type specificity of hormones is achieved by the fact that hormone-sensitive cells possess receptor molecules capable of binding only one type of hormone. Sensitive cells have had to undergo some degree of differentiation before becoming able to respond to the hormone. For this reason, hormones may be developmentally important in enhancing, but never initiating, cell differentiation.

Similar comments apply to the phenomenon of 'induction'. This refers to an interaction between two layers of cells which have come to lie next to each other as a result of the folding movements characteristic of development. The formation of almost every organ in a vertebrate depends on one or more induction steps, and the most likely value of induction in animal development is to delimit more precisely the boundary between two adjacent regions of an embryo which are due to specialize in different ways (Fig. 16). Both embryonic induction and hormone action are probably of value in co-ordinating the timing of development and cell differentiation in different parts of an organism. In spite of an immense amount of work, it has not yet been possible to identify any inducer substance. However, an important point which has emerged from these studies is that the cells which respond to the induction must have acquired a special capacity to do so. This is called 'competence', and means that cells must differentiate in respect of their capacity to respond to an inducer substance.

In conclusion, it seems clear that hormone action, and most cases of induction, contribute to cell differentiation by accentuating or co-ordinating underlying differences which already exist between cells; such differences presumably result from the unequal distribution of cytoplasmic substances previously mentioned.

Interaction of genes and regulatory molecules

Once the molecules which control gene expression have been identified, and the mechanism by which they are distributed within cells is understood, there remains the question of how and when these regulatory molecules interact with genes. One possibility is that genes become progressively repressed in a rather stable way during cell differentiation; regulatory molecules which are formed in the cytoplasm of cells would associate with the appropriate genes and permanently repress their activity in that cell and in all its mitotic progeny. Since it is known that any repression which genes undergo is certainly reversible under the conditions of nuclear transplantation, it is simplest to suppose that such repression is also reversible during normal cell differentiation. The opposite extreme to this view is to suppose that gene regulation is a continuously reversible process. Although there is no strong reason for preferring or excluding any one mechanism of gene regulation, there are some circumstantial reasons for favouring a scheme of the kind shown in Fig. 17. It is assumed that genes are regulated by association with molecules which are themselves gene products and which are synthesized in the cytoplasm of cells. These regulatory molecules are presumed to gain access to the genes whose activity they regulate only once every cell cycle.

FIG. 17. Hypothetical scheme for cyclic reprogramming of gene activity. The determined state of a cell line can be propagated over many cell generations by the cyclic process shown in (a). Under normal conditions daughter cells are determined in the same way as their parent cell. When two sides of a parent cell are exposed to different environmental influences, as in induction, the two daughter nuclei acquire an unequal share of the parent cell's cytoplasmic regulatory molecules; this could lead to differentiation (b). In normal development (c), activated eggs contain unequally distributed cytoplasmic materials, which could cause embryonic cells (= daughter cells of the fertilized egg) to specialize in different ways.

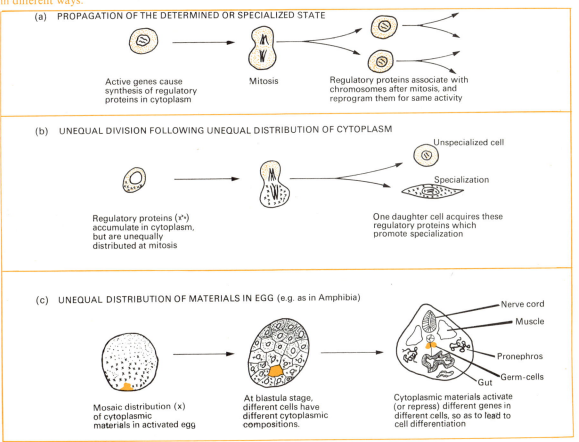

15

Fig. 17 shows the accumulation of regulatory molecules in cell cytoplasm during interphase, and their association with chromosomes at the end of mitosis. Regulatory molecules are assumed to be stably associated with genes during interphase, and released into the cytoplasm as the chromosomes condense for mitosis. The question of which molecules will associate with expanding chromosomes at the end of mitosis would be determined by the relative abundance of those released from chromosomes, and accumulated in the cytoplasm, at the end of the previous interphase. The stability of cell differentiation would be accounted for if active genes cause the accumulation, in the cytoplasm, of molecules which promote their own activity. Thus every daughter cell will differentiate in the same way as its parent cell, unless the cytoplasm is altered, as can happen when cleavage nuclei enter different regions of an egg, or when external influences such as embryonic inducers or hormones cause a change or rearrangement of cell cytoplasm. This proposal is consistent with a number of experimental results that bear on cell differentiation. It is currently useful only because it relates such results to each other, and because it makes certain predictions about the behaviour of regulatory molecules.

Postscript

Many important aspects of cell differentiation have been touched on lightly or not at all in this article. For example, a large amount of current research is devoted to the question of how cells recognize and interact with each other so as to form complex tissues and organs of the right shape and histological composition. The problem of cell differentiation is closely connected with such subjects as the cellular aspects of antibody formation, growth control and cancer. The intention here has been to discuss cell differentiation from the point of view of gene control, for which no apology is made since more is currently known about this than about any other aspect of the subject.

FURTHER READING

General

DAVIDSON, E. H. (1968). *Gene activity in early development*. Academic Press, New York and London.

STEWARD, F. C. (1970). From cultured cells to whole plants: the induction and control of their growth and morphogenesis. *Proc. R. Soc.* B. **175**, 1–30.

WADDINGTON, C. H. (1956). *Principles of development*. George Allen and Unwin, London.

For reference

BROWN, D. D. (1967). The genes for ribosomal RNA and their transcription during amphibian development. *Current topics in Devel. Biol.* **2**, 48–75.

GROSS, P. R. (1968). Biochemistry of differentiation. *A. Rev. Biochem.* **37**, 631–60.

GURDON, J. B. (1968). Transplanted nuclei and cell differentiation. *Scient. Am.* **219** (6), 24–35.

HADORN, E. (1968). Transdetermination in cells. *Scient. Am.* **219** (5), 110–20.

HARRIS, H., WATKINS, J. F., FORD, C. E., and SCHOEFL, I. (1966). Artificial heterokaryons of animal cells from different species. *J. Cell Sci.* **1**, 1–30.

MARKS, P. A., and KOVACH, J. S. (1966). Development of mammalian erythroid cells. *Current topics in Devel. Biol.* **1**, 213–52.

PAUL, J. (1970). DNA masking in mammalian chromatin: a molecular mechanism for determination of cell-type. *Current topics in Devel. Biol.* **5**, 317–52.

See these titles in the Oxford Biology Readers series:
16. *The nucleolus.* E. G. Jordan.
26. *Somatic cell division.* B. John and K. R. Lewis.
33. *Cellular immunology.* J. L. Gowans.
44. *Differentiation and growth in higher plants.* D. H. Northcote.
46. *Metamorphosis.* J. R. Tata.
51. *Development of pattern and form in animals.* L. Wolpert.
61. *Chemical aspects of immunology.* R. R. Porter.
65. *The meiotic mechanism.* B. John and K. R. Lewis.
75. *Transcription of DNA.* A. A. Travers.
79. *The cellular effects of hormones.* P. J. Randle and R. M. Denton.

2

16

Oxford Biology Readers
Edited by J.J. Head and O.E. Lowenstein

The Nucleolus

E.G. Jordan

Oxford University Press, Ely House, London W.1

GLASGOW   NEW YORK   TORONTO   MELBOURNE   WELLINGTON

CAPE TOWN   SALISBURY   IBADAN   NAIROBI   DAR ES SALAAM   LUSAKA   ADDIS ABABA

BOMBAY   CALCUTTA   MADRAS   KARACHI   LAHORE   DACCA

KUALA LUMPUR   SINGAPORE   HONG KONG   TOKYO

E. G. Jordan is Lecturer in Biology at Queen Elizabeth College, London.

PHOTOSET AND PRINTED IN GREAT BRITAIN BY

BAS PRINTERS LIMITED, WALLOP, HAMPSHIRE

The basic information for the regulation of cell activity is contained within the DNA of the *chromosomes*. Some of this information is transcribed into *messenger* RNA (mRNA) molecules which control the synthesis of enzymes and other proteins of the cell. Other information in the DNA does not code for the synthesis of proteins and is not transcribed into a messenger RNA, though it is used for synthesis of other types of RNA. One of these is a vital structural component of ribosomes, the small particles in the cytoplasm which translate mRNA into proteins. Ribosomal RNA (rRNA) is produced by specialized regions of certain chromosomes and the aggregation of it together with the proteins necessary for ribosome structure constitute the *nucleolus*. In other words the nucleolus is a gene (or a sequence of identical genes) together with both its RNA product and the later stages of the conversion of this product into ribonucleoprotein particles, the precursors of ribosomes.

The appearance and disappearance of the nucleolus

The nucleoli are usually the only features which can be distinguished readily in the nucleus of non-dividing cells under the light microscope. They appear as refractile, more or less spherical bodies (Fig. 1). The chromosomes of non-dividing cells are dispersed as thin threads which cannot be resolved with the light microscope. During nuclear division this situation is reversed. The nucleolus disperses and is no longer visible whilst the chromosomes condense into clearly defined thick strands. Exceptions are known and there are stages where both chromosomes and nucleoli are visible at the same time.

Number and size

The number of nucleoli per nucleus may vary from cell to cell but their combined volume per nucleus is more or less constant for a particular cell type. If a cell has two nucleoli when most others of the same kind in the same tissue have only one, then the volume of each would be half that of the single one. There is some evidence that the surface area rather than the volume is the constant feature, the

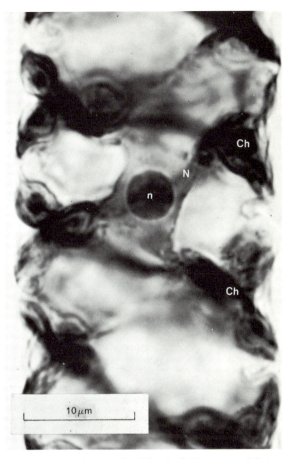

FIG. 1. Light micrograph of fixed whole mount of *Spirogyra*. N, nucleus; n, nucleolus; Ch, chloroplasts with pyrenoids.

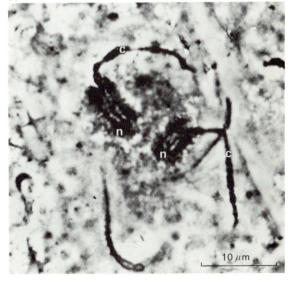

FIG. 2. Prophase stage of *Spirogyra* sp. stained with iron-alum acetocarmine. Two chromosomes, c, have regions which are embedded in the disorganizing nucleoli, n. This stage is directly comparable with that of the electron micrographs, Figs. 5 and 6. (Preparation by M. B. E. Godward.)

3

two nucleoli having a smaller combined volume than the single one. When the nucleus divides the material of the parent nucleolus is divided equally between the two daughter nuclei; during interphase each daughter nucleolus doubles its volume, restoring the pre-division level.

Persistence of nucleoli

When the nucleolus disappears at division its material disperses into the cytoplasm and later collects around the chromosomes to be reformed into nucleoli within the new nuclei. In those organisms where the nucleolar material does not disappear its course can be followed through division. *Spirogyra* is such a case; the nucleolus is partitioned at anaphase into two clumps which move with the chromosomes to the spindle poles where they are incorporated into the new nuclei.

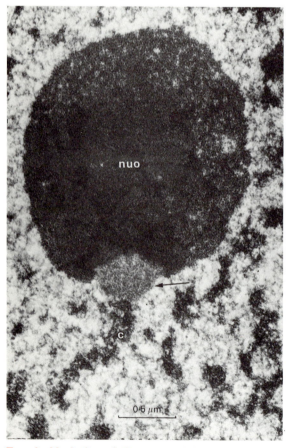

FIG. 3. Nucleolus of dormant artichoke tuber tissue *Helianthus tuberosus*, showing the change in the chromosome, c, as it joins the nucleolus, nuo; arrow, less dense chromatin, the nucleolar organizer.

The nucleolar organizer

In 1931 Heitz showed with *Vicia* sp. that at telophase the nucleoli always formed at the same place on the chromosomes. It is now recognized that particular regions of chromosomes are associated with nucleoli and in some cases actually penetrate them. The prophase of *Spirogyra* is shown in Fig. 2, and parts of two chromosomes can be seen embedded within the nucleoli. These regions are called the *nucleolar organizing regions* of the chromosomes since it is here that the nucleolus is reorganized after nuclear division. Nucleolar organizers have been recognized in many organisms. They have a different appearance from the rest of the chromosome, staining lightly (Fig. 3) and looking thinner than other regions during mitosis (Fig. 4); for this reason they have been called *secondary constrictions*. Other regions of chromosomes have this appearance, for instance the centromeres (the *primary constrictions*), but are not associated with the nucleolus.

Electron microscopy has confirmed this chromosomal location of nucleoli. Figs. 5 and 6 show the association of nucleolar chromosome and nucleolus in *Spirogyra*. In this organism the nucleolar organizer does not look different from the rest of the chromosome. These figures are of prophase stages which is why the chromosomes can be seen as thick condensed strands while the nucleolus has a loose reticular nature.

In animal nucleoli it is not often possible to distinguish any region as the organizing chromosome by observation alone; histochemical techniques have to be employed. For example, Miller obtained the central part of the nucleoli of newt eggs as a circle of droplets on a thread, like a necklace, and found that the thread was digested by the enzyme DNAase, i.e. DNA (chromosomal material) was present in it. The DNA threads were only 3 nm in diameter and would be difficult to identify if embedded in nucleolar material, so it is not surprising that they do not show up in electron micrographs of sectioned intact nucleoli.

The DNAase technique has been tried on plant nucleoli by Chouinard. The darkly staining structures shown in the nucleoli of onion (Fig. 7) were shown in this way to consist of DNA. These are contained within the lightly staining areas which I would equate with the secondary constrictions, described before.

One of the greatest disadvantages of the electron microscope is illustrated by this problem. Areas can often be seen (the lightly stained areas) which behave most significantly; but their chemical composition cannot be ascertained. Answers to such questions can be hoped for as techniques for histochemistry at the electron microscope level improve.

Genetic nature of the nucleolar organizer

An unusual condition was discovered in the cells of a strain of the African clawed toad, *Xenopus laevis*, by Fishberg, Elsdale, and Smith. They found that there was only one of the usual two nucleoli in each cell. The condition was found to be inherited and therefore due to a genetic change, presumably associated with the nucleolar organizer. When the animals were mated, one quarter of the progeny had two nucleoli per nucleus, one half had one and the remaining quarter had none. The last group died four days after hatching, but this was long enough for experiments to be performed on them.

Radioactive RNA precursor molecules were fed to all three types of tadpole. Those with two nucleoli per cell incorporated them into all types of RNA (messenger RNA, transfer RNA, and ribosomal RNA), as did those with one nucleolus per cell; but those with no nucleoli, while they made messenger and transfer RNA at the same rate as the other groups, made no ribosomal RNA at all. This was taken to mean that the genes producing mRNA and tRNA were working normally but the gene(s) producing rRNA were not. With no rRNA, the animals could not synthesize ribosomes, which in turn meant that they could not synthesize proteins and could not, therefore, grow. The growth they did make was at the expense of the rich rRNA endowment of the egg.

Various hypotheses were advanced to explain the inability of the anucleolate tadpoles to synthesize rRNA. The most obvious, perhaps, is that they had lost the gene(s) in the nucleolar organizer which brings about synthesis of rRNA. Others are that an operator gene controlling the rRNA gene(s) had been lost, so that the rRNA gene(s) was permanently 'switched off'; or that the enzyme RNA polymerase, which is necessary for

the transcription of the rRNA gene(s) had been lost.

These possibilities were resolved by Wallace and Birnstiel, using the technique of RNA/DNA

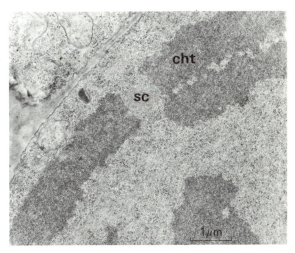

FIG. 4. This electron micrograph shows the difference between the secondary constriction, sc, and the rest of the chromosome, cht, at mitosis in *Vicia faba*. (em by Jean G. Lafontaine.)

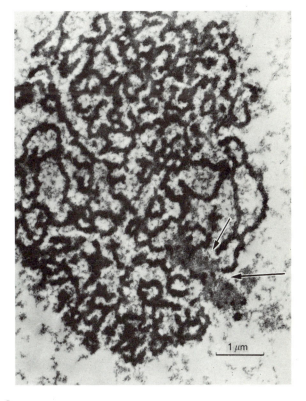

FIG. 5. Prophase nucleolus of *Spirogyra* sp. showing the loose reticular nature and also the connection with the nucleolar chromosome (arrows).

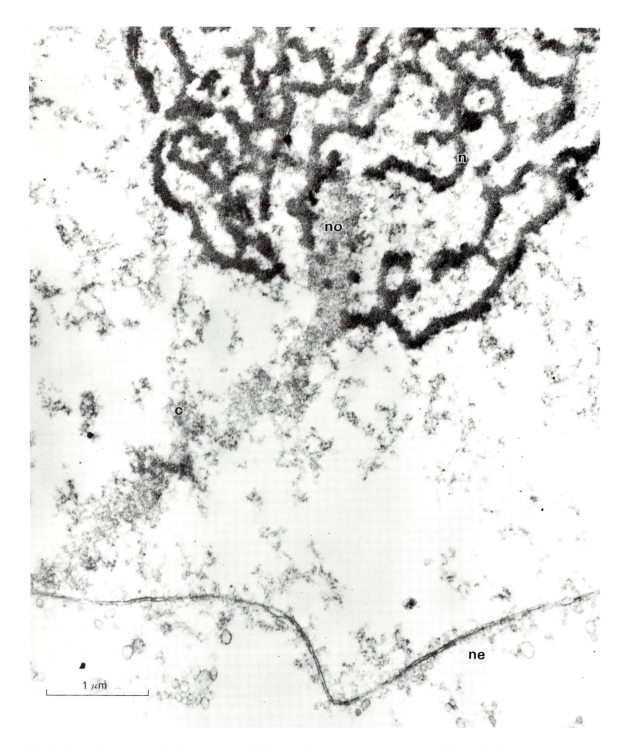

Fig. 6. Part of the same nucleolus as sectioned in Fig. 4. Here a long stretch of one of the nucleolar chromosomes has been sectioned showing the coiled nature of the main chromosome region, c, and the connection of the nucleolar organizing region, no, with the nucleolus, n. ne, nuclear envelope.

6

hybridization (Fig. 8). A DNA sample is denatured by heating, which causes the double helix to unwind and become single strands. This material is incubated with radioactive rRNA, which binds with the DNA only at the site(s) of the rRNA gene(s) where the matching is perfect. The unbound rRNA is washed away and the quantity of bound rRNA—that left covering the rRNA genes of the DNA single strand—is measured; from this the number of rRNA genes can be calculated.

The results were that about 0·3% of all the genes in the nucleus were involved in rRNA synthesis. In the normal tadpole (with both nucleoli present in each nucleus), this represented about 1600 genes. In the mononucleolate animals, 800 genes per nucleus were covered by radioactive rRNA, and in the anucleolate animals, none. Two conclusions can be drawn from this evidence. First, that the anucleolate animals have indeed lost genes

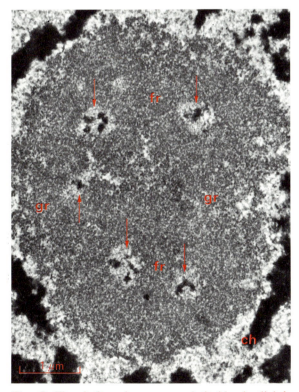

FIG. 7. The dark regions, arrowed, inside the lightly staining areas of the nucleolus have been shown to disappear when digested with DNAase, revealing their chromosomal nature. *Allium cepa*. fr, fibrillar zone; gr, granular zone; ch, chromatin. (em by L. A. Chouinard.)

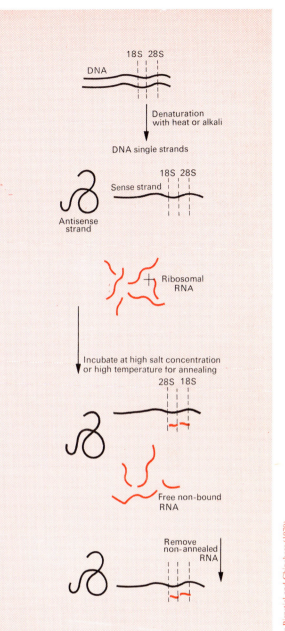

After Birnstiel and Chipchase (1970).

FIG. 8. DNA hybridization technique. Strands of DNA are separated by heat or alkali treatment so as to expose the nucleotide bases. The single strands of DNA are incubated with radioactive ribosomal RNA which binds to the DNA where the sequence of bases is the same, i.e. at the site of ribosomal genes. The number of RNA genes can be calculated from the amount of radioactivity bound to the DNA strand.

(1600 of them) which cause the synthesis of rRNA, for if the explanation were that an operator gene had been lost or that an RNA polymerase had been lost, how could we explain the quantitative relationship of 800 genes missing in the mononucleolate tadpoles and 1600 missing in the anucleolate? Second, the 1600 rRNA genes in the normal diploid tadpole are presumably equally distributed between the two homologous chromosomes which bear the nucleolar organizers, 800 on each. All 800 genes must be very close to each other, in a cluster, since one mutational event has eliminated all of them; if they were scattered throughout the chromosomes, such a change would be impossible.

The obvious inference is that the 800 genes are situated in the secondary constriction, or nucleolar organizer, the region of the chromosome where the nucleolus is formed. It has been possible to observe the deletion of this region microscopically in *Drosophila*, where some flies have been found with shorter organizer regions than others. The number of rRNA genes left on the chromosome is exactly proportional to the length of organizer left.

The nucleolar organizer, then, is a continuous sequence of genes (thought to be identical) which

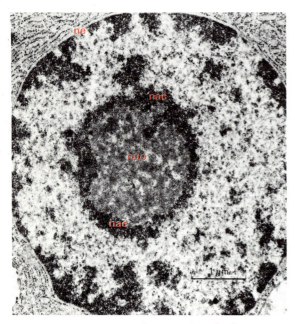

FIG. 9. Nucleolus (nuo) of pancreatic cell of rat showing the surrounding chromatin usually called nucleolar associated chromatin, nac; ne, nuclear envelope.

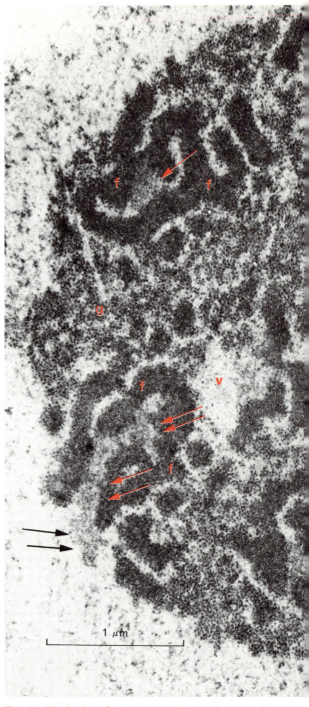

FIG. 10. Nucleolus of *Spirogyra* sp. This is the normal interpha The chromosome (nucleolar organizer) is the most lightly sta stretch can be seen in longitudinal section (double arrows), c section (single arrows). This chromosome is surrounded by the zone, g, characterized by the 15 nm granules. v, nucleolar vacu

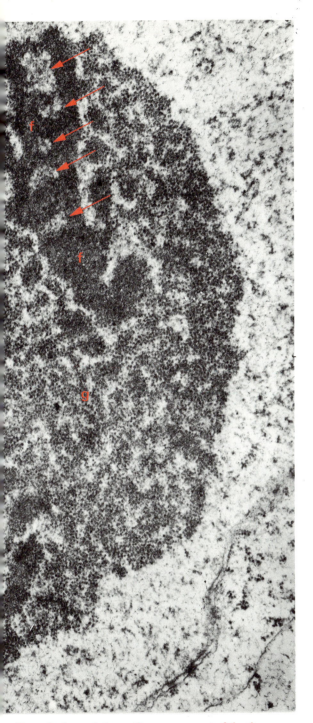

code for rRNA; the nucleolus is the organelle formed round them.

Ultrastructure of the nucleolus

One of the first observations made with the electron microscope was that the nucleolus has no membrane though particularly in animal cells it is often surrounded by condensed chromatin (chromosomal nucleoprotein). This is called *nucleolar associated chromatin* (nac in Fig. 9) and should not be confused with the organizer.

The non-chromosomal part of the nucleolus usually consists of two different zones, the fibrillar and the granular; they are easily distinguished in Fig. 10. The fibrillar zone is dense and composed mainly of fine fibrils 5 nm in diameter which are closely packed and not easily distinguished. The granular zone surrounds the fibrillar and is composed of many 15 nm granules. In plant nucleoli it is usually possible to see the lighter zone which I have shown to be the chromosome region. Sometimes nucleoli contain spaces indistinguishable from the nuclear sap; these are nucleolar vacuoles (Fig. 11).

The arrangement of the zones varies with the state of the cell. In the new nucleolus of a telophase nucleus the fibrillar and chromosomal regions are restricted to one or two areas (Fig. 12), but as we

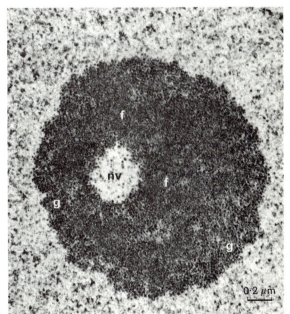

...e of a nucleolus and shows the arrangement of the three zones. ...d follows a meandering course through the nucleolus; a long ... surface of the nucleolus. More usually it is seen in transverse ... region, f, and the rest of the nucleolus consists of the granular

FIG. 11. Nucleolus of carrot root *Daucus carota* showing a nucleolar vacuole, nv; g, granular zone; f, fibrillar zone.

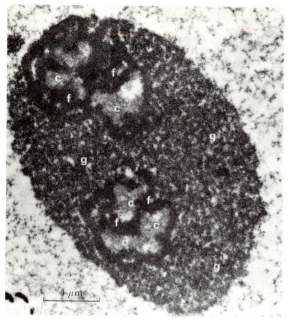

FIG. 12. *Spirogyra* sp. nucleolus at telephase just after reorganization. Note that the organizer zones, c, are clumped into two regions and are not dispersed as at interphase (see FIG. 10). f, fibrillar region; g, granular region.

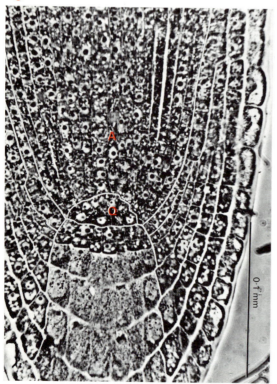

have seen the situation in an active interphase cell is very different (Fig. 11): the chromosomal region is dispersed as small 0·1 μm threads throughout the nucleolus with the fibrillar zone ensheathing it. RNA is synthesized at the chromosome and moves via the fibrillar to the granular zone and thence to the cytoplasm, and the extensive distribution of the chromosome in the interphase nucleolus could be due to the accumulation of the products of synthesis all around it.

To summarize, between mitotic divisions in *Spirogyra*, the nucleolus approximately doubles in size; the chromosome and fibrillar material become more dispersed throughout the nucleolus.

Nucleolar changes associated with metabolism

Great increases in nucleolar size also accompany increased metabolism. In plant root tips there is a group of cells which are more or less inactive (the quiescent centre) surrounded by cells which are in all stages of growth and differentiation (Fig. 13). The growing cells can be seen to have the larger nucleoli. Hyde showed that they differed in volume by a factor of about ten.

At Queen Elizabeth College Dr. Chapman and I have found other nucleolar changes to be associated with metabolic activity. The material we studied was slices of tuber tissue from the Jerusalem artichoke (*Helianthus tuberosus*) previously stored at 4°C. When the slices were shaken in water at 25°C synthesis of nucleic acid, gibberellic acid, and cell wall material commenced, and there was an increase in respiration, protein synthesis, and nucleolar size.

At the same time the zones of the nucleolus changed. At first (0 h dormant tissue, Fig. 14) the nucleolus is a small dense structure with hardly any noticeable granular zone; the chromosomal component is situated outside as a separate structure on the surface. After 3 hours the chromosome region has moved below the surface of the enlarging nucleolus (Fig. 15), and by 24 hours (Fig. 16) a large quantity of granular material has appeared

FIG. 13. Root tip of *Plantago ovata* showing the smaller nucleoli in the quiescent centre, Q, as compared with those in the active cells, A. (Photo by B. B. Hyde.)

FIG. 16. Artichoke tuber nucleolus 24 hr after activation. This nucleolus shows the normal active configuration. The organizer, c, has now moved deep into the body of the nucleolus and is breaking up into smaller regions, s. The fibrillar region, f, surrounds the organizer. A distinct granular zone, g, has now appeared at the periphery of the nucleolus.

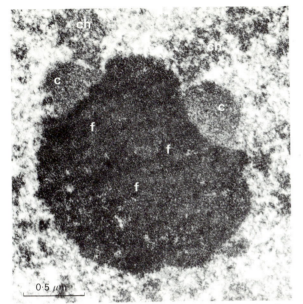

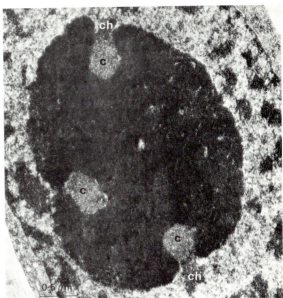

FIG. 14. Nucleolus of dormant artichoke tuber *Helianthus tuberosus* showing the typical 'inactive' appearance. c, nucleolar organizer at the outside of the dense fibrillar region, f. There is no clear granular zone at this stage. ch, chromatin other than the nucleolar organizing region.

FIG. 15. Nucleolus of artichoke tuber tissue 3 hr after activation. (Slices are removed from the cold stored tubers at 4°C and shaken in water at 25°C.) The nucleolar organizer, c, has now moved into the enlarging nucleolus. There is still no clear granular zone at this stage. The connections between the chromatin, ch, and the nucleolar chromatin, c, are particularly clear and show the different appearance of the two in higher plants (see FIG. 6).

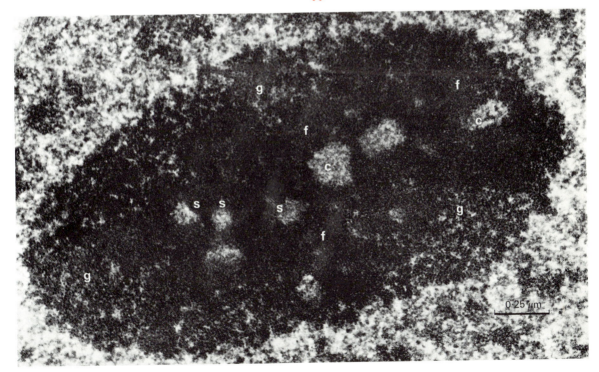

and the chromosome region is deeply embedded in the nucleolus, not as a clump, but as meandering threads about 0·1 μm in diameter. This is the normal appearance in actively growing cells (see also *Spirogyra* nucleolus, Fig. 10). It appears that the activity of the nucleolus is related to the position of the nucleolar organizer relative to it.

FUNCTION OF THE NUCLEOLAR ORGANIZER AND NUCLEOLUS IN MAKING RIBOSOMES

Evidence has come from two independent and different sources, biochemistry and electron microscopy. They agree remarkably well. We will take the biochemical evidence first.

How DNA, mRNA, tRNA, and rRNA co-operate in the synthesis of protein

Every protein in the cell is represented in the genetic code of DNA in the chromosomes. Messenger RNA for a particular protein is formed at the DNA site and then leaves the nucleus, entering the cytoplasm. A ribosome, consisting of about 50% rRNA and 50% protein, becomes attached to one end of the mRNA and works its way along it, joining together amino acids in the order dictated by the code of the mRNA. The amino acids are 'ferried' and held in position while this takes place by transfer RNA. When the ribosome reaches the far end of the mRNA molecule, protein is released and the ribosome is free to become attached to a fresh mRNA molecule; but before this, when it has moved along the mRNA for only a short distance, a second ribosome becomes attached to the mRNA, then others, producing a complex (the polysome) consisting of a mRNA molecule, several ribosomes, and unfinished protein chains of various lengths.

Ribosomes

The major function of the nucleolus so far established is the formation of ribosome precursors, which pass into the cytoplasm. The finished ribosome is about 20 nm in diameter and consists of two subunits called 40S and 60S (S is a Svedberg unit; it refers to sedimentation velocity during centrifugation and is therefore a reflection of molecular size).

Fig. 17 shows how these subunits are made. DNA of the nucleolar organizer is transcribed into rRNA by an RNA polymerase enzyme. Protein molecules join the newly made rRNA (dashes),

which also becomes methylated in two places (dots). The chain is released, after which the excess (non-methylated) rRNA is digested away leaving the 18S and 28S rRNA molecules with their methyl groups and proteins. These coil up forming the 40S and 60S ribosomal subunits. The stages after release of the chain are called *processing*.

How the genes are arranged in the nucleolar organizer

The 18S and 28S RNA molecules of the ribosome must be coded for in the DNA of the nucleolar organizer. It could be that the 18S and 28S genes are arranged in blocks: a block of 18S genes followed by a block of 28S genes; alternatively, 18S and 28S genes might be intermingled in the DNA, and perhaps might even alternate.

Birnsteil showed in two ways that the latter was true. He was able, by centrifugation, to make a preparation of just that portion of the DNA of *Xenopus* which hybridized with rRNA; in other words a preparation of pure rRNA genes. When this was centrifuged again in a caesium chloride density gradient, it formed a single distinct band with a density of $1·724$ g/cm^3. Calculations showed that the density of the 18S gene was $1·713$ g/cm^3 and of the 28S gene, $1·723$ g/cm^3. If blocks of 18S and 28S genes were separate, two distinct bands would form, not the one band of density $1·724$ g/cm^3 which was observed. Thus it seems that 18S and 28S genes are intermingled in the rDNA.

The second way was to chop up the rDNA chain into portions which were small enough to contain only one 18S and one 28S gene. These fragments were incubated with excess 18S rRNA, which bound to them. This increased their density (observed by centrifugation in the caesium chloride gradient) up to a maximum when all the 18S genes on the rDNA were covered by rRNA. Now 28S rRNA was added and the density of the fragments increased again. The 28S site left uncovered by the first treatment (with 18S rRNA) was now being covered; that is to say, the 18S and 28S genes must be adjacent in the DNA.

These experiments also revealed that only 40% of the rDNA gene combines with rRNA during hybridization: DNA other than rDNA forms 60% of the total rDNA gene. This has been called 'excess' RNA gene.

We have seen previously that the nucleolar organizer of *Xenopus* contains about 800 genes. Each is thought to consist of (a) a DNA stretch

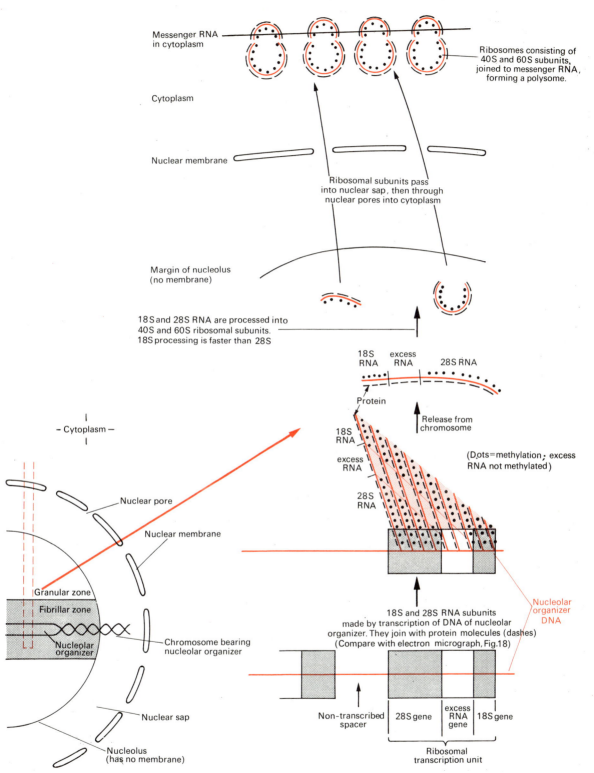

Messenger RNA in cytoplasm

Ribosomes consisting of 40S and 60S subunits, joined to messenger RNA, forming a polysome.

Cytoplasm

Nuclear membrane

Ribosomal subunits pass into nuclear sap, then through nuclear pores into cytoplasm

Margin of nucleolus (no membrane)

18S and 28S RNA are processed into 40S and 60S ribosomal subunits. 18S processing is faster than 28S

18S RNA   excess RNA   28S RNA

Protein

Release from chromosome

18S RNA

excess RNA

(Dots = methylation; excess RNA not methylated)

28S RNA

Nucleolar organizer DNA

18S and 28S RNA subunits made by transcription of DNA of nucleolar organizer. They join with protein molecules (dashes) (Compare with electron micrograph, Fig.18)

Non-transcribed spacer

28S gene   excess RNA gene   18S gene

Ribosomal transcription unit

— Cytoplasm —

Nuclear pore

Nuclear membrane

Granular zone

Fibrillar zone

Nucleolar organizer

Chromosome bearing nucleolar organizer

Nuclear sap

Nucleolus (has no membrane)

FIG. 17. An interpretation of nucleolar function.

13

coding for the ribosomal precursor, and (b) a 'spacer' stretch of DNA which has no known function. Stretch (a) consists of three different DNA sequences, the 18S gene, the 'excess' RNA gene, and the 28S gene.

The 18S rRNA molecule, which gives rise by combination with protein to the small (40S) subunit of the ribosome, is processed faster than the 28S molecule. It probably passes straight out of the nucleolus and nucleus while the 28S RNA becomes a large (60S) subunit after one hour. The 15 nm granules of the granular zone are probably an accumulation of these particles during their maturation in the nucleolus.

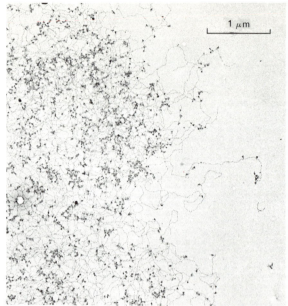

FIG. 19. Granular zone of the nucleolus of *Xenopus laevis* oocyte. No active genes are observed in this zone, but the granules appear with fibrils. (From Miller and Beatty.)

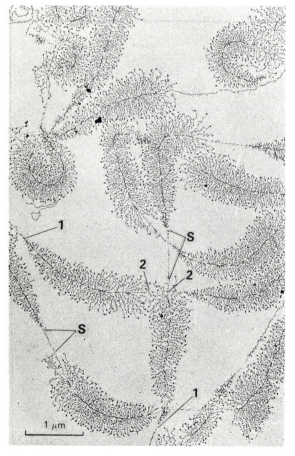

FIG. 18. Ribosomal genes with 'tufts' of RNA being synthesized on them. The shortest molecules are at the beginning of the gene (1), progressively increasing in size as more RNA is copied, continuing up to the end of the gene (2). The non-transcribed or spacer DNA is marked by S. This micrograph is of isolated fibrillar zones from *Xenopus laevis* oocyte nucleoli, spread out, and then stained with uranyl acetate. (From Miller and Beatty.)

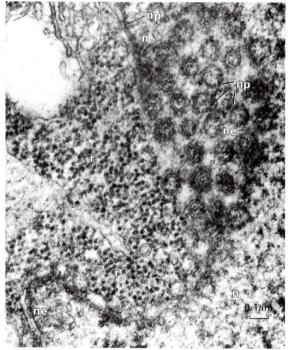

FIG. 20. Oocyte of *Nereis virens* showing the heavy ribosome population, r, of cytoplasm adjacent to the nucleus, n. The nuclear envelope can be seen in both surface section and cross section, ne. np, nuclear pores.

Another RNA molecule is found in the nucleolus which eventually forms part of the larger (60S) ribosome subunit. This is small (5S) RNA. It is made outside the nucleolus and moves in joining the 28S RNA during its maturation in the nucleolus.

Where rRNA is made in the nucleolus

N. and P. Granboulan have shown by autoradiography that [³H]uridine, which is incorporated specifically into RNA, is first located in the fibrillar zone and can be found there only 5 min after application. After 15 minutes or more, it appears over the granular zone. The fibrillar zone, the main constituent of the inactive nucleolus, is thus the site of the RNA synthesis, which fits in well with the close association of this zone with the chromosomal material.

Evidence from electron microscopy

Miller and Beatty have looked at isolated fibrillar zones in newt and *Xenopus* nucleoli and their electron micrographs are strikingly similar to the model built up from biochemical evidence (Fig. 18). The nucleolar DNA is specially prepared so that it is spread out. The strand is a single DNA double helix coated with protein. Along its length are regions of fibrils separated by 'spacer' regions. These fibrils have been shown by enzyme digestion to be composed of RNA and protein, that is to say they are rRNA molecules in process of formation. Their length is slightly shorter than might be expected and they are probably finely folded. Each region of fibrils is about 2·5 µm long; this would code for an RNA of molecular weight about 2·5 million, which is in fact the molecular weight of the precursor rRNA molecule.

Each region of fibrils is thought to be a single gene for precursor rRNA. The fibrils, which increase in length along the gene, are precursor RNA chains bound with protein; they increase in length as the polymerase molecule travels along the rDNA transcribing it into rRNA. The isolated granular zone (Fig. 19) shows many granules and fine fibrils, but no active genes.

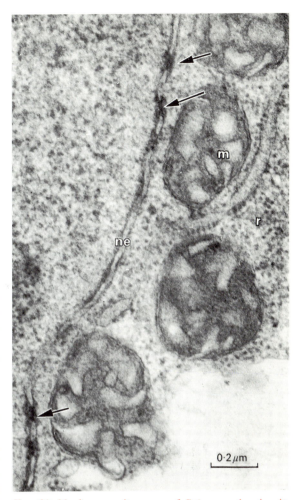

FIG. 21. Nuclear envelope, ne, of *Spirogyra* showing its double nature, and nuclear pores (arrows). The cytoplasm shows ribosomes, r, and mitochondria, m.

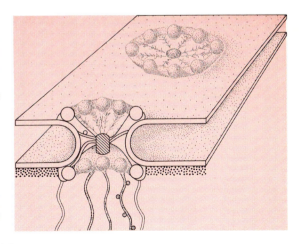

FIG. 22. Diagram of nuclear pore showing the eight subunits and their association with intranuclear fibrils. These fibrils can sometimes be seen connected to the nucleolus. (After Franke and Scheer 1970.)

15

Transport to the cytoplasm

Since ribosomes are not active until they reach the cytoplasm, we might ask, how do they get there? Transport of ribosomal subunits must be happening, but it has been very difficult to observe. In developing oocytes large quantities of ribosomes are made and they accumulate in the cytoplasm (Fig. 20). This would seem to be a useful cell in which to investigate the process. Sometimes clouds of granules can be seen in the cytoplasm apparently streaming away from the nuclear envelope. The nucleus is surrounded by a double membrane which has regions of apparent discontinuity called pores (Fig. 21); these are the most likely place for ribosome subunits to leave the nucleus.

The nuclear pore (Fig. 22) is a complex structure consisting of an annulus composed of eight subunits and a central granule. Fibrils from the subunits radiate into the nucleus; these are sometimes connected to the nucleolus. Perhaps the transport of the nucleolar particles involves a degree of unwinding as they pass through the nuclear pores. Balbiani ring granules, another granular product of the nucleus, are known to change their shape as they 'squeeze' through the central region of the pore. The mRNA must also cross the nuclear membrane on its way to the cytoplasm. More research into the structure of the nuclear pore and its part in the passage of these materials from the nucleus is needed.

The nucleolus and mRNA

Some early studies with RNA precursors indicated that mRNA passed from the genes to the cytoplasm via the nucleolus. Sidebottom and Harris inactivated the nucleolus with a microbeam of ultraviolet light, and this stopped the normal passage of mRNA from the nucleus to the cytoplasm. Other experiments have also shown that a functional nucleolus is required before transport of mRNA from the nucleus can be effected.

In summary we can say that the nucleolus is the accumulation of protein and rRNA at the site of the rRNA genes. The ribosome precursors thus produced are transported to the cytoplasm and assembled into functional ribosomes. Since the activity of the nucleolus is basic to protein synthesis and therefore to all metabolism and growth, this organelle clearly forms part of the key to our understanding of cellular activity.

FURTHER READING

General

BIRNSTIEL, M. (1967). The nucleolus in cell metabolism. *A. Rev. Pl. Physiol.* **18**, 25–58.

—— and CHIPCHASE, M. (1970). The nucleolus: pacemaker of the cell. *Science Journal* **6**, 41–8.

HARRIS, H. (1968). *Nucleus and cytoplasm.* Clarendon Press, Oxford.

For reference

BERNHARD, W. and GRANBOULAN, N. (1968). Electron microscopy of the nucleolus in vertebrate cells. In *Ultrastructure in biological systems*, Vol. 3 (eds. A. J. Dalton and F. Haguenau). Academic Press, New York.

BUSCH, H. and SMETANA, K. (1970). *The nucleolus.* Academic Press, New York.

GODWARD, M. B. E. and JORDAN, E. G. (1965). Electron microscopy of the nucleolus of *Spirogyra britannica* and *Spirogyra ellipsospora*. *Jl R. microsc. Soc.* **84**, 347–60.

HAY, E. D. (1968). Structure and function of the nucleolus in developing cells. In *Ultrastructure in biological systems*, Vol. 3 (eds. A. J. Dalton and F. Haguenau). Academic Press, New York.

HYDE, B. B. (1967). Changes in nucleolar ultrastructure associated with differentiation in the root tip. *J. Ultrastruct. Res.* **18**, 25–54.

JORDAN, E. G. and CHAPMAN, J. M. (1971). Ultrastructural changes in the nucleoli of Jerusalem artichoke (*Helianthus tuberosus*) tuber discs. *J. exp. Bot.* **22**, 627–34.

LAFONTAINE, J. G. (1968). Structural components of the nucleolus in mitotic plant cells. In *Ultrastructure in biological systems*, Vol. 3 (eds. A. J. Dalton and F. Haguenau). Academic Press, New York.

MILLER, O. L. and BEATTY, B. R. (1969). Nucleolar structure and function. In *Handbook of molecular cytology* (ed. Lima-De-Faria). North-Holland.

PERRY, R. P. (1969). Nucleoli the cellular sites of ribosome production. In *Handbook of molecular cytology* (ed. Lima-De-Faria). North-Holland.

13

Oxford Biology Readers
Edited by J.J. Head and O.E. Lowenstein

The Origin of Life

J. D. Bernal and Ann Synge

Oxford University Press, Ely House, London W.1

GLASGOW NEW YORK TORONTO MELBOURNE WELLINGTON
CAPE TOWN SALISBURY IBADAN NAIROBI DAR ES SALAAM LUSAKA ADDIS ABABA
BOMBAY CALCUTTA MADRAS KARACHI LAHORE DACCA
KUALA LUMPUR SINGAPORE HONG KONG TOKYO

ISBN 0 19 914113 4

PHOTOSET AND PRINTED IN GREAT BRITAIN BY
BAS PRINTERS LIMITED, WALLOP, HAMPSHIRE

J. D. Bernal, F.R.S., who died last year, was Professor of Physics at Birkbeck College, London, from 1937 to 1963, when he became its first Professor of Crystallography. Among his many publications are: *The physical basis of life* (Routledge 1939), and *The origin of life* (Weidenfeld and Nicolson 1967).

Ann Synge has translated much of the work that has been written in Russian on the origin of life, especially that of A. I. Oparin. She is a schoolteacher.

This Reader is, in essence, a digest of Chapters 4–9 of a book of the same name by the late J. D. Bernal, published by Weidenfeld and Nicolson in 1967. He had asked me to collaborate with him in preparing it but his health failed and I have done the work alone. Thus, although most of the ideas and a large proportion of the actual text are his, I am responsible for the reader as such. I have therefore felt free to include some supplementary material, in particular the section on the further evolution of metabolism and that on coacervates, which, I thought, would be helpful to a student meeting the subject for the first time.

Norwich, March 1971   Ann Synge

Introduction

It is not possible to give an authoritative or comprehensive account of the origin of life. What is given here is mostly the product of over thirty years of thought on the subject by a physicist. The physical and biological sciences cannot be strictly differentiated but the older scientists were right to regard living things as being fundamentally different from non-living ones. They were wrong in not understanding that the difference is a material, not a spiritual one.

Even in the nineteenth century it was not possible to produce a realistic scheme of the origin of life because the essential biochemical knowledge about the way living things function did not yet exist. It was therefore appropriate that, when the question was again brought into the forum of scientific discussion in the 1920s, it should be done by two biochemists, Oparin in the Soviet Union and Haldane in Britain. Most of the work on the subject since that time has been based on their ideas.

Metabolism and reproduction are characteristic of all living things and as long as we deem that these can be traced back we are dealing with the evolution of life, not with its origins. However, these processes can hardly have appeared *de novo*; if we accept the idea of continuity they must have been preceded by simpler processes and it is these we wish to discuss here. It must be admitted that the discussion of the origin of life is still very tentative; it is speculative science and not by any means observational or experimental science, though observation and experiment are used in its study.

We can see that before there could have been organic evolution in the Darwinian sense there must have been chemical evolution, longer and with changes which were, perhaps, slower than those which have occurred since the first cells began to evolve towards the multitude of organisms we see today.

It seems natural to divide the history of this evolution into three sections. First, the formation of the small molecules such as amino acids, organic bases, and monosaccharides which form the basis of all living things. Second, the formation of large polymers of these smaller organic groups, each with its own definite structure. Third, the integration of the polymers to form organisms.

2

FROM ATOMS TO MONOMERS

It is now generally held that the planets, including Earth, were originally formed by the agglomeration of dust particles. This primitive dust must have consisted mainly of the most common materials in the surroundings of the Sun—silicates, particularly magnesium silicate, and also metallic iron containing nickel. This dust has largely, but not quite, disappeared from our solar system (Fig. 1), but some of it functions as nuclei for meteors. The particles are most abundant in the noctilucent clouds that are sometimes seen in high latitudes long after the Sun has set. Here each silicate or iron particle seems to be surrounded by a layer of ice. There are also gases in the solar system and in the outer parts, at least, some of them must have condensed on the dust particles. They include simple organic compounds such as cyanogen (C_2N_2) and hydrocarbons such as methane (CH_4) and may have condensed to form less volatile organic compounds such as are found in the carbon-containing meteorites or shooting stars which fall on Earth from interplanetary space. It is at least reasonable to suppose that the silicate and iron particles with their skins of organic compounds might have come together to form first the smaller bodies known as asteroids, then the larger planets. We know that the major planets, Jupiter and Saturn, and probably Uranus as well, are composed largely of hydrogen, methane, and ammonia, and also of ice. There is a strong presumption that it was while the organic compounds were dispersed on the dust particles (and so had a very large surface area) that the chemical reactions giving rise to the first intermediate-sized molecules took place. It is, however, a controversial presumption. Many people think that the essential monomers of which living things are built were synthesized from simpler compounds in the atmosphere and oceans of Earth.

Whether or not the Earth contained organic compounds similar to those found on meteorites at the time of its formation, they could easily have arisen there later by the condensation of smaller molecules such as methane, carbon dioxide, and ammonia. In 1953 Miller passed electric discharges through mixtures of methane, ammonia, hydrogen, and water vapour and found that amino acids were formed as well as other compounds. Similar experiments have since been carried out using other mixtures of gases and other sources of energy such as ultraviolet radiation. Many of the

FIG. 1. Horsehead nebula in Orion (1300 light years from Earth); the dark mass of the horsehead is a cloud of dust and gas obscuring more distant stars. (Photo from the Hale Observatories.)

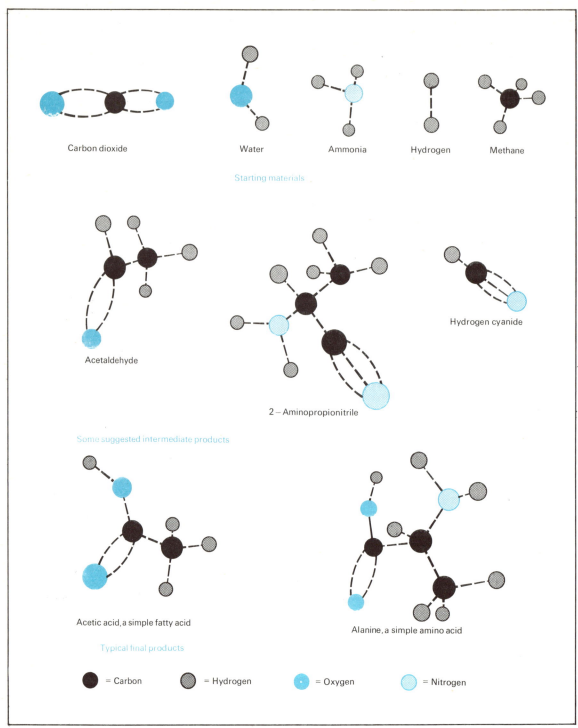

Carbon dioxide Water Ammonia Hydrogen Methane

Starting materials

Acetaldehyde

2 – Aminopropionitrile

Hydrogen cyanide

Some suggested intermediate products

Acetic acid, a simple fatty acid

Alanine, a simple amino acid

Typical final products

● = Carbon   ● = Hydrogen   ● = Oxygen   ● = Nitrogen

FIG. 2. Models of some of the molecules which might have been concerned in the synthesis of amino acids and fatty acids in the primitive atmosphere. In the top row are the starting products and in the bottom row examples of finished products. In the middle row are some suggested intermediates.

compounds synthesized in this way (Fig. 2) are of special importance for life and Miller's experiments were crucial in showing how such compounds, which are now only made by living things, could have arisen on Earth before the origin of life.

The frequency with which the specific life-forming molecules are produced in such experiments can hardly be accidental but must depend on the particular stability of these molecules. The limited number of monomers taking part in biochemistry has long been noted by biologists and Wald has written:

> It turns out that about 29 organic molecules are enough to introduce the bare essentials. They include glucose, the major product of photosynthesis and major source of metabolic energy and hydrogen; fats as a principal storage form of metabolic energy; phosphatides as a means of circulating lipids in aqueous media and for their remarkable structure-forming proclivities; then the 20 amino acids from which all proteins, including all enzymes, are derived. Five nitrogenous bases (adenine, guanine, cytosine, uracil, thymine), together with ribose or its simple derivative desoxyribose, and phosphoric acid, form all the nucleic acids, both RNA and DNA. These 29 molecules give students a first entry into the structures of proteins and nucleic acids, the coding of genetic information, the structures of enzymes, the composition and general properties of cell structures, and bring them to a point from which they can begin to explore the complexities of energy metabolism. That this is not the whole of biochemistry goes without saying; the extraordinary thing is that it makes so good a start. Yet this alphabet of biochemistry is hardly longer than our verbal alphabet.

The reasons for this restricted list are not clear. Experiment has shown that they are particularly stable, though reactive compounds, but it may be that they are the residue of a far longer list which appeared at various stages in the origin of life and which has been reduced by a process of natural selection of molecules. In any case it is clear that a number of molecules which are now only formed by living things are likely to be synthesized in the neighbourhood of any star where there is enough suitable gas and dust to act as raw material and where the temperature is not too hot for them to exist. This synthesis would take place along normal chemical lines, not involving any special biological principles, but would determine the kinds of molecule which would later form living things. It is the solution of these compounds in the waters of the oceans of Earth which is often referred to as the 'primitive soup'.

FROM MONOMERS TO POLYMERS

This stage, beginning with the primitive soup and ending with the large, ordered polymers, the proteins and nucleic acids, and at least some of the processes by which the former are produced by means of the latter, is the most difficult to understand. We can begin by considering how the monomers became sufficiently concentrated to polymerize and go on to see how they could actually have polymerized, but when we come to the study of how order was introduced into the polymers and handed on we run into serious difficulties.

Concentration

There are several ways in which concentration could have occurred. At present we sometimes find a very thin layer of organic material floating on the surface of the ocean. This is derived from the bodies of dead marine organisms but, in earlier times, the organic compounds of the primitive soup could have formed such a layer. Surface layers of this sort inevitably drift with the wind and are concentrated in the foam of breaking waves. The foam may also be piled up in estuaries and adsorbed on the clay there. Thus the most dilute solutions of organic material may end up as concentrates on the estuarine mud.

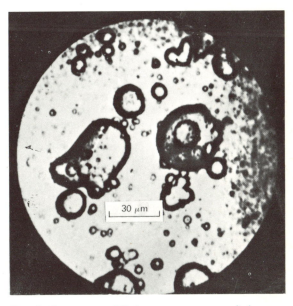

FIG. 3. Coacervate with three components, gelatin, gum arabic, and RNA. (Photo by A. I. Oparin.)

5

These concentrates will contain many different sorts of compound but there is a process, well known as segregation in mineralogy, which causes like elements and compounds to be brought together in the crust of the Earth. Thus, mercury only constitutes 5 parts in 10 000 000 of the crust of the Earth but some ores contain as much as 7% of mercury in the form of the sulphide. It has also been shown that polymerization occurs in dilute (0·02M) solutions of amino acids in the presence of cyanogen derivatives such as produced in Miller's experiments, especially in the presence of ultraviolet radiation.

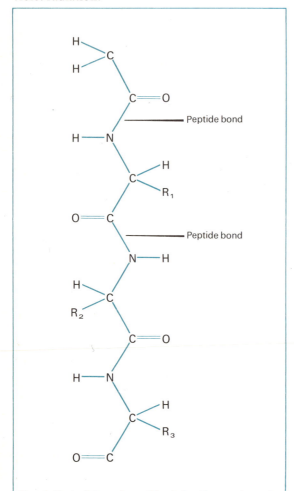

FIG. 4. Part of the polypeptide chain of a protein molecule. The chain is formed by combination of the amino group (NH_2) of one amino acid with the carboxyl group (COOH) of the next. This is the peptide bond. The side chains project in various directions and the molecules are irregularly twisted.

Polymerization

It is usually held that this took place in the cold but amino acids have been made to form polymers similar to proteins at high temperatures and pressures such as occur in lavas and volcanic ash. However, polymerization is now known to be an extremely common process in nature and the existence of such abiogenically formed organo-metallic compounds as the Ziegler catalysts used in making polythene (polyethylene —$(CH_2)_n$) would provide the means of building straight-line polymers of a great variety of compositions but not necessarily of a very complicated character, because the operations of such catalysts are likely to be of a serial kind, producing the same combinations again and again.

Coacervates

It may be that at this stage the concentration of large molecules became great enough for coacervation to take place. A coacervate is formed from dilute solutions of two substances each having large polymeric molecules of opposite charges. The two most commonly studied are gelatin and gum arabic. At particular concentrations these separate out into two phases, sol and gel. Each phase contains both polymers but in different concentrations. However, both sol and gel are

Photo by George Rodger, Magnum.

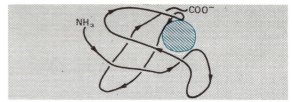

FIG. 5. Model of a molecule of myoglobin. The haem group is the dark flat piece on the upper right.

6

mainly composed of water and so can accommodate large amounts of other molecules in solution. As may be seen from Fig. 3 it is even possible to produce complex coacervates, one gel droplet within another. If the droplets contain enzymes they can carry out within themselves a number of reactions and form a model for the biochemistry of the cell.

Oparin, who has worked continuously on the origin of life for longer than anyone else, attaches great importance to coacervates in the evolution of ordered polymers and he and his colleagues have carried out a number of experiments with systems of enzyme-containing coacervate droplets. On the other hand, it is hard to see how coacervates could have been formed in the absence of a considerable concentration of identical or similar molecules and this implies a mechanism for their formation which is not likely to have existed at so early a stage.

Ultraviolet radiations

It is generally held that until the evolution of a photosynthetic mechanism dependent on chlorophyll there was little or no free oxygen in the atmosphere. This meant that short-wave ultraviolet radiations from the Sun could reach the surface of the Earth. As proteins and nucleic acids are liable to break down when irradiated with such radiations they must have been formed at some place where it could not reach them, probably under a considerable layer of water. At present we are shielded from these radiations by a layer of ozone which is formed by these same radiations from the free oxygen of the air at a high altitude. This ozone screen has allowed living things to come to the surface of the water and even on to the land.

Evolution of functional molecules

The types of polymer most characteristic of living things are the proteins and nucleic acids. Both are heteropolymers, the proteins being made up of some 20 different kinds of amino acid and the two kinds of nucleic acid of 5 different nucleotide monomers. As we know them today these polymers consist of very long chains of monomers folded and arranged in a definite order characteristic of the organism producing them (Figs. 4 and 5). However, we may be sure that in the earliest forms of life such polymers as may have been present were

smaller, and maybe the order of the monomers in them was less strictly determined.

The main characteristic of reactions occurring in living things is that they are carried out by stages in a relatively dilute water-based medium with the help of catalysts. These include the enzymes which, in present-day organisms, are always proteins. Their chemical function is to promote the building up and breaking down of molecules, involving energy-liberating steps, some of which might occur spontaneously in any case, but at a much slower rate than they do in life. Simpler 'proto-enzymes' must have preceded the

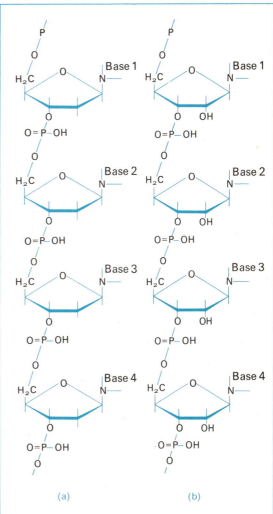

(a) (b)

Fig. 6. (a) DNA polynucleotide (b) RNA polynucleotide. Notice the absence, in the DNA molecule, of the hydroxyl group in the sugar.

7

enzymes of today. It is not hard to form some idea of what they may have been. The enzyme catalase speeds up the breakdown of hydrogen peroxide into water and oxygen 20 000 times. Cobalt ions only speed it up 10 times. When cobalt forms complexes with organic compounds they are more effective catalysts than inorganic cobalt. Several other inorganic ions have catalytic properties which can be augmented by forming complexes with organic molecules, especially porphyrins, and it is easy to imagine that at least some of the enzymes found in living things were originally formed in this way.

Another characteristic function carried out in the tissues of living things is energy transfer, that is the supplying of energy in unit quantities in reactions which are carried out in stages by a series of enzymes, as in the Krebs di- and tricarboxylic acid cycle. The compounds which perform this function are by no means as specific as enzymes. One of the commonest of these compounds is adenosine triphosphate (ATP). In this case the energy is transferred by means of the phosphate radical and, although this is bound to an organic group, it has been shown in the laboratory that even inorganic phosphates have some energy-transferring properties which are

increased by associating the phosphate radical with organic groups.

The third characteristic activity carried out in living things is the recording and transferring of information; this is carried out by the nucleic acids, desoxyribonucleic acid (DNA) and ribonucleic acid (RNA) (Figs. 6 and 7). In this case we cannot point to a possible prototype. It is of some interest, however, that adenosine is present both in the ubiquitous energy-transferring substance ATP and in the nucleic acids. This suggests that the triple monomer (nitrogenous base + sugar + phosphate) existed as such and polymerized at an early date

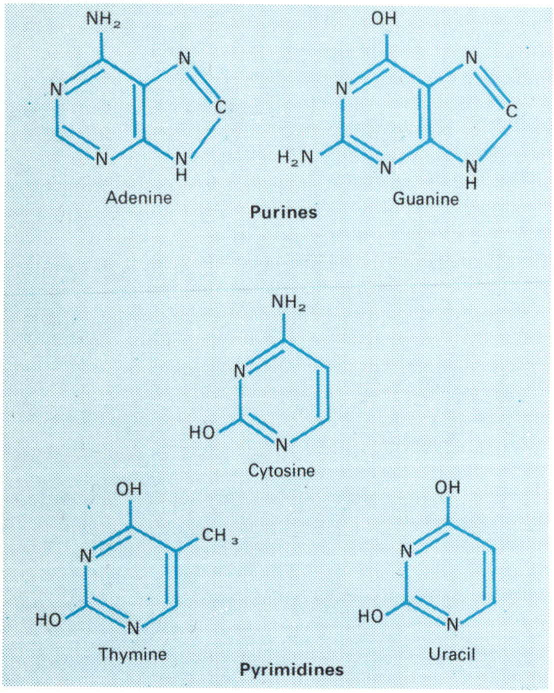

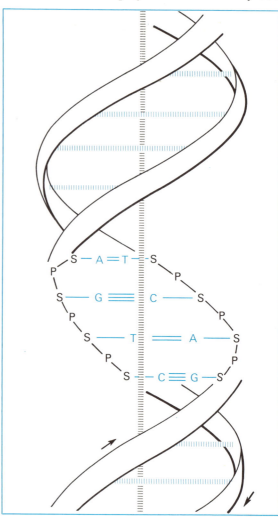

FIG. 8. The DNA double helix. Note that the chain is formed by alternate sugar (S) and phosphate (P) groupings and that the bases project inwards, towards the axis of the helix, to link up with their appropriate partners.

NH₂

Adenine

OH

Guanine

Purines

NH₂

Cytosine

OH

CH₃

Thymine

OH

Uracil

Pyrimidines

FIG. 7. The nitrogenous bases of DNA and RNA.

8

into a nucleic acid. This polymerization was the origin of self-reproducing molecules.

Ordered copying of polymers

The repeated production of polymers with their monomers arranged in the same definite order is now seen as the central problem of the origin of life. In existing living things it is carried out by the nucleic acids. DNA exists in the form of two chains of nucleotides twisted together in a double helix like 2-ply wool (Fig. 8). There are 4 different nucleotides, each consisting of a nitrogenous base, a sugar, and a phosphate group. The sugar and phosphate groups form the chains and the bases stick inwards towards those of the other chain in the helix. The two chains can only be formed when each base on one chain is opposite the right base on the other. Thus adenine on one chain must always be opposite thymine on the other and guanine on one must be opposite cytosine on the other. It is supposed that new molecules of DNA are formed by the unwinding of such a double helix. Each nucleotide in each strand then collects from the cell contents its appropriate partner and these new nucleotides join together to form new chains (Fig. 9). Ribonucleic acids can be built on DNA in the same way only the sugar involved is ribose instead of the desoxyribose of DNA; further the base which acts as a partner to adenine is uracil and not thymine (Fig. 7). We thus have a mechanism for producing large numbers of identical molecules of RNA. Now RNA is concerned in the production of proteins and the order of the bases in the RNA determines the order of the amino acids in the protein produced under its influence, so we have a mechanism for producing large numbers of identical protein molecules too. As many of the proteins are catalysts, this arrangement makes sure that the reactions catalysed by them will occur smoothly again and again. This is a very sophisticated mechanism and it is logical to assume that it was preceded by a simpler one. What we have to do is to get some idea of the minimum complexity with which a process of this sort could start and then to find out how it could have developed the complexity which it shows today. This problem has not been solved.

Life before cells

It has been suggested that some sort of natural selection of polymers took place, analogous to the natural selection which plays so important a part

in the evolution of living things. This might have occurred in the coacervate droplets, if they existed early enough. It seems likely that prototypes of proteins and nucleic acids could have developed in the primitive soup; but whether the interaction of polymerized molecules which gave rise to the nucleic acid-induced synthesis of proteins which we know today took place before or after the origin of the first living things is a controversial point. However, the microstructure of the organelles and membranes of all organisms is such that they could only be assembled from a large number of identical or nearly identical units. This implies a fairly highly developed mechanism for their synthesis. If such a mechanism existed, it raises the question of whether life, in the sense of a metabolizing and self-reproducing system, may not have existed before there were any separate organisms. So we may see Earth just before the occurrence of separate organisms as being much as it is now, apart from the absence of oxygen, though somewhere on the mud of the estuaries there would be 'sub-vital' areas with complicated and inter-locking sets of chemical reactions taking place. These reactions would be limited by the rate of diffusion of active molecules and would

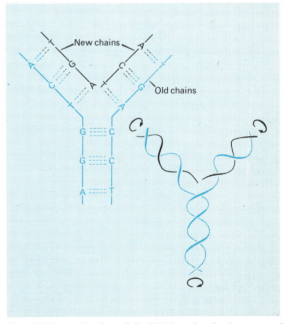

FIG. 9. The replication of the DNA molecule. A suggested mechanism for duplicating the strands without first separating them.

9

take place more slowly than they now do in living things, because the catalysts and other compounds taking part would be less well adapted to their purpose. There would be no separate organisms and this would also imply the absence of movement and the limitation of communication between one part and another to purely chemical movements and interactions dependent on diffusion.

FROM POLYMERS TO CELLS

The transition from uniform sub-vital areas to cells seems logically and historically to be effected by membranes. The essential feature of the cell membrane or plasmalemma is the unit membrane which consists of a double layer of phospholipid molecules lying between two layers of protein molecules (Figs. 10 (a) and (b)). The lipid layers are of uniform thickness, and this implies either a mechanism for making fatty acid chains of a uniform length or one for folding them in loops of a uniform length. This does not present great problems. There are signs of regularity in the hydrocarbon chains found on meteorites and polyethylene crystallizes in layers determined by the length of the loops which form spontaneously in the molecule. These lengths vary with the temperature at which crystallization occurs.

The importance of unit membranes has three main aspects. First, they cut a bit of protoplasm off from the rest so that it can vary and reproduce independently and so become subject to natural selection. Second, they divide the cells up into compartments and, being semipermeable and able

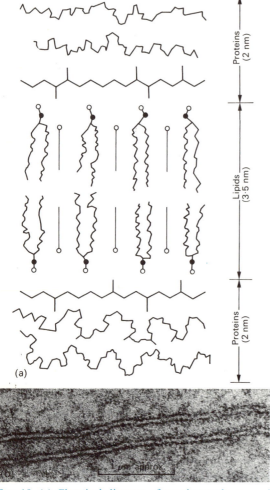

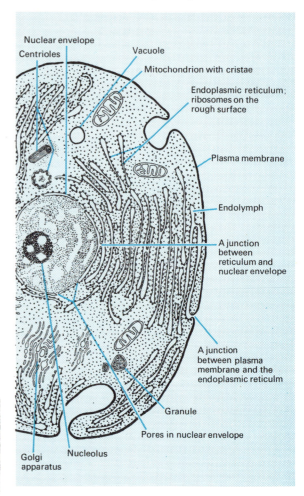

FIG. 10. (a) Chemical diagram of a unit membrane and (b) two membranes as seen with an electron microscope.

FIG. 11. Diagram of generalized eukaryotic cell to show the relationship between the various structures.

10

to transfer molecules from one side to the other, they can control the composition of the liquids in each. Third, they provide anchorage for enzymes and structures so that they are held in a fixed position with regard to one another. This makes it possible for their activities to be co-ordinated.

Cells

The discovery, in the early nineteenth century, that all the larger animals and plants are made of cells and cell products and that all cells are produced by other cells, was a great advance in biology, but it is becoming clear that even unicellular organisms have a complicated structure and must have a long history behind them.

Existing single-celled organisms may be classified as eukaryotes, having a distinct nucleus which undergoes mitosis, and prokaryotes which have no such nucleus. The prokaryotes include the mycoplasmas (see Fig. 15), the bacteria, and the blue-green algae. They seem to be more primitive than the eukaryotes (Fig. 11) though they are by no means simple. They are already cells with a considerable history behind them.

The organelles

Organelles can be classified in two groups, the particulate and the membranous. The most important particulate organelles are the microsomes, the microtubules (Fig. 12), and cilia, and certain contractile fibres such as the myosin of muscle. Ribosomes are found in all cells—though those of prokaryotes are different from those of eukaryotes (Fig. 13). It is here that protein synthesis is carried out under complex direction from the DNA of the nucleus and perhaps other DNA-containing structures. Microtubules are more complicated in their structure and are also found in both

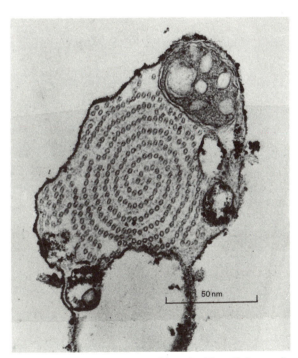

FIG. 12. Microtubules arranged in a double coiled array in the exopod of the prozoan *Actinosphaerium*. (Photo by K. R. Porter.)

FIG. 13. Diagram showing the results of an experiment in which RNA extract from various organisms was driven electrophoretically through a gel which acted as a molecular seive. The rate of movement depends on the size of the molecule but is not directly proportional to it. The size of the molecules is expressed in Svedberg units which

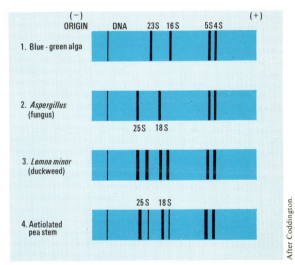

are related to molecular weight (S value $\sim$ R.M.M. $\frac{2}{3}$). There is an approximately linear relationship between log S and the distance travelled per unit time. Note that the RNA extracts from all the organisms showed the DNA line as a contaminant in the preparation (very large molecules) and the 5S and 4S soluble RNA lines (comparatively small molecules). In addition that from the one prokaryote (blue-green algae) shows only the 23S and 16S lines while that from the eukaryote without chloroplasts (*Aspergillus*) shows only the 25S and 18S ribosomal RNA lines. The eukaryote with fully developed chloroplasts (*Lemna minor*), on the other hand, shows both ribosomal RNA pairs (25S and 18S, 23S and 16S), while that in which the chloroplasts are undeveloped (aetiolated pea stems) shows the 25S and 18S pair and traces of the 23S and 16S pair.

prokaryotic and eukaryotic cells. They seem to be built of simple protein molecules arranged as a cylinder, and the more complicated cilia and flagella appear to be derived from them.

The membranous organelles seem to be part of the general membranous structure of the cell. They are composed of the same unit membrane that is found in the cell wall, the nuclear membrane and the endoplasmic reticulum. The most important are the mitochondria and the chloroplasts which are the organelles which build up the energy stores of the cell. Both have an outer and an inner membrane, the outer one being, apparently, continuous with the endoplasmic reticulum, like the membranes derived from the host cell which are sometimes found covering particles of the larger viruses in infected cells. The inner membrane seems to belong to the organelle itself and shows a characteristic folded pattern (Figs. 14 (a) and (b)). They have their own DNA and their own RNA which is different from that of the rest of the cell and more like that of a prokaryote (Fig. 13). They also have all the components of a protein-synthesizing mechanism and, although there is as yet little direct evidence that it actually synthesizes proteins, it is known that some of the characteristics of the mitochondrion may be inherited otherwise than via the DNA of the nucleus.

Multiple origin of cells

Many of the features of the organelles suggest that the transition from sub-vital areas to cells may not have been direct but that there may have been a stage of independent organelle life, and that only at a later stage were the free-living organelles grouped together to form the complex that we now call a cell. The transformation from organelle life to cell life may, indeed, not have occurred all at once but by stages over a long period.

Mycoplasmas and viruses

Nobody has ever found a free-living organelle, but some of the organelles found in cells not only show a certain independence of the genetic apparatus but may also be hard to distinguish from prokaryotic cells. For instance, it is sometimes hard to tell a blue-green alga from a plastid, or a mycoplasma from a mitochondrion. Also, although it is now generally accepted that viruses are not primitive organisms, one may hope to learn something about the hypothetical independent organelle from the real dependent virus.

Mycoplasmas are the simplest organisms we know (Fig. 15). They have no proper plasmalemma, only a unit membrane separating the contents of the cell from the environment. Their only internal organelles, apart from a few special granules, are their ribosomes (which are like those of bacteria) and their strands of DNA which take the form of a loop (also like those of bacteria only smaller). They do not show the same infolding of their cell membrane as eukaryotic cells, perhaps because they are so small that they have an adequate surface–volume ratio without such folding. Some are saprophytes but most are parasites and may have simplified their structure accordingly. Even so we may accept them as marking a degree of simplification possible without losing their independence as living organisms.

We can go still further in our search for simplicity and consider the viruses. The simplest of

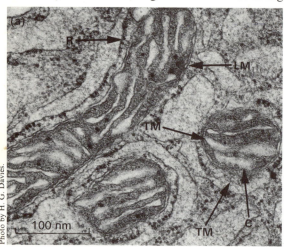

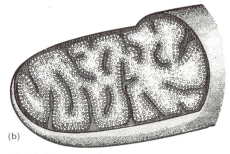

(b)

FIG. 14. (a) Electron micrograph of a plasma cell showing mitochondria. LM, longitudinal section through a mitochondrion; TM, transverse section through a mitochondrion; C, crista; R, ribosomes. (b) Stereogram reconstruction of a mitochondrion.

them contain only a specific protein and a specific nucleic acid. The function of the protein is not yet fully known but it is shed before the nucleic acid enters the host. Once inside, the nucleic acid begins to replicate. When it has done so a sufficient number of times the host cell begins to synthesize molecules of virus protein according to the specification laid down by the virus nucleic acid. This protein then aggregates itself round each of the nucleic acid chains. An interesting feature of this multiplication of viruses is the fact that the protein molecules formed assemble themselves into quite complicated structures without help from the nucleic acids. Empty shells of viruses are often found, having the same shape

and packing into the same kind of protein crystals as normal virus particles, but without the nucleic acid (Fig. 16). Virus particles may even be dis-aggregated chemically, the protein being separated from the nucleic acid. The nucleic acid may then be destroyed by ribonuclease but the protein will still reaggregate to form the original structure *in vitro*. This may well provide a clue for the formation from protein molecules of many of the complicated structures of the cell. Such self-assembly can occur even without viruses, for example in the transition from globular to fibrous insulin. It implies that the properties of biological structures can be, so to speak, built into the unit molecules of which they are made.

Minimum requirements for an organism
In his last contribution to the study of the origin of life, Haldane produced the following specification for a simple organism:

Suppose, then, that we have reduced the amount of RNA needed to specify a protein to a minimum—what protein would be needed? If our culture medium included amino acids, ribose, the four bases, and a source of high-energy phosphate, the following reactions would have to be carried out:
First, formation of nucleotides.
Secondly, coupling of nucleotides to form chains.
Thirdly, combination of amino acids with ATP or some related substance.
Fourthly, coupling of these amino acids to form a peptide chain.

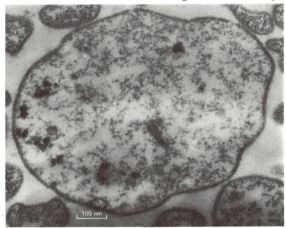

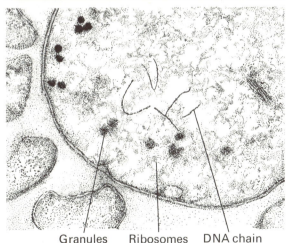

Granules   Ribosomes   DNA chain

FIG. 15. Electron micrograph of *Mycoplasma gallisepticum* showing the special granules, the ribosomes, and the threads of DNA. Note that there is no infolding of the outer membrane such as is seen in larger organisms. (Photo by Prof. R. W. Horne.)

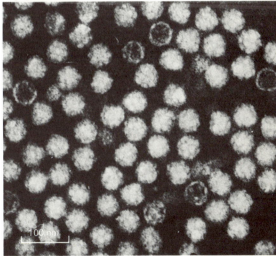

FIG. 16. Particles of turnip yellow mosaic virus. The solid-looking particles contain RNA; the empty-looking ones do not. (Photo by John Innes Institute.)

In existing organisms, these reactions are catalysed by different enzymes. Even in the simplest organisms, it is desirable that enzymes should not only be efficient but specific, that is to say, should catalyse a limited series of reactions. Otherwise, the control of metabolism would be impossible. I want to suggest that the initial organism may have consisted of one so-called 'gene' of RNA specifying just one enzyme, a very generalized phospho-kinase, which could catalyse all the above reactions.

The further evolution of metabolism

If the first cells were formed in an environment containing organic compounds but little or no free oxygen, we may suppose that their metabolism was something like that of anaerobic saprophytes

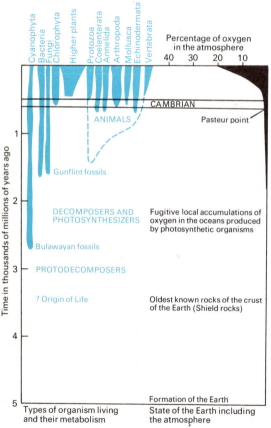

FIG. 17. Diagram correlating the fossil record with other events on the Earth. The black area represents the percentage of free oxygen in the atmosphere. The rest of the atmosphere has probably always consisted largely of nitrogen though various other gases have occurred in different amounts at different times. The Pasteur point is the point at which oxygen reached 1 per cent of the atmosphere which is the point at which aerobic respiration can become more efficient than anaerobic.

living today. This is a primitive form of metabolism which can occur, though not at all effectively, in simple solutions without any cellular organization. Our supposition is borne out by the presence of the appropriate set of enzymes in all living things today, though in the higher animals and plants the reactions concerned are linked to more recently evolved chains and cycles of reactions, e.g. aerobic respiration.

We suppose that the primitive organisms consumed the organic substances of the primitive soup and broke them down to waste products such as carbon dioxide and alcohol by means of their enzymes faster than they were synthesized spontaneously. Thus food became scarce and those organisms which contained enzymes which enabled them to use the energy of sunlight in their metabolism were at an advantage. Several such mechanisms arose and can be seen to this day in the metabolism of such lowly organisms as the sulphur bacteria. All involved the use of light to mobilize the hydrogen from some compound such as hydrogen sulphide and use it to reduce carbon dioxide by stages to some useful compound such as glucose. The most effective mechanism involves chlorophyll, which mobilizes the hydrogen of water, the oxygen being given off into the air. This requires a very high degree of spatial organization of catalysts and could not possibly have evolved before the appearance of fairly complicated membranous organelles.

At this stage many forms of life which could not use the energy of light, or used it inefficiently, must have been wiped out. However, some lived on and became parasites, living at the expense of the organic material built up by the successful photosynthesizers. All these organisms were anaerobic but, as oxygen accumulated, first in the water and then in the air, some of them, both photosynthetic and non-photosynthetic, developed mechanisms for using oxygen to oxidize their waste products completely to water and carbon dioxide. This made them so much more efficient in the use of food that, again, they must have eliminated many rival forms which had not developed these mechanisms.

These two crises of metabolism probably account for the great uniformity of the metabolism of the more advanced forms of life compared with the metabolic diversity of the bacteria which may be regarded as survivors from pre-chlorophyll days.

DATING THE ORIGIN OF LIFE

Fig. 17 is an attempt to summarize what is known and guessed of the relative order of the events concerned with the origin of life and to fix their dates within the time available.

The earliest date of 5 000 000 000 (5×10^9) years ago is the date when it is believed that the solar system originated. The various bodies in it gradually differentiated out and about $4 \cdot 5 \times 10^9$ years ago the Earth had roughly the same mass and composition as it has now. None of the events which we have discussed in connection with the origin of life has left any detectable trace

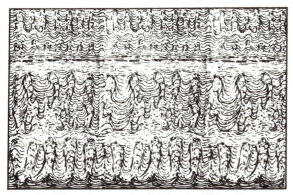

FIG. 18. Biogenic deposits in Pre-Cambrian limestone, probably at least $2 \cdot 7 \times 10^9$ years old. (Drawn from a photograph in Rutten 1962.)

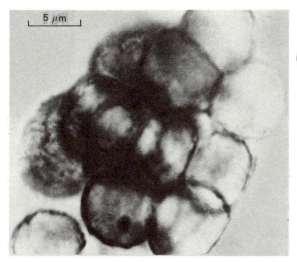

FIG. 19. Micrograph of a single colony of primitive algal plants. Some of the earliest fossils we know. About $1 \cdot 6 \times 10^9$ years old. (Photo by Prof. E. S. Barghoorn in *Science*, **119**.)

till we come to the evolution of photosynthesis. The presence of oxygen in the air led to oxidation of the minerals of the crust of the Earth and the newer rocks are brownish and reddish from the presence of ferric compounds in them, while the older ones are blackish or greenish from the presence of ferrous compounds. Unfortunately difficulties have arisen in trying to date the first appearance of an oxidizing atmosphere in this way. From this and other evidence, though, it seems that at some time between 2×10^9 and 1×10^9 years ago oxygen appeared in the air in considerable quantities; however, photosynthesis began much earlier (about 3×10^9 years ago). All the oxygen liberated by the earliest photosynthesizers was taken up in the water and in various chemical oxidations.

The earliest sediments believed to have been formed by living things were found in Rhodesia and are dated at least $2 \cdot 7 \times 10^9$ years ago (Fig. 18). The organisms which excreted them must have been anaerobic and although they are referred to as 'algal' there is very little reason to think that they were photosynthesizers. The oldest actual fossils known were found in Canada. They are called the 'Gunflint' fossils (Fig. 19) and there are several different kinds, some algae and some fungi. There is some confusion about their date but it is probably about $1 \cdot 6 \times 10^9$ years ago. There are also some very striking algal limestone deposits in the Sahara (Fig. 20). They were probably laid down between 2×10^9 and 1×10^9 years ago too; but their main interest lies in the fact that similar formations were still being laid down as late as the Ordivician (that is later than $0 \cdot 5 \times 10^9$ years ago) and so must belong

FIG. 20. Reef of Conophyton in the Pharusian Pre-Cambrian of the Hoggar, central Sahara. Probably between 2 and 1×10^9 years old. (Photo by Dr. M. Gravelle.)

to the period when there was a considerable concentration of oxygen in the air.

Although the fossil record is poor before the beginning of the Cambrian (0.6×10^9) years ago living things must have been evolving, because at that time a rich assembly of recognizable fossils appears containing representatives of most of the phyla now seen on Earth. It must be assumed that something happened at about this date which led to the formation of organisms with calcareous or horny shells which left recognizable fossils while the earlier soft-bodied organisms did not. It has been suggested that this change was connected with the increasing concentration of oxygen in the air, brought about by the proliferation of photosynthetic organisms.

Stage	Years ago (millions)	Atmosphere and hydrosphere	Sources of free energy	Location of life or pre-life	Forms of chemical evolution	Forms of structural evolution
III (From molecule to cell)	800 Base of Cambrian	As at present with variable oxygen content	Visible sunlight (ultraviolet cut off by high ozone layer)	In sea and on land In algal reefs	Chemical evolution slowing down	Evolution of plant and animal phyla
		Formation of oceans Continental drift		In surface waters		Eukaryotes with nuclei, single- and multi-celled organisms
II (From monomers to polymers)	2400 (gunflint algae)	Oxygen concentration growing			Oxidation replacing fermentation	Folded membranes in mitochondria and other organelles, flagellae Folded membranes in chloroplasts Prokaryotes, i.e. bacteria
	3000	Ozone near surface	Ultraviolet light	Under surfaces of pools	**Photosynthesis** yielding O_2 yielding S	**Separate organisms** Membrane formation
	4000	Formation of hydrogen Anoxic secondary atmosphere N_2, CO_2	Radioactivity? and local heating	Estuarine muds and proto-soils Sub-vital areas Primary soup	Lipids Polymerization Nucleic acid and protein synthesis Carbohydrates Metal-containing proto-enzymes	Coacervates
I (From atoms to monomers)	5000	Simple molecules containing H, C, N, and O atoms		On the surfaces of planets, planetesimals, asteroids, cosmic dust	High-energy phosphates as proto-coenzymes	

Fig. 21.

FURTHER READING

General

BERNAL, J. D. (1967). *The origin of life.* Weidenfeld and Nicolson, London.

FOX, S. W. (ed.) (1965). *The origin of prebiological systems; and of their molecular structure.* Academic Press, New York.

OPARIN, A. I. (1961). *Life, its nature, origin, and development.* Oliver and Boyd, Edinburgh.

RUTTEN, M. G. (1962). *Geological aspects of the origin of life on earth.* Elsevier, Amsterdam.

For reference

HALDANE, J. B. S. (1954). The origins of life. *New Biol.* **16**, 12–27.

KLUYVER, A. J. and VAN NIEL, C. B. (1956). *The microbe's contribution to biology.* Harvard University Press.

OPARIN, A. I. (ed.) (1959). *Proceedings of the First International Symposium on the origin of life on the earth.* Pergamon Press, Oxford.

WALD, G. (1964). The origins of life. *Proc. natn. Acad. Sci. U.S.A.* **52**, 595–611.

See these titles in the Oxford Biology Readers series:

5. *The Euglenoids.* G. F. Leedale.
9. *Photosynthesis.* C. P. Whittingham.
16. *The Nucleolus.* E. G. Jordan.
19. *Mitochondria.* J. B. Chappel and S. C. Rees.
26. *Somatic cell division.* B. John and K. R. Lewis.
34. *Protein Structure.* D. C. Phillips and A. C. T. North.
43. *Symbiosis of algae with invertebrates.* D. C. Smith.
50. *Fine structure, origin and evolution of the algae.* D. A. Greenwood.

13

Oxford Biology Readers
Edited by J.J. Head and O.E. Lowenstein

11

Homology,
An Unsolved
Problem

Sir Gavin de Beer

Oxford University Press, Ely House, London W.1

GLASGOW   NEW YORK   TORONTO   MELBOURNE   WELLINGTON
CAPE TOWN   SALISBURY   IBADAN   NAIROBI   DAR ES SALAAM   LUSAKA   ADDIS ABABA
BOMBAY   CALCUTTA   MADRAS   KARACHI   LAHORE   DACCA
KUALA LUMPUR   SINGAPORE   HONG KONG   TOKYO

ISBN 0 19 914111 8

Sir Gavin de Beer, F.R.S., who is now retired, was formerly Professor of Embryology in the University of London and Director of the British Museum (Natural History). He is author of a large number of books on biology, embryology, and evolution which include *Introduction to experimental embryology* (Clarendon Press 1926), *Development of the vertebrate skull* (Clarendon Press 1937), *Embryos and ancestors* (Clarendon Press, third edition 1962), *Charles Darwin* (Nelson 1963), and *Atlas of evolution* (Nelson 1964).

Figs. 1 and 7 drawn by Derek Whiteley.

PHOTOSET AND PRINTED IN GREAT BRITAIN BY
BAS PRINTERS LIMITED, WALLOP, HAMPSHIRE

1. The concept of homology

The term *homology* is derived from the Greek *homologia* which means 'agreement', and is applied to corresponding organs and structures of plants and of animals which show 'agreement' in their fundamental plan of structure, as for example the leaf of an oak tree with the leaf of an ash tree, or the right forelimb of a dog with the right forelimb of a horse. Richard Owen introduced the term into biological language in 1843 to express similarities in basic structure found between organs of animals which he considered to be more fundamentally similar than others.

The basis of such similarity and its fundamental nature was for Owen, as for other anatomists of the Transcendental School who considered ideas that grouped facts to be more important than the facts themselves, that such organs corresponded to their representatives in a hypothetical 'archetype', a primeval pattern which was regarded as a sort of blueprint on which groups of similar animals had been created. This concept was pre-Darwinian and pre-evolutionary. The way to define an archetype was to make an abstraction of all the similarities that could be found in common in a group of animals, paying no attention to the variations which individuals and populations showed. The archetype was therefore nothing but a metaphysical concept, and views such as these were held widely by anatomists towards the end of the eighteenth century, and especially by those of the German school of *Naturphilosophie*.

As it turned out, Owen was right in basing homology and homologous organs, or homologues, on their structure regardless of their function. An organ is homologous with another because of what it *is*, not because of what it *does*. Homologous organs are the 'same' organs however modified in detailed form and in the function that they carry out. The forelimb of a horse is homologous with the wing of a bat, although the former serves for locomotion on land and the latter for flight in the air. Homology is therefore to be distinguished sharply from *analogy*, the term applied by Owen to structures that perform similar functions but do not correspond to the same representative in the archetype. The wings of an insect serve the same function as the wings of a bird and are analogous to them, not homologous with them. The entire science of comparative anatomy is concerned with the recognition of homologous organs in different groups of organisms, plants and animals, and their distinction from analogous organs.

Like other people, Owen had predecessors in his way of thinking, and the earliest was Aristotle who may be said to have founded comparative anatomy in his *Historia animalium*, when he wrote: 'There are living beings such that all the parts of one recall the corresponding parts of others'; forelimb of quadruped, wing of bird, fin of fish. Aristotle, who based his views largely on external comparative anatomy, did not carry the analysis very deep, and it had to wait two thousand years before further progress was made. But Aristotle also did something else, which is reflected in the views of Owen and many of his contemporaries; like Plato, Aristotle believed that absolute reality resided not in a thing itself, but in the idea or essence of a thing, and this metaphysical notion is evident in the views of Owen. It was killed stone dead for biology by T. H. Huxley who, in his *Theory of the vertebrate skull*, published in 1858, showed that the Transcendental idea of the skull being only a variation on the theme of a vertebra, so that the skull 'was' modified vertebrae, was as absurd and untenable as the converse idea that vertebrae were modifications of the skull. From then on, transcendental anatomy was killed in England, and facts counted more than idealistic hypotheses and concepts.

To return to the development of comparative anatomy, Pierre Belon (1517–74) published a figure of the skeletons of a man and of a bird, showing that the bones corresponded, bone for bone. Felix Vicq d'Azyr (1748–93) made an important contribution when he analysed the correspondence of such structures and organs in great detail, by paying particular attention to their bones, joints, blood-vessels, ligaments, muscles, glands, etc., thereby establishing the 'correspondence' on a broad anatomical basis. Goethe (1749–1832) was so deeply interested in the correspondence of form that he coined the term morphology, the science of shape, to express the meaning of comparative anatomy; but he was so imbued with the idealistic conceptions of interpretation in anatomy that, on seeing a sheep skull broken into three rings, lying on the ground in the Jewish cemetery in Venice, he immediately concluded that the skull was only modified vertebrae, a blunder, as has been seen above. His recognition that parts of flowers are essentially modifications of leaves has more to be said for it.

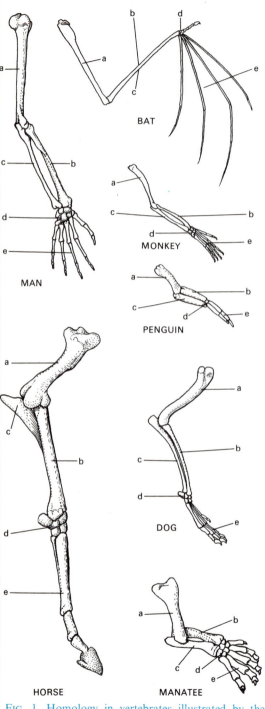

FIG. 1. Homology in vertebrates illustrated by the bones of the forelimb. (a) Humerus (b) radius (c) ulna (d) carpals (wrist) (e) metacarpals (fingers). (After de Beer 1964.)

MAN

BAT

MONKEY

PENGUIN

DOG

HORSE

MANATEE

Mention must also be made of Étienne Geoffroy-Saint-Hilaire (1772–1844) whose obsession with unity of type led him to believe that *all* animals were built on the same plan of structure, a view in the tradition from Aristotle to Owen, which was shattered by Cuvier (1769–1832) who contended that there were four plans of structure in animals. Geoffroy-Saint-Hilaire did, however, put forward a criterion in comparative anatomy: 'the only general principle that can be applied is given by the position, the relations, and the dependences of the parts, that is to say by what I name and include under the term *connections*.' This was an extension of Vicq d'Azyr's work, and is still the way in which a comparative anatomist studies the morphology of organs to satisfy himself that they are, or are not, what is called homologous.

Darwin's bombshell of evolution, which burst in 1859, had a profound effect on the concept of the explanation of homology, but without touching the criteria by which it is established. At one stroke, it was obvious that metaphysical 'archetypes' do not exist, and that homology between organs is based on their correspondence with representatives in a common ancestor of the organisms being compared, from which they were descended in evolution. 'What can be more curious,' asked Darwin, 'than that the hand of a man, formed for grasping, that of a mole for digging, the leg of the horse, the paddle of the porpoise, and the wing of the bat, should be all constructed on the same pattern, and should include similar bones, in the same relative positions?' In the 6th edition of the *Origin of species* (1872) he went on to quote Sir William Flower: 'We may call this conformity to type, without getting much nearer to an explanation of the phenomenon, but is it not powerfully suggestive of true relationship, of inheritance from a common ancestor?'

In other words, it is homologous organs that provide evidence of affinity between organisms that have undergone descent with modification from a common ancestor, i.e. evolution. Furthermore, since evolution is the explanation of the 'agreement' between homologous organs, their study, if they are hard parts susceptible of fossilization, is not restricted to the morphology of living organisms, but the entire range of palaeontology is available for it. So, provided with a cast-iron explanation in terms of affinity, of

4

inheritance in evolution from a common ancestor, it looked as if the concept of homology was at last soundly based and presented no more problems of principle; however, as will be seen below, it unfortunately does.

2. Homology in plants: leaves and flowers

The leaf of a land plant is a lateral appendage of the stem, morphologically different from the stem, with, typically, a bud in the axil between the leaf-base and the stem. The leaf contains plastids with chlorophyll and is therefore green; a foliage leaf is exposed to sunlight with the energy of which the chloroplast performs the chemical reactions of photosynthesis. Foliage leaves can differ widely in detailed shape, from the needles of conifers to the stalked undivided blades of lilies, the indented leaf of the oak, the subdivided compound leaf made up of leaflets of the pea. The whole leaf, or a leaflet, can be modified into a tendril of a climbing plant as in the vine, ending in adhesive discs as in Virginia creeper. In the fly-catching sundew, the leaf bears tentacles that secrete a sticky substance that catches the fly, digests it, and then absorbs it. Leaves can also be modified into scales and bracts, but the most interesting modification is into floral leaves.

The flower of an angiosperm typically consists of four concentric whorls of elements. The frond or foliage leaf of a fern shows in its simplest form that it is a sporophyll: it forms and bears spores on its under surface. The innermost whorl of the elements of a flower is formed by the carpellary leaves, the carpels, which usually grow together to form an enclosed chamber, the ovary, surmounted by its style and stigma; but the carpels betray their sporophyll nature by the fact that they produce spores. These spores which develop into embryo-sacs, are contained within the ovules or future seed-coats. As sexual dimorphism, with its great genetical selective advantages, affects the flowering plant (the sporophyte), the spores produced by the carpels are sedentary macro-spores, which is why the carpels are regarded as the female elements in the flower.

The second whorl of floral elements consists of the stamens, thin stalked structures ending in anthers which produce pollen-sacs containing the pollen grains which are microspores, adapted to travel and dispersal to find the macrospores, which is why the stamens are regarded as male sporophylls. The third whorl is made up of petals, which show clear similarity to the structure of foliage leaves in spite of the fact that they may be of different colours. These colours attract insects, an adaptation to the pollination of flowers by insects which increases the chances of cross-pollination of the stigma of the flower of one plant by the pollen of a different plant. As Darwin noticed, no flower that is wind-pollinated, like those in catkins, has coloured petals. The evolution of the modification of floral leaves into petals that attract insects and of insects that pollinate flowers is a striking result of the fact that both flowering plants and insects evolved at the same time, in the late Mesozoic era, each thereby contributing to the rapid evolution of the other.

The evolutionary derivation of the parts of a flower from the unspecialized leaves of an ancestor is supported by the facts that in some Cycads,

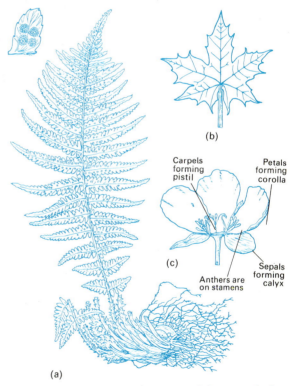

FIG. 2. Homology in plants illustrated by sporophylls, foliage, and floral leaves. (a) Fern sporophyte, showing frond or sporophyll bearing sporangia on its under surface. (b) Foliage leaf of maple. (c) Floral leaves (flower) of *Paeonia*. (After E. Strasburger (1921) *Textbook of botany*, Macmillan.)

5

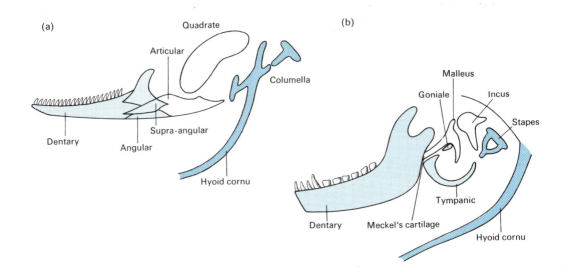

Fig. 3. Homology in vertebrates illustrated by the hinge of the lower jaw in reptiles (a); and the ear-ossicles (stapes) in mammals (b). (After E. S. Goodrich (1930) *Studies on structure of vertebrates*, Macmillan.)

the most primitive gymnosperms living, the carpels are simple sporophylls, like foliage leaves bearing ovules, and that in the Magnoliaceae, the most primitive living angiosperms, the stamens are often broad sporophylls, bearing their spores (pollen grains) on their under surface.

3. Homology in animals: the ear ossicles
In reptiles the hinge between the upper and lower jaws is the joint and articulation between two bones: the quadrate of the upper jaw and the articular of the lower jaw. The quadrate abuts against the side of the auditory capsule by its otic process. Both quadrate and articular are cartilage-bones, preformed in cartilage which then becomes ossified. The reptilian lower jaw also contains a number of membrane-bones, ossifications without cartilaginous precursors, such as the dentary in front, which bears the teeth, the angular and supra-angular behind situated laterally to the articular, the pre-articular and coronoid on the inner side of the jaw. Some fossil reptiles show even more bones.

The jaws are part of the 1st visceral or mandibular arch which is separated from the 2nd or hyoid arch by the tympanic cavity, derived from the 1st visceral pouch, and connected with the throat by the eustachian tube. In the hyoid arch, the uppermost skeletal element is the columella auris, cartilage-bone, a rod conveying vibrations of sound from the tympanic membrane on which sound waves impinge, to the fenestra ovalis of the auditory capsule where the vibrations are imparted to the lymph fluid which stimulates the sense organs of hearing. As the tympanic cavity lies between the 1st (or mandibular) arch and the 2nd (or hyoid) arch, the quadrate and articular bones project into the tympanic cavity from in front, and the columella auris from behind, and the latter is able to vibrate in an open space instead of in thick tissue.

In mammals the conditions at first sight seem to be very different, because the lower jaw consists of a single bone, the dentary, from which an uprising extension articulates with the fossa of a membrane-bone of the brain case, the squamosal. The hinge of the lower jaw in mammals is therefore different from that in reptiles. When the question is asked what has happened in mammals to the old hinge bones of the reptiles, the answer is sensational. These bones have become inserted between the columella auris and the tympanic membrane and are known as the incus and malleus respectively, while the columella, now

6

called the stapes, continues to fit into the fenestra ovalis, receiving the vibrations from the incus which in turn receives them from the malleus impinging on the tympanic membrane. The leverage which these bones can exert on one another makes the transmission of vibrations more sensitive. So there is a chain of three ear ossicles in mammals, and between two of them, the incus and the malleus, is the old hinge joint of the lower jaw of reptiles.

The other bones of the reptilian lower jaw have also changed their functions and their names. The angular in mammals has become the tympanic bone which surrounds and protects the tympanic cavity; the pre-articular (also called goniale) becomes attached to the front of the malleus; the coronoid and supra-angular disappear.

The important point to notice in these changes is the perfect morphological correspondence between the conditions in reptiles and in mammals. All the elements that are cartilage-bones in the former are so also in the latter: the same is true of the membrane-bones and their relative positions correspond exactly. This correspondence also extends to minute details. The columella in reptiles is frequently pierced by a hole through which the stapedial artery passes; this is constant for the stapes of mammals, and is the reason why it is called the 'stirrup'. The lateral head vein runs back medially to the quadrate in reptiles and to the incus in mammals. The facial nerve passes out of the brain case and runs backwards on the median side of the quadrate in reptiles and of the incus in mammals. The nerve passes above the tympanic cavity on the outer side of the stapedial artery and gives off a branch, the chorda tympani, which runs forwards above the tympanic cavity and then down on the median side of the lower jaw elements, articular or malleus, in exactly the same way in reptiles and in mammals.

Minute morphological analysis of the conditions in reptiles and in mammals, carried out on embryonic and on adult material, proves beyond possibility of error that the reptilian quadrate, articular, and columella are respectively homologous with the mammalian incus, malleus, and stapes. This is a good example of the detailed 'correspondence' looked for by Vicq d'Azyr and of the 'connections' sought by Geoffroy-Saint-Hilaire. What makes this study even more significant is that the results of comparative anatomy are confirmed by those of palaeontology, for there are fossil reptiles that show advances towards the mammalian condition, and the superseding of the quadrate-articular hinge of the lower jaw by the squamosal-dentary articulation. All this evolution took place without any functional discontinuity. It is a sobering thought that every man carries in his ear ossicles the homologue of the lower jaw hinge of his reptilian ancestors. This is one of the most demonstrative examples of how comparative anatomy can determine homology of structures inherited from common ancestors in evolution.

4. Conservative effects of homology

The courses taken by certain nerves and blood vessels in adult mammals are determined by the structure of their embryos which repeat the embryonic conditions of the ancestors' embryos. The recurrent laryngeal nerve is an example of how the topology of homologous structures determines some curious anomalies in adult anatomy. The recurrent laryngeal nerve is a branch of the vagus nerve which in fishes has four branchial branches, each of which passes down behind visceral pouches 3, 4, 5, and 6, and runs forwards ventrally but on the median side of the arterial arches that also run down behind those visceral pouches which, in fishes, are pierced as gill-slits.

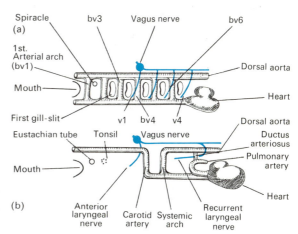

FIG. 4. Morphology of the arterial arches and the vagus nerve in (a) dogfish; (b) rabbit. bv 1, 3, 4, 6, blood-vessels running in the 1st, 3rd, 4th, and 6th visceral arches; v, vagus nerve; v1, 4, 1st, 4th branch of vagus. (After de Beer 1966.)

In mammals these arterial arches are reduced in number by the disappearance of arches 1, 2, and 5. The 3rd or carotid, the 4th or systemic aorta, and the 6th or pulmonary persist. The systemic aorta persists only on the left side where there is still the old connection between the aorta and the pulmonary artery by means of the ductus arteriosus, which is of great importance to the embryo when still in the uterus where respiration is carried out by the placenta. At birth respiration immediately becomes pulmonary, and the ductus arteriosus closes up and becomes nothing but a ligament. But the old 4th branchial branch of the vagus, now called the recurrent laryngeal nerve still loops round the remains of the ductus arteriosus, remnant of the old 6th arterial arch.

In early stages of development, the heart lies far forward, in the neck, and the laryngeal nerve does not have far to go to innervate the larynx. But as development proceeds, the heart and the arterial arches are drawn back into the thorax. This is why the recurrent laryngeal nerve on the left side, after running backwards and looping round the ductus arteriosus, then runs forwards again to innervate the muscles of the larynx. In man, this course of the nerve is several inches longer than it need have been in the adult if it went straight to the larynx from the point where the nerve emerges from the skull. In the giraffe its course must be several feet longer. The explanation is the homology between the mammalian ductus arteriosus and the 6th arterial arch of the fish, which is respected in descendant forms, resulting in apparently anomalous conditions.

5. The displacement of homologous structures
There is no doubt whatever that the forelimb in the newt and the lizard and the arm of man are strictly homologous, inherited with modification from the pectoral fin of fishes 500 million years ago. They have identical elbow and wrist joints and their hands end in five fingers. The bones and muscles that they contain also correspond. But a minute examination of their comparative anatomy reveals the astonishing fact that they do not occupy the same positions in the body. The limbs of vertebrates are always formed from material that is contributed from several adjacent segments of the trunk. So, in the newt the forelimb is formed from trunk segments 2, 3, 4, and 5; in the lizard from 6, 7, 8, and 9; in man from

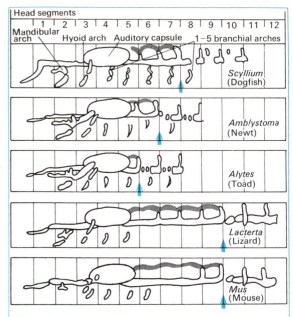

FIG. 5. Homologous structures can change their position: the occipital arch (▲). The arch is the hindmost component of the developing brain-case. (After de Beer 1937.)

trunk segments 13 to 18 inclusive. This can be determined embryologically by observing the contributions made by the segmental muscles to the muscles of the limbs, and anatomically in the adult by studying the ordinal numbers of the spinal nerves that make up the brachial plexus and innervate the muscles of the limbs, because ventral motor nerve roots are always 'faithful' to the muscle plates of their own segments. So the undoubted homology of the forelimb in newt, lizard, and man does not imply that they occupy identical positions in the body. They have shifted in position during evolution.

There is no shifting in position during embryonic development, but what has happened is that in the course of evolution, transposition has occurred; new adjacent segments further back in the trunk have been drawn into contribution to the formation of the limb, and segments further forward, which previously contributed, cease to do so. The limb is a pattern which has been transposed over the long axis of the vertebrate body, like a tune that can be transposed over the keys, as E. S. Goodrich showed.

Another example of the same phenomenon is the position of the occipital arch of the skull which marks its hind end and the hinder limit of the head. In sharks the occipital arch is in the septum separating the 7th from the 8th segment, counting from the front of the head. In newts it is between the 6th and 7th segments, in frogs between the 5th and 6th; in reptiles and mammals between the 9th and 10th segments. Man therefore has 9 segments in his head. Without forfeiting its homology, the occipital arch has been transposed up or down the segments of the body.

These examples illustrate the important principle of the *pattern* which is where the problem of homology lies, not in identity of position in the body. A completely independent but comparable case is the shifting in position of the pattern of teeth in mammals. Extensions of the pattern of typical premolar teeth to teeth immediately in front of or behind them can be seen in related species, but do not always affect the 'same' teeth if 'sameness' means identical numerical position of the teeth in the jaws.

The realization that homologous organs conform to a pattern is valuable, and will appear again below.

6. Serial homology

Serial homology is really a misnomer, because it is not concerned with tracing organs in different organisms to their representatives in a common ancestor, that is to say with evolution, but with the similarity between organs repeated along the anteroposterior axis of one and the same organism. Such organisms are those that show metameric segmentation, orderly repetition of parts. Examples of serial homology include the parapodia of marine annelid worms, in which one pair corresponds to each segment of the body, and the segmental nerve roots, muscle plates, and ribs of vertebrates. Aristotle was interested in the correspondence between forelimbs and hindlimbs; in mammals for instance, each shows one bone in the upper arm and thigh, two bones in the forearm and shank, several bones in the wrist and ankle, and several more in hand and foot each of which ends in five digits, fingers or toes. This is not real homology, as forelimb and hindlimb cannot be traced back to any ancestor with a single pair of limbs, At most it might be said that there had been reduplication of a pattern. At the hands of Transcendental anatomists, serial homology has led to abject nonsense, such as attempts to claim serial homology for the soft palate of a mammal's mouth cavity and the diaphragm.

One aspect of serial homology may have an indirect bearing on homology. In the paired limbs of arthropods, one pair of which corresponds to each segment, the limbs near the mouth serve, not for locomotion or respiration, but for feeding, and are modified into 'mouth-parts'. Higher crustacea have a pair of mandibles, 2 pairs of maxillae, and 3 pairs of maxillipeds, followed posteriorly by the series of swimming, walking, or respiratory limbs. Insofar as these mouth-parts really are serially homologous with ordinary paired limbs, it is possible to argue that in the ancestor the mouth-parts were ordinary limbs that have become modified. In a sense the homology of the elements of a flower with foliage leaves is a kind of serial homology, because a plant grows in height.

7. Latent homology

The concept of homology which refers organs to a representative in a common ancestor concerns itself with homologous organs as visible phenotypic structures, but it is more than possible that the criterion is over-exacting in insisting that the representative structure must be visible in the common ancestor. This suspicion arises from many sources. One example is the Titanotheres, extinct mammals, in many lineages of which knobs appeared on the head as soon as they reached a certain size. It is difficult to deny the homology between these knobs, but they cannot be referred to anything visible in a common ancestor. It must be inferred that these separate lineages inherited a trait, as a result of which each lineage would have exhibited the structure as soon as a limiting factor was removed, in this case presumably insufficient size.

On a simpler level, there is the problem of recurrent mutations. In the fruit fly *Drosophila* there have been repeated mutations from the normal red eye to white. It is difficult to rule out the possibility of a common inheritance of a tendency for this mutation to occur, even if the common ancestor did not have white eyes.

More complex is the problem of spiral cleavage. This is a very precise set of manoeuvres by which the fertilized egg is cleaved. First, four sub-equal cells are formed by the first two cleavage divisions, but after that there is a sort of quadrille as successive quartets of smaller cells are formed above the four original cells, by division spindles

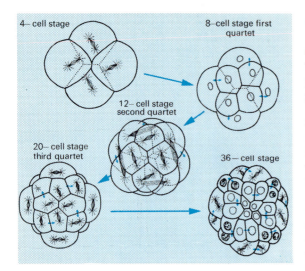

FIG. 6. Spiral cleavage of the egg, found in polyclad Turbellaria, Nemertinea, marine Annelida, Mollusca (other than Cephalopoda), showing the determinate succession and orientation of cell-divisions; probably a sign of latent homology. (After E. J. W. Barrington (1967) *Invertebrate structure and function*, Nelson.)

orientated first half-left, then half-right, then half-left again, with the precision of a drill, so that at the 64-cell stage it is possible to work out a cell lineage with indications of the prospective fates of the cells and their progeny in subsequent development. Spiral cleavage occurs in polyclad turbellarians, nemertines, marine annelids, and molluscs other than Cephalopoda. It surely indicates a general affinity between the different groups in which it is found, because it is difficult to see how such a complicated mechanism could have been evolved separately in each group, and this affinity is supported by other embryological and morphological considerations. But did the common ancestor of all these groups itself develop by spiral cleavage? It is impossible to say and difficult to assert, because in many species of these groups it does not occur. Development by means of 'polar lobes', as found in the annelid *Sabellaria* and in the mollusc *Dentalium*, present an analogous problem, because so many of their close relatives do not. These are perhaps other cases of latent homology, and another may be that of the evolution of social instincts in wasps, bees, and ants, because in each of these groups there are examples of solitary and social species, and although they are all Hymenoptera it is impossible to believe that

their common ancestor had social instincts.

Latent homology therefore conveys the impression that beneath the homology of phenotypes, there is a genetically based homology which provides some evidence of affinity between the groups that show it.

8. Homology and functional change

It is one of the definitions of homology that homologous organs can vary in the functions that they perform. There are several proofs of this, of which one of the simplest is the case of muscles and electric organs in fishes. Every time that a muscle contracts there is an output of electromotive force which in normal cases is so slight that it exerts no effect. In some fishes, however, the muscles of certain parts of the body are modified to produce electric organs which are batteries of muscles, insulated in series, which can make electric discharges powerful enough to deter predators and to kill prey. As it was difficult to imagine how these specializations arose by natural selection, and what advantages could have been conferred by initial stages of such specialization, Darwin warned that 'it would be extremely bold to maintain that no serviceable transitions are possible by which these organs might have been gradually developed'. This prophecy has been fully verified by H. W. Lissmann, who showed that weak electric discharges given off by the muscles of certain fishes function in a manner analogous to radar and provide the fish with information of the proximity of other objects, by reflection of the electric waves and their perception by the sense organs of the lateral line of the fish. This is not so much a change of function as the exaggeration of a function with the result that it serves a different purpose.

The original method of feeding of the primitive vertebrates was by the production of a ciliary current of water directed towards the mouth, wafting in particles of food. But the water must then pass out through the gill-slits, and to prevent the loss of food particles with it, an endostyle was present, as in amphioxus, consisting of bands of cilia along the floor of the throat and bands of mucus-secreting cells, the mucus of which catches up the food particles like a moving flypaper. True vertebrates feed by means of predatory jaws, and lampreys by means of a specialized sucking mouth and rasping tongue; but the larval form of the lamprey, the ammocoete, preserves the ciliary method of feeding, with an endostyle, and when it metamorphoses into the adult lamprey, the endostyle closes up, and from the opening that connected it with the floor of the throat there develops the thyroid gland. In all other vertebrates the thyroid develops in the same way from a downgrowth of the floor of the throat, partially homologous therefore with the endostyle.

Other ductless glands in vertebrates probably have a comparable origin: the pineal gland of mammals is homologous with the pineal eye of lower vertebrates which, even in the lizard embryo, still has the layer of melanin pigment. The thymus may be homologous with the excretory organs, the nephridia which, like those of amphioxus, develop from the epidermis of the dorsal end of the gill-slits.

Perhaps the most striking example yet found of function change is that of the hinder pair of wings in Diptera, common flies, which are modified into little rods, the halteres, which vibrate so fast that they serve as gyroscopic organs, as J. W. S. Pringle showed.

9. Non-homology

Just as morphology can provide proof that certain organs and structures are homologous, it can also show that others are not. In the hyoid arch of sharks, the cartilaginous skeletal elements, hyomandibular above and ceratohyal below, articulate with one another on the median side of the afferent hyoidean artery, laterally to which are the cartilaginous hyal rays some of which are joined together at their bases forming dorsal and ventral pseudohyoid bars. These stiffen the edge of the hyoid arch in the front wall of the 1st gill-slit.

In skates, the lower cartilaginous element of the hyoid arch is lateral to the afferent hyoidean artery, and therefore cannot be homologous with the ceratohyal cartilage of the sharks, but is the ventral pseudohyoid bar. In *Rhynchobatus*, a form somewhat intermediate between sharks and skates, an intermediate condition is found with a reduced ceratohyal cartilage in addition to the ventral pseudohyoid bar.

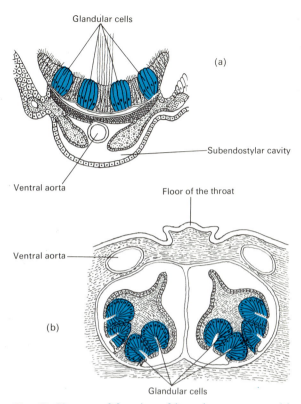

FIG. 7. Changes of function of homologous organs. (a) Transverse section through the endostyle of amphioxus showing the four tracts of ciliary and of glandular cells; (b) through the endostyle of the ammocoete larva of the lamprey shortly before metamorphosis when the thyroid gland will develop from the endostyle. (After de Beer 1966.)

FIG. 8. (a) Dorsal view of a dipteran fly showing the hind wing converted into a gyroscopic organ (haltere). (b) Enlarged view of the haltere. (Courtesy of J. W. S. Pringle.)

11

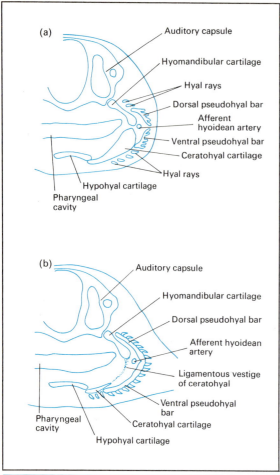

(a)

Auditory capsule

Hyomandibular cartilage

Hyal rays

Dorsal pseudohyal bar

Afferent hyoidean artery

Ventral pseudohyal bar

Ceratohyal cartilage

Hyal rays

Hypohyal cartilage

Pharyngeal cavity

(b)

Auditory capsule

Hyomandibular cartilage

Dorsal pseudohyal bar

Afferent hyoidean artery

Ligamentous vestige of ceratohyal

Ventral pseudohyal bar

Ceratohyal cartilage

Pharyngeal cavity

Hypohyal cartilage

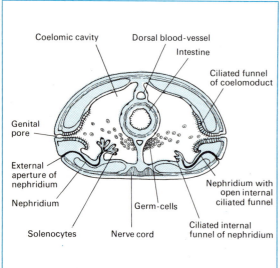

Coelomic cavity

Dorsal blood-vessel

Intestine

Ciliated funnel of coelomoduct

Genital pore

External aperture of nephridium

Nephridium

Solenocytes

Germ-cells

Nerve cord

Nephridium with open internal ciliated funnel

Ciliated internal funnel of nephridium

One of the most fundamental facts in the morphology of animals is the body cavity or coelom, in all above the evolutionary level of the flatworms. This cavity serves primarily for the formation and temporary storage of the germ cells, which are eventually evacuated into the surrounding medium by means of tubes known as coelomoducts. A completely different set of tubes, nephridia, serve to void excretory products out of the body, and these are found even in the flatworms which have no coelom. Where a coelomic cavity exists the nephridia project into it.

The distinction between the two sets of tubes was made clear 75 years ago by E. S. Goodrich, who showed that nephridia are developed inwards from the ectoderm, whereas coelomoducts are developed outwards from the mesodermal wall of the coelomic cavity. Both kinds of tubes can be seen in annelid worms. The difficulty in interpreting these two kinds of structure in any given form is partly that both nephridia and coelomoducts may end internally in the coelomic cavity with open funnels surrounded by beating cilia which create a current of fluid serving to expel their content whether excretory products or germ cells. Another difficulty is that in some animals the function of the nephridia has been taken over completely by the coelomoducts and the nephridia disappear, as for example in molluscs and in vertebrates. This is why man has a urinogenital system, while annelids have separate excretory and genital systems. Nephridia and coelomoducts are not homologous at all. Even today, this fundamental fact of morphology is often ignored by continental European zoologists.

10. Homology and embryology
Since every organ and structure in any organism has come into existence only as a result of embryonic development, it is natural to look to embryology for evidence on homologous structures. At late stages of development, when

FIG. 9. Non-homology, demonstrated by morphological relations of the afferent hyoidean artery in the hyoid arch of sharks and skates, as seen in transverse sections of heads of (a) shark, and (b) skate. (After de Beer (1932), *Q. Jl microsc. Sci.*)

FIG. 10. Non-homology of nephridia and coelomostomes, shown by the separate presence of both pairs of structures in annelid worms. Solenocytes are blind internal endings of primitive nephridia. (After Goodrich 1946.)

morphological relations between structures are established, such studies may yield valuable results, as in the case of the ear ossicles mentioned above. But at very early stages, such research leads to disappointment.

Progress in early embryology has made such strides that there are two levels on which the relations of homology and embryology can be studied. The first concerns the correspondence of places of origin of homologous structures in the fertilized egg or young embryo of related species, because the prospective fates of portions of embryos are now well known, and they can be plotted back on to the egg. In many cases it can be observed and proved by experiment, e.g. on the mollusc *Dentalium* and in tunicates, that there are extensive translocations and movements of these 'organ-forming substances'. Structures as obviously homologous as the alimentary canal in all vertebrates can be formed from the roof of the embryonic gut cavity (sharks), floor (lampreys, newts), roof and floor (frogs), or from the lower layer of the embryonic disc, the blastoderm, that floats on the top of heavily yolked eggs (reptiles, birds). It does not seem to matter where in the egg or the embryo the living substance out of which homologous organs are formed comes from. Therefore, *correspondence between homologous structures cannot be pressed back to similarity of position of the cells of the embryo or the parts of the egg out of which these structures are ultimately differentiated.*

The same conclusion arises from cases of larval divergence. *Polygordius* is a primitive worm. In one species, *P. lacteus*, the trunk of the future worm develops all coiled up inside the body of the trochophore larva; in *P. neapolitanus* the trunk develops outside the larva as a worm-like extension of it. In spite of these developmental differences, the fully formed worms of the two species are practically indistinguishable. G. Fryer has drawn attention to the remarkable differences in larval structure and development between two bivalve molluscs, *Mutela* and *Unio*, the adults of which are so similar.

The other level of embryology at which the relations between it and homology can be studied is the induction of tissues to undergo differentiation, as a result of diffusion of substances from a master structure called an *organizer*. It was found by Hans Spemann that the dorsal lip of the blastopore of a newt embryo at the gastrula stage, has the power, when grafted anywhere into the body of another embryo, of inducing the tissues by which it by accident finds itself surrounded to differentiate into all the structures characteristic of a vertebrate embryo: notochord, segmental muscle plates, kidney tubules, spinal cord, brain with eyes, etc. If they had been left undisturbed these tissues would have differentiated into very different structures. This is another proof that the quality of a structure is not dependent on the place of origin of the material out of which it is formed.

It was a problem to know why the lens of the vertebrate eye, which develops from the epidermis overlying the optic cup, should develop exactly in the 'right' place, and fit into the optic cup so perfectly, until it was discovered that the optic cup is itself an organizer which induces the epidermis to differentiate into a tailor-made lens. At least, this is what it does in the common frog, *Rana fusca*, in the embryo of which, if the optic cup is cut out, no lens develops at all. But in the closely related edible frog, *Rana esculenta*, the optic cup can be cut out from the embryo, and the lens develops all the same. It cannot be doubted that the lenses of these two species of frog are homologous, yet they differ completely in the mechanism by which determination and differentiation are brought about.

This is no isolated example. In true vertebrates, the spinal cord and brain develop as a result of induction by the underlying organizer; but in the 'tadpole larva' of the tunicates, which has a 'spinal cord' like the vertebrates, it differentiates without any underlying organizer at all. All this shows that *homologous structures can owe their origin and stimulus to differentiate to different organizer–induction processes without forfeiting their homology.*

Attention must now be paid to the germ layers. It was discovered a hundred and fifty years ago by C. Pander and K. E. von Baer that the fertilized eggs of all animals above the jelly-fish give rise to layers of tissue, three in number: ectoderm, endoderm, and mesoderm, which become folded up in different ways. It was then found that in general, ectoderm gives rise to epidermis, nervous system, sense organs, and nephridia; endoderm to the alimentary canal and its derivatives (in vertebrates: thyroid, lungs, liver, pancreas, appendix, urinary bladder); mesoderm to dermis, connective tissue, cartilage, bone, muscles, germ cells, coelomoducts or genital ducts, and also to kidneys where nephridia have been lost.

Very soon, this generalization became a dogma, and it was held that homologous organs *must* always arise from the 'correct' germ layer. This position was first shaken when experiments involving extirpation of the neural crest (from which nerve cells arise) in newt embryos also resulted in absence of cartilages of the jaws and other visceral arches. It was morphological heresy to think that cartilage could arise from ectoderm. The orthodox view was that no valid conclusions could be drawn from experimentally mutilated embryos. It therefore became necessary to demonstrate the facts from the study of embryos on which no experiments had been performed, and this is what I did in 1947.

In newt eggs, ectodermal tissues arise from the upper superficial part of the egg which is black, because of the presence of innumerable small melanin granules, which persist for a long time in the cells derived from it, and indicate their ectodermal origin. On the other hand, endodermal and mesodermal cells contain small globules of yolk which betray their origin. By means of these natural indicators I was able to show that not only the cartilages of the jaws and visceral arches consist of cells containing the tell-tale melanin granules, but also the osteoblasts of the dermal bones of the skull (frontal, parietal), and the odontoblasts in the papillae which secrete the dentine that composes the body of the teeth. Enamel had always been regarded as an ectodermal product, formed from the stomodaeal epidermis which grows in and lines the front of the mouth cavity. But enamel can be formed from ectodermal stomodaeal cells (with melanin granules), or from endodermal cells (with yolk globules), according to where the tooth rudiments are, for they act as enamel organizers.

So the imagined embryological specific monopoly of the germ layers and what they invariably give rise to was shattered. This had an effect outside zoology, for the old dogma had cut so deep that even malignant cancers used to be classified according to the germ layer of origin of the tissues in which they arose.

It is therefore necessary to give the lie direct to the entry on 'Homology' in the glossary by W. S. Dallas which Darwin most unfortunately appended to the 6th edition of the *Origin of Species*. It defines homology as 'That relation between parts which results from their development from corresponding embryonic parts.' This is just what homology is *not*. The real situation was well defined by E. B. Wilson in 1894, when he pointed out that 'Embryological development does not in itself afford at present any absolute criterion whatever for the determination of homology . . . . comparative anatomy, not comparative embryology, is the primary standard for the study of homologies.'

As if this were not enough, there are also the processes in regeneration and asexual reproduction, whereby organs are replaced or new individuals differentiated. Such cases of morphogenesis differ completely from the sequence of events in embryonic development from the egg. In many cases, as can be seen most strikingly in nemertine regeneration, and polyzoan and ascidian asexual reproduction, no respect whatever is paid to the germ layers from which the structures of the new organism are made.

Before leaving embryology, there is a further aspect of the subject that is worth consideration. It is sometimes called sexual homology, and it refers to the correspondence between organs of the genital system that have undergone different

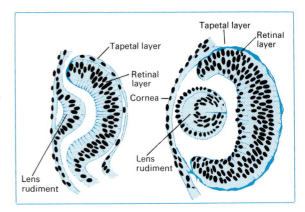

FIG. 11. Induction of the lens in Amphibia by organizer-action of the eye-cup. The eye-cup develops as an outgrowth from the brain, but the lens develops from the epidermis overlying the mouth of the eye-cup. In *Rana fusca*, the common frog, this is brought about by diffusion of an inducing substance from the eye-cup acting on the overlying epidermis, exactly in the right place. This is proved by the fact that if the eye-cup is removed, no lens is formed. But in *Rana esculenta*, the edible frog, removal of the eye-cup does not prevent the formation of a lens, which therefore owes its origin to a different mechanism. (After Huxley and de Beer 1934.)

development in the two sexes. For instance, the testis corresponds to the ovary; the scrotum containing the testes of the male corresponds to the labia majora of the vulva of the female, and the correspondence is made even more obvious in abnormal cases where the ovaries undergo 'descent' like testes, and pass by the canal of Nuck into the labia. The penis corresponds to the clitoris which, although diminutive, also contains erectile tissue. Part of the prostate corresponds to the uterus, and this fact can be made use of in certain pathological cases where enlargement of the prostate can be treated by sex hormones.

Are these corresponding organs homologous? Not in the strict sense, since it is not possible to refer them to a single representative in a common ancestor, which in vertebrates was certainly not hermaphrodite. They are the result of divergent embryonic development consequent on sexually dimorphic differentiation, due in part to genes and in part to sex hormones. The rudiments from which they have developed are homologous.

11. Homology and genetics

Because homology implies community of descent from a representative structure in a common ancestor it might be thought that genetics would provide the key to the problem of homology. This is where the worst shock of all is encountered.

It was seen in the section on latent homology that the theory of homology is concerned with homologous structures as phenotypes; but as their essence is hereditary descent from a common ancestor, it is natural to investigate the question how homology applies to genotypes. But what can be made of a gene such as that in certain fowls which not only controls the formation of a crest of feathers, but also brings about a cerebral hernia, with upswelling of the skull in the form of a knob, to accomodate it? There is no homology whatever between these two conditions. What is more, the feather crest character controlled by this gene is dominant in the wild-type gene complex, but the cerebral hernia is recessive. In the Japanese silky fowl, the gene complex suppresses the formation of the cerebral hernia altogether, while the production of the feather crest is unaffected.

Another example is that of the gene 'antenna' which, in *Drosophila*, controls the production of an extra antenna instead of an eye, structures that are not homologous. This phenomenon recalls that known as heteromorphosis in regeneration where the organ regenerated is different from that which was lost or amputated: e.g. a leg instead of an antenna, or an antenna instead of a stalked eye, described by H. Przibram and by H. W. Lissmann and A. Wolsky.

Cases are known where identical, 'homologous' genes (as can be proved by breeding experiments) control characters which can be shown to have evolved independently. *Triphaena comes* is a moth which on the mainland of Britain is grey, but in the Orkneys and the Hebrides has dark races. But as E. B. Ford showed, the manner in which the genetic control of the dark races was built up was quite different in the Orcadian and Hebridean forms, which means that the dark colour in the two is not homologous.

What all this means is that *characters controlled by identical genes are not necessarily homologous*.

The converse is no less instructive. In *Drosophila* there is a gene, 'eyeless', which deprives its possessor of eyes. It is a recessive character, which is important because it means that when its effect is produced, the fly has inherited the 'eyeless' allele from both parents, and no normal eye-controlling allele is present. If a stock of individuals pure (homozygous) for the 'eyeless' gene is inbred for many generations, there is high mortality as would be expected from the adverse effects of natural selection acting on a gene with such lethal effects. But eventually, flies appear in the offspring possessing normal eyes. It can easily be shown that the 'eyeless' gene has not changed, because when one of these phenotypically eye-possessing but genotypically homozygous 'eyeless' flies is mated with the original wild stock, i.e., the 'eyeless' gene is put back into the original gene complex, the virulent effects of the 'eyeless' gene reappear. What has happened during the inbreeding is that all the other pairs of alleles making up the gene complex have been reshuffled until a gene complex has been produced that prevents the phenotypic manifestation of the 'eyeless' allele. Other genes must therefore deputize for the absent normal gene that controls the formation of eyes. But why should they, and by what mechanism? Nobody can deny that the restored eyes that develop in genetically 'eyeless' stocks are homologous with the original normal eyes. Therefore, *homologous structures need not be controlled by identical genes*, and *homology of phenotypes does not imply similarity of genotypes*.

15

12. Revision

It is now clear that the pride with which it was assumed that the inheritance of homologous structures from a common ancestor explained homology was misplaced; for such inheritance cannot be ascribed to identity of genes. The attempt to find 'homologous' genes, except in closely related species, has been given up as hopeless. As S. C. Harland said: 'The genes, as a manifestation of which the character develops, must be continually changing .... we are able to see how organs such as the eye, which are common to all vertebrate animals, preserve their essential similarity in structure or function, though the genes responsible for the organ must have become wholly altered during the evolutionary process'.

But if it is true that through the genetic code, genes code for enzymes that synthesize proteins which are responsible (in a manner still unknown in embryology) for the differentiation of the various parts in their normal manner, what mechanism can it be that results in the production of homologous organs, the same 'patterns', in spite of their *not* being controlled by the same genes? I asked this question in 1938, and it has not been answered.

It is useless to speculate on any explanation in the absence of facts. But attention may be drawn to the work of T. M. Sonneborn (1970) on 'Gene action in development', in which he describes results obtained by him on the unicellular protozoon *Paramecium*, which show that although the 'pattern' of the cortex of that organism must be the result of genetic action, parts of that cortical pattern are necessary for the development of cortical structures at the next cell division. To the question 'Is the whole of development encoded in DNA (that is to say, in the genes)?' the answer in *Paramecium* is 'No'. Whether this is applicable to 'patterns' in higher organisms, and whether homologous structures are controlled by non-DNA mechanisms awaits further research.

FURTHER READING

References arranged conformably with the sections in this article.

1. DARWIN, C. R. (1872). *The origin of species*, 6th edn. (World's Classics edn.) Oxford University Press, London.
   GHISELIN, M. T. (1969). *The triumph of the Darwinian method*. University of California Press, Berkeley and Los Angeles.

HUXLEY, T. H. (1858). On the theory of the vertebrate skull. *Proc. R. Soc.* **9**, 381.
   OWEN, R. (1843). *Lectures on comparative anatomy and physiology of invertebrate Animals*. London.
2. CRONQUIST, A. (1968). *The evolution and classification of flowering plants*. Nelson, London.
3. DE BEER, Sir GAVIN (1964). *Atlas of evolution*. Nelson, London.
4. DE BEER, Sir GAVIN (1966). *Vertebrate zoology*. Sidgwick and Jackson, London.
5. BUTLER, P. M. (1937). Studies of mammalian dentition. *Proc. zool. Soc. Lond.* B, 103.
   DE BEER, G. R. (1937). *The development of the vertebrate skull*. Clarendon Press, Oxford.
   GOODRICH, E. S. (1913). Metameric segmentation and homology. *Q. Jl microsc. Sci.* **59**, 227.
7. HUXLEY, J. S. (1932). *Problems of relative growth*. Methuen, London.
8. LISSMANN, H. W. (1951). Continuous electric signals from the tail of a fish. *Nature, Lond.* **167**, 201.
   PRINGLE, J. W. S. (1948). The gyroscopic mechanism of the halteres of Diptera. *Phil. Trans. R. Soc.* B **233**, 347.
9. GOODRICH, E. S. (1946). The study of nephridia and genital ducts since 1895. *Q. Jl microsc. Sci.* **86**, 115.
10. DE BEER, G. R. (1947). The differentiation of neural crest cells into visceral cartilages and odontoblasts. *Proc. R. Soc.* B **134**, 377.
    —— (1962). *Embryos and ancestors*. Clarendon Press, Oxford.
    HUXLEY, J. S. and DE BEER, G. R. (1934). *The elements of experimental embryology*. Cambridge University Press
11. FORD, E. B. (1965). *Mendelism and evolution*. Methuen, London.
    MORGAN, T. H. (1929). The variability of eyeless. *Publs Carnegie Instn.* **399**, 139.
12. DE BEER, G. R. (1938). Embryology and evolution. In *Evolution*: *Essays presented to E. S. Goodrich* (ed. G. R. de Beer). Clarendon Press, Oxford.
    HARLAND, S. C. (1936). The genetical conception of the species. *Biol. Rev.* **11**, 83.
    SONNEBORN, T. M. (1970). Gene action in development. *Proc. R. Soc.* B **176**, 347.

11

Adaptation

Sir Gavin de Beer

Oxford University Press, Ely House, London W.1

GLASGOW NEW YORK TORONTO MELBOURNE WELLINGTON

CAPE TOWN SALISBURY IBADAN NAIROBI DAR ES SALAAM LUSAKA ADDIS ABABA

BOMBAY CALCUTTA MADRAS KARACHI LAHORE DACCA

KUALA LUMPUR SINGAPORE HONG KONG TOKYO

© Oxford University Press 1972

ISBN 0 19 914132 0

Sir Gavin de Beer, F.R.S., who is now retired, was formerly Professor of Embryology in the University of London and Director of the British Museum (Natural History). He is author of a large number of books on biology, embryology and evolution which include *Introduction to experimental embryology* (Clarendon Press 1926), *Development of the vertebrate skull* (Clarendon Press 1937), *Embryos and ancestors* (Clarendon Press, third edition 1962), *Charles Darwin* (Nelson 1963) and *Atlas of evolution* (Nelson 1964).

Figures 6 and 7 drawn by D. A. A. Whiteley.

PHOTOSET AND PRINTED IN GREAT BRITAIN BY

BAS PRINTERS LIMITED, WALLOP, HAMPSHIRE

1. Definitions

The word adaptation has any one of three biological meanings, which must be distinguished. *First*, it is applied to the great general principle, the most important in biology, that every living organism inherits the necessary structures and functions by means of which it interacts with the factors of the environment in which it lives and carries out the essential processes of life. These processes are: assimilation of new matter, reproduction, and capture of energy.

Assimilation, at the molecular level, means the synthesis by means of enzymes of chemical substances different from those of the organism into others similar to itself. At the level of the organism it means intake by diffusion in water or in air of simple inorganic substances or decaying organic substances, or ingestion into the body of organic substances.

Reproduction, at the molecular level, means the replication of deoxyribonucleic acids. At the level of the organism it means fission, simple or multiple, either of single cells, or division of whole multicellular individuals, either extremely unequally or subequally.

Capture of energy involves either chemosynthesis, as in the case of bacteria which derive their energy from the oxidation of iron salts, sulphur salts, ammonia to nitrites or these to nitrates; or photosynthesis making use of the energy of sunlight, based on light-absorption by metal complexes for which a pigment is necessary, as in blue-green algae or green plants; or absorption of high-energy compounds from decaying organisms by saprophytic bacteria or fungi, or from living organisms by parasites; or capture, ingestion, and digestion of living plants or animals (and therefore ultimately of plants) and use of their high-energy compounds by combustion with oxygen, as in animals. Every living organism is *adapted* to carry out these three processes.

Secondly, the word adaptation is used for the particular inherited structure or function which 'adapts' the organism to its environment. All organs (except really abortive organs) and instincts are adaptations of one kind or another.

Thirdly, the word adaptation is used to denote the personal and non-hereditary acclimatization or habituation which enables an organism to overcome the effects of changed and even harmful environmental factors, when it is exposed to habitats different from that in which it and its lineage developed. This is 'exogenous' adaptation, to use Waddington's term. A man from temperate regions protects himself from excessive solar radiation in the tropics by increasing the pigmentation in his skin. A man from sea-level acclimatizes himself to living at high altitudes by increas-

2

ing the number of his red blood corpuscles, by means of which he avoids the oxygen-lack that would otherwise ensue from decreased oxygen pressure in the atmosphere. Although these adaptations are not inherited, the capacity to develop them is.

By gradually increased small doses, an organism can tolerate concentrations of toxic substances which, if administered straight off, would be lethal. Crime stories have made use of the possibility that a man can in this way habituate himself to high intake of arsenic. He then invites to lunch with him a man whom he wishes to eliminate; they eat identically the same meal, but with different consequences. More important is the application of the principle of exogenous adaptation to the 'training' of bacteria, or other unicellular organisms, to antibiotic substances, as Sir Cyril Hinshelwood showed. Some cases give the appearance of being inherited, because all the susceptible organisms are killed off and replaced by resistant organisms. Training is not inherited.

Learning has been defined by W. H. Thorpe as adaptive changes in individual behaviour, as a result of experience or instruction. At its simplest, it is habituation; at a higher level, a conditioned reflex; then trial-and-error, and finally conscious, insight learning as in man, which has to be started from scratch at each and every generation.

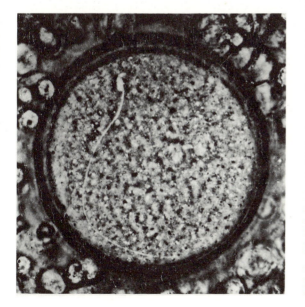

2. History

As might have been expected, the obvious facts of adaptation did not escape the notice of ancient Greeks. A remarkable example of this is in the words put into Socrates's mouth by Xenophon in his *Memorabilia Socratis*, 400 B.C.: 'Because our sight is a delicate thing, it had been shuttered with eyelids . . . eyelashes have been made to grow as a screen; and our foreheads have been fringed with eyebrows to prevent damage . . . The front teeth of all animals are adapted for cutting, whereas the molars are adapted for masticating . . . the mouth through which the things that animals like are admitted, is situated close to the eyes and nose, whereas the outlets for excrement, which is disagreeable, are directed as far as possible from the sense-organs.' (Transl. H. Tredennick, Penguin Books 1970.) Wonderful observation.

Socrates, of course, attributed organs that are 'useful for some end', to design and not to chance. Chance was exactly what Lucretius invoked in his *De rerum natura*, c. 50 B.C., to explain living organisms themselves, and the adaptations of cunning, prowess, or speed, by which their lives have been protected and preserved.

The piety exacted by Holy Church prevented any such ideas from being entertained for eighteen centuries, until the genius of French philosophical speculation pierced the obscurantist fog. As L. G. Crocker has shown, Julien Offroy de La Mettrie, in his *L'homme machine*, 1747, held that design is neither metaphysical nor providential, but 'springs from the dynamics of matter in action, free from the finalism of extra-natural intelligence.' Similar emancipation from the dogma of providence is found in the works of Pierre-Louis Moreau de Maupertuis in 1749, and of Denis Diderot in 1751.

But the Rev. William Paley, in his *Natural theology*, 1802, turned to God for help in stemming the tide of the French ungodly. The lens of the eye in fishes is more spherical than that of the eye in land animals, which shows that each eye is *adapted* to the refractive index of the medium in

FIG. 1. The differentiation between the sexes of germ-cells, and between the individuals that produce them, is the culminating stage of an adaptation of incalculable importance in evolution because it confers on the species the possibility of variation and therefore of adaptability. The egg, containing yolk, is too large to be mobile; the sperm, a nucleus with a flagellum, is mobile and finds and penetrates the egg. Here a mouse sperm has just entered an egg. (Photo J. Smiles.)

which the animal lives. Down-feathers for warmth, webbed feet for propulsion in water, poison-fangs in snakes for defence, pouches for the young in marsupials, the long tongue of the woodpecker, the life-history of mistletoe; some adaptations are even anticipatory, like migration in birds, to avoid a cold season that has not yet arrived. 'The marks of design are too strong to be gotten over. Design must have a designer. That designer must have been a person. That person is God.' Even the alternation of day and night were *designed* to suit man's periods of work and sleep. The rest of Paley's book is concerned with special pleading to demonstrate the beneficent background to evil, disease, pain, suffering, and death, helped by T. R. Malthus's argument that death is necessary to prevent overpopulation.

After Paley's pathetic and uninformed outburst, which shows the mental climate that Darwin had to overcome and which supplied him with a catalogue of adaptations for which he discovered a natural explanation, it is a relief to return to a biologist who knew something of what he was talking about, when he, Georges Cuvier, in 1817, stressed the importance of the 'combination in an animal of organs in such order that they may be in relation to the part which the animal has to play in its place in nature.' This was what Cuvier called 'the conditions of existence', and it implied adaptation; but he still required supernatural causes to explain it.

So matters remained until 1859 when Darwin showed how blind natural selection of heritable variations was the cause of adaptation.

FIG. 2. When Palaeolithic Man made this flint biface, he had an end in view, a model to copy, and a purpose which preceded his starting to work on the flint and aimed at producing an implement to fulfil a desired purpose. Such a purpose can be called teleological, in contrast to the accidental, teleonomic purposes, fulfilled by the organs of plants and animals, produced by variation and natural selection. (Photo British Museum (Natural History).)

3. Teleology versus teleonomy

In 1875, the Rev. Asa Gray congratulated Darwin because he had restored teleological purpose to biology. Gray was Darwin's foremost supporter in the United States for his principle of natural selection, but he coupled this with the view that the heritable variations on which selection worked were favourably designed, with which, of course, Darwin could not agree. But Gray's position serves as an anvil on which to hammer out and analyse the meaning of teleology and of purpose.

Gray wanted to preserve the idea of divine or providential guidance, fulfilment of design, towards a predetermined end. This is an old story, articulated by Aristotle in his *De partibus animalium* c. 330 B.C., when he wrote that each of the parts of the body, like every other instrument, is for the sake of some purpose. Darwin, who showed that design has no part to play in adaptation or evolution, never formulated his views in a general principle; this was done by a great but forgotten Englishman: William Kingdon Clifford, in 1875.

The word *purpose* has been used in a sense to which it is, perhaps, worth while to call attention. Adaptation of means to an end may be provided in two ways that we at present know of: by processes of natural selection, and by the agency of an intelligence in which an image or idea of the end preceded the use of the means. In both cases the existence of the adaptation is accounted for by the necessity or utility of the end. It seems to me convenient to use the word purpose as meaning generally the end to which certain means are adapted, both in these cases and in any other that may hereafter become known, provided only that the adaptation is accounted for by the necessity or utility of the end. And there seems no objection to the use of the phrase 'final cause' in this wider sense, if it is to be kept at all. The word 'design' might then be kept for the special case of adaptation by an intelligence. And we may then say that, since the process of natural selection has been understood, *purpose* has ceased to suggest design to instructed people, except in cases where the agency of man is independently probable.

Here, it must suffice to point out that it was only after the evolution of man, with his memory, experience, and language that objective evidence of final causes appeared on earth. When man made a flint implement, a trap for game, a canoe, a house, sowed seed, or domesticated an animal, he did it with a conscious end in view, when, as Clifford wrote, 'the idea of the end preceded the use of the means.' But when natural selection originates or improves an adaptation in a living organism, it acts blindly and by chance. It remained only for Colin S. Pittendrigh to facilitate discussion of this all-important topic by reserving the word *teleological* for cases involving final causes, where the idea of the end precedes the use of the means, and the word *teleonomic* for cases where the results of blind chance produce a structure or a function which perform some purpose useful to their possessor, but was never foreseen. As will be seen below, the origins of adaptations provide many examples of sequences of unpredictable events.

4. Adaptability

As conditions of the environment have always changed, and always will, the only passport to survival for an organism is adaptability. This implies a sufficient supply of heritable variations out of which natural selection can produce adaptations, and this in turn necessitates a mechanism of genes, their infrequent mutation, but their exchange at the start of each generation and permutations between their possible recombinations: in other words, chromosomes, and exchange of genetic material which is the essential principle of 'sex'. Exchange of genetic material and sex are the adaptations which confer adaptability. Mutations are unimportant for short-term 'purposes', though they are the ultimate source of heritable variation which segregation and recombination of genes multiply to an astronomical extent in the course of time.

There are organisms such as blue-green algae, in which no exchange of genes is known at all. They are the oldest known organisms (about 2000 million years old – the age of the earth is about 4500 million years), and there can be no doubt that they have remained as they are precisely because they have no mechanism for gene-exchange, or have lost it, and with it, the power of abundant variation. In such a population, if 10 mutations occur, the number of new variations produced (provided that they are not lethal) will be 10. But if 10 mutations occur in a population possessed of a gene-exchange mechanism, the number of new variations will be 2^{10}, or 1024. When Mendel went out of his way to provide Darwin with the solution of his great problem of an adequate supply of heritable variation, he pointed out that if two individuals differ by only 7 pairs of genes, the number of new

variations which they will produce as a result of segregation at germ-cell formation, and random recombination at fertilization (provided that they leave sufficient offspring), will be 16 384, distributed between 2187 different genetic types.

Without this genetic mechanism that produces abundant heritable variation, there is nothing for natural selection to work on to produce and improve variation and adaptation (except for occasional mutations, which are nearly always disadvantageous). The result is inevitable extinction from inadaptability, which overtakes inbreeding because it makes stocks homozygous for more and more of their genes and thereby reduces variance.

Some biologists have thought that asexual reproduction was older and more primitive than sexual reproduction because asexual reproduction by repeated fission, not preceded each time by exchange of genetic material, is common in protists, plants, and animals; and also occurs as budding in plants and some sessile animals; as fission in flatworms, some annelid worms, and even acorn-worms which are Chordates; and as parthenogenesis in water-shrimps and insects (which is only very specialized sexual reproduction without a male). This is completely untrue for the simple reasons that, in the first place, if it had not been for exchange of genetic material (i.e. sexual reproduction), there would have been no adaptation or evolution at all; and in the second place asexual reproduction is itself only an adaptation to rapid increase in numbers of a population under temporary and stable optimal conditions; but no adaptation or evolution takes place in such populations undergoing asexual reproduction.

As a result of his observations and experiments on orchids, pin-eyed and thrum-eyed primroses, and many other plants, Darwin demonstrated that flowering plants are adapted to be crossed 'at least occasionally' by pollen from a different plant. But this evidence must also apply to the very earliest stage in the evolution of living organisms, when they were little more than molecules of deoxyribonucleic acid and protein enzymes, surrounded by a cell membrane. Unless they already possessed genes, which are portions of deoxyribonucleic acids, and pooled their resources by one organism fusing with another organism and mixing their genes, no variation or adaptation would have been possible to overcome the changes in conditions which must have occurred, and there would have been no organic evolution at all. For this reason, it must be concluded that from the very start, the first organisms had genes and exchanged them. Bacteria do exchange genes, in an incomplete way. All other organisms exchange genes, except for those like blue-green algae which have remained as they are.

The evolution of sex has been a long series of adaptations. At the most primitive stage, exchange of genetic material was simply brought about by two cells, germ-cells we can call them, meeting and fusing. But as evolution proceeded and particularly when the fused or fertilized 'zygote' had to grow and develop to re-establish the 'adult' body, one type of germ-cell accumulated food-matter, yolk, and increased in volume to contain the supplies which the zygote would live on during its development. That germ-cell was then incapable of moving but this did not matter because the other type of germ-cell kept the

(Photo British Museum (Natural History).)

FIG. 3. Adaptation in the woodpecker. Two toes pointing backwards, with which it clings firmly to the bark of the tree, stiff-feathers which prop it firmly up, a strong, stout beak which can chisel holes in the bark, very strong neck muscles that can make the beak work as a hammer, and a very long protrusible tongue that can reach the grubs under the bark, all enable the woodpecker to obtain its food. (Photo British Museum (Natural History).)

power of movement, by flagella, cilia, or amoeboid movement, and was able to find and meet the other. The mobile germ-cell became the sperm, and the immobile germ-cell the egg. This differentiation between germ-cells as a result of a physiological 'division of labour' has been pressed back to the organisms that produce them; sperm-producers have become males, and egg-producers females. Every other sexual difference in anatomy, physiology, and psychology is a direct result of the differentiation between mobile and immobile germ-cells and the vital necessity that they shall meet. This is the ultimate reason why there are men and women.

5. The fitness of the environment

In his *Notebooks on transmutation of species*, Darwin sometimes attributed 'fitness' to the environment instead of to the organism as in the notes: 'non-adaptation of circumstances' which leads to extinction, and 'The constitution [of the organism] being hereditary and fixed certain physical changes [in the environment] at last become unfit [for the organism] the animal cannot change quick enough and perishes.' This line of approach was taken up by L. J. Henderson in what has been called 'his golden book'.

There are a number of physical and chemical qualities about the material of the earth's surface which are commonly taken for granted, but which are of the highest significance for the very existence of living organisms as we know them. Whatever may have been the primitive conditions when life first appeared (no free oxygen, reducing chemical situation, strong ultraviolet light), the present properties of water, carbon dioxide, and oxygen are such that they constitute a unique set of conditions without which life could not continue.

The properties of water are a case in point. Its chemical inertness, great stability, and unrivalled qualities as a solvent have made it possible for organisms to carry out their essential life-processes. The power of water to dissolve substances is seen in the enormous variety of elements and salts in the sea, and the enormous quantities of them that the sea contains. As lower organisms live chiefly in the sea, their body-fluids are related to its contents, not only as regards the chemical substances contained in them, but in their osmotic relation to sea-water. The extent and variety of ionization in water far surpass those of any other solvent, and its dielectric constant is almost the highest known, which means that the degree of electrolytic dissociation in water is also very high and keeps the ions separate and electrically charged, so that they do not unite to form electrolytically neutral molecules. This is of great importance for osmotic regulation and for the maintenance of homeostasis in body-fluids, particularly in the most highly differentiated forms because these fluids, which constitute the internal environment, are solutions and suspensions in water.

This does not end the unique properties of water. It has a very high specific heat, and the chief effect of this in the environment of organisms is the maintenance of a nearly constant temperature in the oceans, and the moderation of extremes of temperature on dry land. Within the organism itself, the production of heat leads to little change of temperature. The high latent heat of evaporation of water, the highest known, provides a mechanism of heat-loss and an essential process for temperature regulation in higher vertebrates. The high latent heat of melting of water and its very high freezing-point buffer extreme falls of temperature in seas and lakes. The unique property of having its maximum density at 4°C means that ice floats on water, and life can continue beneath it. Yet another thermal property of water is its rate of conductivity of heat, which is very high for a liquid. This brings about a rapid equal distribution of heat in the body-fluids and plays an important part in temperature regulation.

There are not wanting other properties of the physical universe without which life, as we know it, would be impossible. The most obvious of these is the range of temperature within which solar radiation heats the earth, which depends on the stage of stellar evolution of the sun, and its distance.

While the properties of water tend to make the conditions of the aquatic environment uniform, the conditions on dry land can be extremely diverse, so that its surface is subdivided into countless habitats and micro-habitats. A principle put forward by C. S. Elton acquires its significance here: as a result of their power of movement, animals migrate, sometimes on a large scale, as in birds, insects, and fishes, but mostly on a very small irregular and individual scale, with the result that in addition to the natural selection of variant animals, there is selection of habitats by animals. Marmots choose burrows for hibernation and monotremes for aestivation; small animals in danger of water-loss choose shady nooks where they escape from desiccation; insects choose their food-plants, and the plants in which they lay eggs. Such migration increases the possibilities of adaptation. In most cases the animals will select those environments to which they are already best adapted; but sometimes they may select environments to which they are by unpredictable chance pre-adapted.

6. General and special adaptation

To say that an organism is adapted to its environment is a gratuitous platitude, for if it were not adapted it would not live or be an organism. Nevertheless, there are a number of general adaptations that are so common that they are

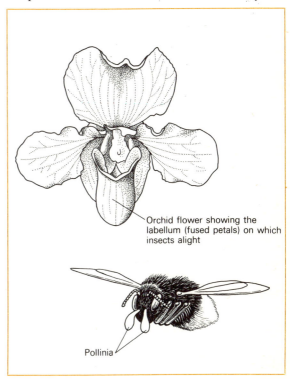

Orchid flower showing the labellum (fused petals) on which insects alight

Pollinia

FIG. 4. Orchids have evolved a complex adaptation ensuring cross-pollination with the help of insects that visit their flowers to obtain nectar. The anthers are modified into pollinia which become detached from the flower and attached to the insect, by means of a viscous fluid that rapidly sets hard. The bee flies away with them and pollinates the next flower that it visits.

taken for granted, although they are very significant. In a land plant, roots, stem, branches, and leaves are all adaptive in their normal environment, and if the function of any one of them is interfered with (no water for the roots, stem blown down, no light for the leaves), the plant dies. A fish breathes with its gills oxygen dissolved in the water; on dry land the fish dies. A man breathes gaseous oxygen through his lungs; submerged in water a man dies. In warm-blooded animals, heat-loss is relatively greater in those of small size, because the ratio between surface (through which heat is lost) and volume (in which heat is produced) is greater the smaller the size. This is why there is a limit to the geographical extension of the ranges of small birds and mammals towards polar regions, a generalization known as Bergmann's law.

General adaptations characterize all members of large taxa such as land-plants, fishes, mammals, etc. Special adaptations are found in small taxa, and they are extremely numerous. Only two examples will be given here. The woodpecker has five structural features, not found elsewhere except in related species, that are directly concerned with the mode of life which it has evolved. Two of its toes are turned backwards, which enable it to take a firm grip on the bark of the tree to which it clings; its tail-feathers are very stiff, and as they are propped against the tree-trunk they give the bird a firm stance for the chiselling that it performs with its beak; this beak is particularly strong and long; very strong neck-muscles give the beak the power of a hammer as it rains blows on the bark; the tongue is so long that when retracted it is coiled up inside the back of the head, but when extended it can reach the grubs under the bark.

The sundew plant, *Drosera*, bears on the upper surface of its leaves a large number of fine tentacles, multicellular hairs each of which ends in a gland that secretes a sticky substance. When a fly alights on any of the tentacles, neighbouring tentacles bend towards it, and between them the sticky substance holds the fly fast until it dies. The glands of the tentacles then secrete an enzyme which dissolves most of the tissues of the fly, which are then absorbed into the leaf through the same glands on the tentacles. That this extraordinary specialization is adaptive is proved by the facts that sundew plants artificially 'fed' on flies or small pieces of meat, compared with plants not so 'fed', have larger and more numerous leaves, higher flower-stalks, and more numerous seed-

capsules; and also by the fact that the sundew plant has few roots, which in ordinary plants are the organs through which nitrogenous food-material is absorbed in solution in water. This is why sundew plants can live on very poor soil, for they have evolved a novel method of nitrogen-supply from animal material.

There remain co-adaptations, where structural and functional adaptations involve the efficient co-operation of two individuals. The simplest and most widespread example of this is the external genital organs of the two sexes in higher animals. The most remarkable cases, however, involve a plant on the one hand and an animal on the other. Darwin noticed that no plant pollinated by wind

FIG. 5. The adaptation of Batesian mimicry has been made by females of the edible butterfly, *Papilio dardanus* (left-hand column of four forms). The females mimic different, unpalatable species, and derive immunity from their being shunned as food by birds that have learnt the distaste of their models (right-hand column). The males are not mimics. In the first row, left, *Papilio dardanus* var. *hippocoon* mimics *Amauris niavius*. Second row: *P. dardanus* var. *cenea* mimics *Amauris echeria*. Third row: *P. dardanus* var. *niobe* mimics *Bematistes tellus*. Fourth row: *P. dardanus* var. *planemoides* mimics *Acraea alcippe*. The mimics must not be too numerous relative to the models, or the birds would fail to learn their lesson. This is why females of *P. dardanus* show polymorphism and spread the risk between several models. (Photo British Museum (Natural History).)

has a coloured corolla. Colour and scent in flowering plants are adaptations that attract insects which visit the flower for the nectar and pollen which they contain, and, in visiting other flowers, pollinate them with pollen from the previous flower. If this was on a different plant, the effect is to provide a mechanism for cross-pollination and the exchange of genetic material, the advantages of which have been described above. In a similar way, the colour of some fruits attracts birds, which eat them and pass the seeds out with their droppings, thereby securing dissemination over a wide area, another advantage.

Probably the best example of plant–insect co-adaptation is that in orchids and moths, described by Darwin who also showed that all departures of flowers from the radial type into bilaterally symmetrical forms 'is governed in relation to insects'. Flowering plants and insects both evolved and underwent adaptive radiation in the late Mesozoic Era, each as a result of co-adaptation with the other.

7. Adaptation and polymorphism
Polymorphism is the occurrence together in the same habitat of two or more discontinuous forms of a species, each being maintained by natural selection, as E. B. Ford showed. One of the best-known examples is the Batesian mimetic butterfly, *Papilio dardanus*, a non-poisonous palatable species on which birds prey. The females, the more important sex from the point of view of reproduction, can protect themselves by evolving striking resemblances to non-palatable and poisonous butterflies of other species and genera, shunned by birds as soon as they have learnt to associate their bad taste with the particular pattern. Mimicry is therefore an adaptation. Thus, *P. dardanus* var. *hippocoon* mimics the model *Amauris niavius* and gains protection thereby; but only under certain numerical conditions: the model must be markedly more numerous than the mimic, or birds would not learn to shun them. At Entebbe the mimics are only 4 per cent of the models, and the mimicry is perfect. At Nairobi the mimics are more common than the models, their variability is great, and the mimicry imperfect. The protection given by Batesian mimicry therefore depends on the mimics not being too numerous, because there is little adaptive value in resembling a distasteful species which birds have not learnt to shun, as happens when many of the pattern that they eat are palatable.

Papilio dardanus gets over this difficulty by another adaptation. Other female forms than *hippocoon* mimic other genera and species of distasteful butterflies: var. *cenea* mimics *Amauris echeria*; var. *niobe* mimics *Bematistes tellus*; var. *planemoides* mimics *Acraea alcippe*. In this way, *P. dardanus* spreads the risks of its mimicry between several unpalatable models, with each of which an adaptive numerical balance is automatically achieved without affecting the others. So, polymorphism itself can be adaptive.

Man is polymorphic in his blood-groups, of which several are known. The commonest is the ABO system, as a result of which every human being belongs to one or other of groups A, B, AB, or O, a fact of utmost importance for blood-transfusion. It is clear that the proportions in which individuals of different populations show these blood-groups are adaptive, for they are relatively constant over a great number of years in each separate lineage of population, are not maintained by mutation, and must therefore be the result of selection. But we do not know what for. In this shameful state of ignorance, all that can be said is that the expectation of life in white men is greatest in group O and least in group B; the position in women is exactly reversed. A high proportion of cases of cancer of the stomach occur in individuals of group A; duodenal ulcer is commoner in group O. That the blood-groups are old-established characters is shown by the fact that groups A and O are found in chimpanzees and group B in orang-utans, which means that these polymorphisms have been preserved by natural selection for between 5 and 10 million years. What a field for research.

8. Imperfections of adaptation
For a pious divine like William Paley, who considered adaptations to be the result of divine design, there could not be any question of their imperfection. But Darwin realized clearly as early as 1837 in his first *Notebook on transmutation of species*, that the adaptation of organisms to the environments in which they evolved can be proved to be imperfect: 'A race of domestic animals made from influence in one country is permanent in another. Good argument for species not being closely adapted.' He had himself seen that some species of plants, imported from England into St. Helena, succeeded in their new

habitat better than indigenous species, which means that the adaptation of the latter was not as good as that of the former. If placental mammals had got into Australia, no marsupials would have survived at all.

From a general point of view, it is obvious that if adaptation were perfect, it could not be improved, and this would mean that no evolution was possible, since evolution is nothing but the result of improvement of adaptation. Examples of imperfect adaptation are not hard to find.

The Rhesus-positive blood-group is controlled by a gene with dominant effect. If a Rhesus-positive man has a child by a Rhesus-negative woman, the child in the womb will be Rhesus-positive in every case if the father was homozygous for this gene; in half the cases if the father was heterozygous. A Rhesus-positive foetus in a Rhesus-negative mother produces antigens which pass through the placenta into the maternal circulation and stimulate the immunological production of antibodies in her. These cross back again into the foetus, where they cause haemolytic disease of the newborn in an appreciable number of cases. This is a lamentable example of imperfect adaptation, due to the fact that physiological

FIG. 6. The Huia-bird, *Heteralocha acutirostris*, has not proceeded as far on the way of excessive adaptation as the tape-worm, but too far to survive changes in its environment. The male with a strong beak but not long enough to reach insect grubs, and the female with a long beak but not strong enough to chisel holes, were committed to their beetle-larvae food. Deforestation deprived them of this, and being too specialized to become re-adapted to different food-habits, they became extinct.

adaptations to viviparity, over 200 million years old in mammals, have not become co-adapted to the immunological mechanism. The Rhesus blood-group is found in the Rhesus monkey, whose common ancestor with man lived about 40 million years ago, since when man's lineage has possessed this blood-group. But the trouble is not as old as that. It would not occur if the entire population was Rhesus-negative, as in the original Neolithic Mediterranean race of which the Basques are the least modified surviving representatives; nor in a population entirely Rhesus-positive, as the Indo-european Celtic peoples were when they invaded Western Europe 5000 years ago. It was then that a mixed Rhesus-positive and Rhesus-negative population was produced by interbreeding, and the imperfection was made manifest.

Immunological reactions may have other deleterious effects, such as allergy and anaphylactic shock, as a result of injection of proteins, or ingestion of eggs, apples, shellfish, or other substances to which the subject has been sensitized.

9. Excesses of adaptation

Adaptations serve for the survival of the individual in the environmental circumstances in which they arose; when these circumstances change, those particular adaptations may become death-traps. George G. Simpson showed this in the evolution of horses. At the start of the Tertiary Era, the habitat of the ancestral *Eohippus* was marshy ground with leafy vegetation, favouring animals with many toes and short-crowned teeth adapted to browsing. These were the conditions under which, with increase in size, *Eohippus* evolved into *Orohippus*, *Epihippus*, and *Mesohippus*, all many-toed browsers. But conditions then changed in Oligocene times, and leafy plants were replaced by grasses with flint in their leaves, the chewing of which wore down the molar teeth. A variation in these teeth led to high-crowned grinding teeth with transverse ridges of dentine, enamel, and bone, making them into grindstones, and, most important, provided with open ('persistent') pulp-cavities which enabled them to grow indefinitely and make good the results of abrasion. Along the new line of many-toed grazers there evolved *Miohippus*, *Parahippus*, and *Merychippus*. But the other descendants of *Mesohippus*, which had carried on the previous trend of many-toed browsers, *Anchitherium*, *Hypohippus*, and others, paid the

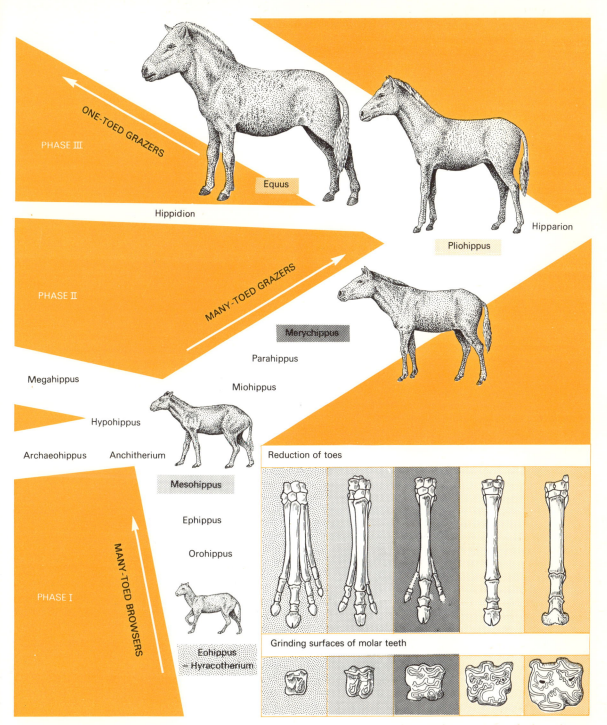

PHASE III

ONE-TOED GRAZERS

Equus

Hippidion

Hipparion

Pliohippus

PHASE II

MANY-TOED GRAZERS

Merychippus

Parahippus

Miohippus

Megahippus

Hypohippus

Archaeohippus   Anchitherium

Mesohippus

Ephippus

Orohippus

MANY-TOED BROWSERS

PHASE I

Eohippus
= Hyracotherium

Reduction of toes

Grinding surfaces of molar teeth

FIG. 7. The evolution of horses, from the beginning of the Tertiary era, has been a zig-zag course of adaptation to changing environmental conditions, determined by natural selection. Phase I: swampy ground with luxuriant leaf vegetation favoured animals with many toes and short-crowned browsing teeth. Phase II: substitution of flinty grass for leafy vegetation favoured animals with high-crowned teeth with persistent growing pulps. Phase III: hard ground favoured animals with feet reduced to single toes, or hoofs.

11

FIG. 8. Parasitism is an extreme form of adaptation, from which there is no possible recovery and survival if conditions change. This is dodder, *Cuscuta europaea*, a convolvulus that attaches itself to other plants and derives its nourishment from them. It has lost all its chlorophyll as it no longer needs to perform photosynthesis, but it produces flowers and propagates its species. (Photo British Museum (Natural History).)

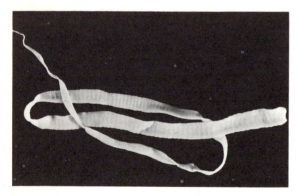

FIG. 9. The tape-worm of man, *Taenia solium*, one of the most extreme animal parasites. Living in the intestine of its host, it has lost its alimentary system, sense-organs, and means of locomotion, and is reduced to a chain of segments (proglottides) containing little but ovaries and testes which perform self-fertilization and give rise to larvae that penetrate a secondary host, normally eaten by the primary host in which the parasite develops into the tape-worm. (Photo British Museum (Natural History).)

penalty of excess of adaptation for eating leaves and not being adapted to eating grass.

A new environmental change set in during the Pliocene period: the ground became dry and hard, and successful adaptation to this was the specialization of the middle or third digit of hand and foot, the expansion of its terminal digit into a hoof, carrying the weight of the animal without any bearing on the 'pad', the specialization of the wrist and ankle joints making them spring-footed with the fetlock–pastern joint, greatly increasing speed, and with reduction of the lateral digits to splint-bones. These were the adaptations with which from *Merychippus* there evolved *Pliohippus* and *Equus*, as one-toed grazers. Meanwhile, other descendants of *Merychippus*, such as *Hipparion*, carried to the extreme the adaptation of the molar teeth, but remained three-toed, and this disqualified them from survival in the prevailing environmental conditions.

These facts show that with changing environmental conditions, the teleonomy of adaptations must also change if the organisms are to survive. They also show how untenable are the notions of programme-fulfilment and orthogenesis, supposed evolution in straight lines, propelled from the past. Horses evolved along an unpredictable zig-zag course, showing that however valuable adaptations may have been in conferring survival under previous conditions, they are lethal in many cases when conditions change.

This inexorable necessity of adaptability in the face of continually changing conditions is and was the cause of most extinction as may be illustrated by a recent example, the Huia-bird, *Heteralocha acutirostris*, of New Zealand. It lived on beetle larvae and wingless insects under the bark of trees and in decayed timber. The male had a fairly long, stout beak with which it chiselled holes. The female had a very long, more slender beak, with which it could reach to the bottom of a boring and take the food, but not strong enough itself to chisel or bore holes. Co-operation between a pair of mated Huia-birds was therefore essential for either of them to get food. With progressive deforestation, the birds' food-supply diminished, and as they were too excessively adapted to their mode of life, and unable to become re-adapted to exploit other food-supplies they became extinct, the last record of them alive being in 1907.

Excessive adaptation is not the only direct

cause of extinction. The Irish elk had antlers of such enormous size that they impeded its progress through forest regions. At earlier stages, when the antlers were not so large, they served a purpose for sexual selection in the competition between males for mates. But their growth was connected with an allometric growth-mechanism, as a result of which increase in size of the animal automatically brought about a disproportionate increased size of the antlers.

All cases of parasitism are potentially excessive adaptations, for the survival of the parasite, always degenerate, is entirely dependent on that of its host. Some cases of symbiosis fall into this category, as when the unicellular alga *Chlamydomonas* lives inside the tissues of the flatworm *Convoluta*, each benefitting the other by its different metabolism, until conditions arise under which the worm digests the alga.

10. The starting-points of adaptations
Henri Milne Edwards once remarked that nature was prodigal in variation, but niggardly in innovation, and this fact is the background to Ernst Mayr's conclusion, that when looking for the origins of adaptational and evolutionary novelties, one finds them in the modification in form and function of pre-existing characters. The hinder pair of wings of dipteran insects have been reduced to small rod-shaped bodies (the halteres) which function as gyroscopic organs, and account for the peculiarities of dipteran flight, as J. W. S. Pringle showed. The poison apparatus of venomous snakes is made of fangs which are only teeth curled into the form of hypodermic needles, and poison glands which are only modified salivary glands.

The sonar adaptation of bats, which enables them to perceive the presence of insect food in flight and of solid obstacles and objects, depends on the production of ultrasonic sounds by the larynx and the perception of the echoes from solid objects by extension of sound-perceptive areas of the skin. Whales also use sonar adaptations, which are all the more important for them because in the sea their senses of smell, taste, touch, and sight are unavailing, and unless they had a mechanism of directional sound-perception males would not find females. This is achieved by the auditory isolation of the ears from the rest of the body, by means of extensions of the air-containing cavities of the middle ear, and filling of these cavities with albuminous blubber-oil foam, which makes a very efficient sound-insulator. The ears are thus sensitive only to waves directed normally into their ear-plugs which stimulate the eardrums, as F. C. Fraser and G. E. Purves showed.

Whales have other adaptations which make it possible for them to dive to great depths in the sea without suffering from 'caisson disease' or bubbles of nitrogen in the blood. This danger is avoided by the multiplication of blood-vessels in contact with the spaces of the middle ear, because the oily albuminous foam that they contain is an absorbent of nitrogen six times more efficient than blood. When a whale breaks surface and 'spouts', it is believed that the matter so excreted is the exhalation of the nitrogen-containing foam. It has not yet been possible to collect a pure specimen of 'spout'. In all these cases it is easy to see that the advantage to the organism accrued from the very start of the adaptation; but also that, structurally, nothing in them was radically 'new'. The whole field of biology is wide open for research on origins of adaptations.

Lucien Cuénot coined the term pre-adaptation to cover cases where the organs and functions that enable an organism to live in a new and different environment are already present and pre-established in the old environment, and therefore owe nothing to the action of the ecological factors of the new environment. The cases are those of wide tolerance of salt and fresh water, humidity, heat, cold, desiccation, etc. As each case is known only *ex post facto*, the concept has no predictive value. Nevertheless, Elton's principle of selection of habitats by animals must in some cases lead to the successful testing of characters that had up till then never been tried. It could be said that air-breathing vertebrates were pre-adapted in the fish ancestor that had a hind pair of gill-pouches modified into air-containing vascular 'lungs', like the crossopterygian ancestors. It might even be argued that the entire plant and animal kingdoms were pre-adapted in the original first organisms because they had genes and the mechanism for the exchange of genetic material. Cuénot coupled his pre-adaptation with mutation (which is blind), as was fashionable among French biologists who did not fully accept natural selection (and still do not); but even he was obliged to concede that only those cases that satisfy natural selection in the new environment could survive. A striking example will be given in the next section.

13

11. The cause of adaptation

Biologists are fortunate in that research has at least enabled them to give an unequivocal answer to the question what causes adaptation, although it bristles with supplementary questions: it is natural selection, or trial and error of heritable variations. It is astonishing that distinguished biologists like William Bateson and Hugo de Vries could for one moment have thought that new varieties and species could arise as nothing but the results of mutations, which are nearly all deleterious in the conditions under which they occur, and do not 'tie in' the organism to its environment. They had no idea of ecology or of the fact that organisms do not evolve in a vacuum, but in the most intimate causal relations with all the factors of their environment. The experiments of Sir Ronald Fisher show conclusively that it is *natural selection* that controls the start, speed, direction, and stop of evolution, and that mutation *in the short term* plays no or only very little part in evolution at all. Further, that evolution is nothing but a by-product of improvement of adaptation.

Biologists are further fortunate in that they have a cast-iron example of how an adaptation has arisen, under man's eyes, in the last hundred years. It is in the Peppered Moth, *Biston betularia*, which was mottled grey in colour, matching the lichens on the barks of trees so perfectly that the moths were difficult to see, not only by man, but also by birds that prey on them. This protective colouration was therefore adaptive.

About the middle of the nineteenth century, a mutation occurred which turned moths black. Such melanic mutations are common, as for example in blackbirds, and in general, they confer increased physiological viability. But in the Peppered Moth the melanic mutations conferred nothing but ecological disaster, for the black moths were conspicuous on the lichen-covered trees, and birds fed on them with alacrity. The possessors of this mutant gene were therefore constantly eliminated. But, as is very common with mutations, this one recurred again and again, showing, by the way, how blind and unadaptive mutation is.

Fig. 10. The Peppered Moth, *Biston betularia*, in its natural surroundings on lichen-covered trees, is almost invisible. The melanic variety, *carbonaria*, which resulted from a mutation, is so conspicuous on such a background that it is quickly eliminated by bird-predators. (Photo British Museum (Natural History).)

Fig. 11. The Industrial Revolution, which killed the lichens and covered trees with soot, provided a background on which the *carbonaria* variety of the Peppered Moth was protected from predation and prospered, whereas the original *betularia* became conspicuous and eliminated by bird predators. (Photo British Museum (Natural History).)

Then, an unpredictable event happened: the Industrial Revolution, as a result of which the pollution of the air by soot, coal-dust, and dirt killed the lichens and blackened the barks of trees in industrial areas. In these areas it was now the turn of the melanic moths, var. *carbonaria*, to benefit from the protection of natural selection on the blackened, lichen-less trees, and for the original var. *betularia* to suffer the rigour of predation by birds. H. B. D. Kettlewell was able to quantify the results of predation, and to establish actuarial tables of expectation of life of the two varieties of moths in natural and in industrial regions. This has made it possible to measure the selection-pressures at work. If one wished, one could say that the melanic variety of the Peppered Moth was pre-adapted to the Industrial Revolution, for all that this might mean. But here, then, is a gene, whose effects were deleterious when it mutated, but which, because of unpredictable changed conditions, has come to confer advantage. Most mutations *become* recessive through selection of gene complexes that hide their effects, but can, under changed circumstances *become* dominant, again as a result of selection of gene complexes.

Unpredictability of environmental changes is not all the story in the origin and cause of adaptation. There is also what I have called the 'bonus principle'. In the mammalian eye, the low-threshold rods, stimulated even in dim light, are most sensitive to blue light. The cones, most sensitive to reddish light, confer acuteness of vision in bright light, with all the more precision when they are innervated one nerve-fibre to one cone, so that the light stimuli are perceived separately from very small areas of the retina. In each of these two functions, seeing more efficiently in the dark by the rods, and seeing more accurately in the light with the cones, improvement results from the very start of selection in those two directions. But when a certain degree of improvement in these two directions has been reached in one and the same eye, a mechanism has been automatically produced in which the different elements of the retina are differentially sensitive to light of different wavelengths. The result is colour vision as E. N. Willmer showed, an unpredictable function for which there never was any direct selection at all.

12. Revision

What is the quality which natural selection 'selects'? In a word, 'fitness', but that is only a word. Two obstacles must be overcome in order to understand the problem properly. One is Herbert Spencer's tautological nonsense of 'survival of the fittest', a circular argument which has done much harm, for it has lent itself to the criticism 'who are the fittest? Those who survive'; and 'which are those that survive? The fittest.' The other is the difference which some have tried to establish between 'Darwinian selection' for survival, and 'reproductive selection'. There is no such difference, for Darwin used the word 'survival' as shorthand for 'propagate their kind in larger numbers than the less well-adapted' (*Origin of species*, World's Classics edn., p. 88.). So most biologists would settle for *net reproductive advantage* for what natural selection confers on successful candidates in the struggle for existence. T. Dobzhansky puts it in a more genetical context when he says that individuals favoured by natural selection contribute a larger proportion of genes to the gene-pool of the next generation.

And yet, this will not quite do. The successful male bird of paradise leaves more offspring than his less extravagantly decorated rivals, but survives only because in his habitat there are no predators. In plants, as J. B. S. Haldane pointed out, there are genes that affect the speed at which the pollen-tubes grow down the style and fertilize the eggs, although the genes in question do not improve adaptation to the environment. It is therefore necessary to add the qualification that natural selection confers net reproductive advantage (net, because the maximum is not the optimum, which is production of offspring that reach their own reproductive stage), provided that this selection is exerted against physico-chemical factors of the environment, and organisms of other species, but *not* against individuals of the same species. In the cases of birds of paradise and pollen-tube race-winners, variations have *cheated* natural selection, but they will not do so for long.

This is the most important problem in biology. We know the general answer: natural selection produces and improves fitness. But we know so little about what is selected, and how. Populations living in nature are known to have very varied genotypes, as a result of gene-recombinations, from which heritable variations arise; but the phenotypes of these populations are remarkably uniform. Why is this? Is their uniformity due in part to non-hereditary habituation? Has Dobzhansky explained it when he stresses that there is no one-to-one relation between a gene and a trait, that evolution does not consist of independent changes of organs or traits; but that what changes is the genetic system and the developmental system that rests on it? Is this also why organs can be homologous in spite of the genes controlling them being different? But how?

Far from the old idea of 'Nature red in tooth and claw', the whole problem is mostly on the biochemical level, in physiology, embryology, and immunology, through genetics and ecology. And how little is known.

FURTHER READING

General
BARRINGTON, E. J. W. (1968). *The chemical basis of physiological regulation*. Scott Foresman, Glenview, Ill.
DE BEER, Sir Gavin (1964). *Atlas of evolution*. Nelson, London.
—(1969). Genetics and prehistory. In *Streams of culture*, p. 173. Lippincott, Philadelphia and New York.
DARLINGTON, C. D. (1958). *Evolution of genetic systems*. Oliver and Boyd, Edinburgh.
ELTON, C. S. (1930). *Animal ecology and evolution*. Clarendon Press, Oxford.
HALDANE, J. B. S. (1932). *The causes of evolution*. Longmans, London.
HENDERSON, L. J. (1913, repr. 1958). *The fitness of the environment*. Beacon Press, Boston, Mass.
HUXLEY, Sir Julian (1963). *Evolution, the modern synthesis*. Allen and Unwin, London.
MAYR, E. (1964). *Animal species and evolution*. Harvard University Press, Cambridge, Mass.
PITTENDRIGH, C. S. (1958). Adaptation, natural selection, and behavior. *Behavior and evolution*. Yale University Press, New Haven.
WADDINGTON, C. H. (1953). The evolution of adaptations. *Endeavour*, **12**.

For reference
DE BEER, Sir Gavin (1963). *Charles Darwin*. Nelson, London.
—(1964). Mendel, Darwin and Fisher. *Notes and Records of the Royal Society of London*, vol. 19, p. 192.
COLEMAN, W. (1964). *Georges Cuvier*. Harvard University Press, Cambridge, Mass.
CROCKER, L. G. (1959). Diderot and eighteenth century French transformism. *Forerunners of Darwin*, p. 114. Johns Hopkins Press, Baltimore.
CUÉNOT, L. (1925). *L'adaptation*. Doin, Paris.

CUVIER, G. (1817). *Le règne animal*. Deterville, Paris.
DARWIN, Charles (1872, repr. 1963). *The origin of species*. World's Classics edn. with Introduction by Sir Gavin de Beer. Oxford University Press, London.
—(1875). *Insectivorous plants*. Murray, London.
—(1876). *Effects of cross and self-fertilization in the vegetable kingdom*. Murray, London.
—(1960). Notebooks on transmutation of species, edited by Sir Gavin de Beer. *Bull. Brit. Mus.* (Nat. Hist.), Historical Series, **2**.
DOBZHANSKY, T. (1959). The evolution of genes and genes in evolution. *Cold Spring Harb. Symp. quant. Biol.* **24**, 27.
—(1962). *Mankind evolving*. Yale University Press, New Haven.
FISHER, R. A. (1930). *The genetical theory of natural selection*. Clarendon Press, Oxford.
—(1932). The bearing of genetics on theories of evolution. *Sci. Prog.* **27**, 2.
—(1958). The discontinuous inheritance. *The Listener*, 17 July. London.
FORD, E. B. (1964). *Ecological genetics*. Methuen, London.
FRASER, F. C. and PURVES, G. E. (1960). *Bull. Brit. Mus.* (*Nat. Hist.*) *Zool.* **7**, 1.
GRAY, A. (1876, repr. 1963). *Darwiniana*. Harvard University Press, Cambridge, Mass.
HINSHELWOOD, Sir Cyril (1956). *The chemical kinetics of the bacterial cell*. Clarendon Press, Oxford.
KETTLEWELL, H. B. D. (1958). Industrial melanism in the Lepidoptera and its contribution to our knowledge of evolution. *Proc. X. Int. Congr. Entomol.* **2**, 831.
MAYR, E. (1959). Where are we? *Cold Spring Harb. Symp. quant. Biol.* **27**, 1.
MEDAWAR, Sir Peter (1957). *The uniqueness of the individual*. Basic Books, New York.
—(1960). *The future of man*. Methuen, London.
MOURANT, A. E. (1954). *The distribution of the human blood groups*. Blackwell, Oxford.
PRINGLE, J. W. S. (1948). *Phil. Trans. R. Soc.* B **233**, 347.
SIMPSON, G. (1951). *Horses*. Oxford University Press, New York.
—(1953). *The major features of evolution*. Columbia University Press, New York.
THORPE, W. H. (1956). *Learning and instinct in animals*. Methuen, London.
WILLMER, E. N. (1949). *Proc. Linn. Soc. Lond.* **161**, 97.

See these titles in the Oxford Biology Readers series:
  3. *The mysterious origin of flowering plants*. K. R. Sporne.
 11. *Homology: an unsolved problem*. Sir Gavin de Beer.
 13. *The origin of life*. J. D. Bernal and A. Synge.
 18. *The origin of chordates*. Q. Bone.
 55. *Evolution studied by observation and experiment*. E. B. Ford.

22

Oxford Biology Readers
Edited by J. J. Head and O. E. Lowenstein

Evolution Studied by Observation and Experiment

E. B. Ford

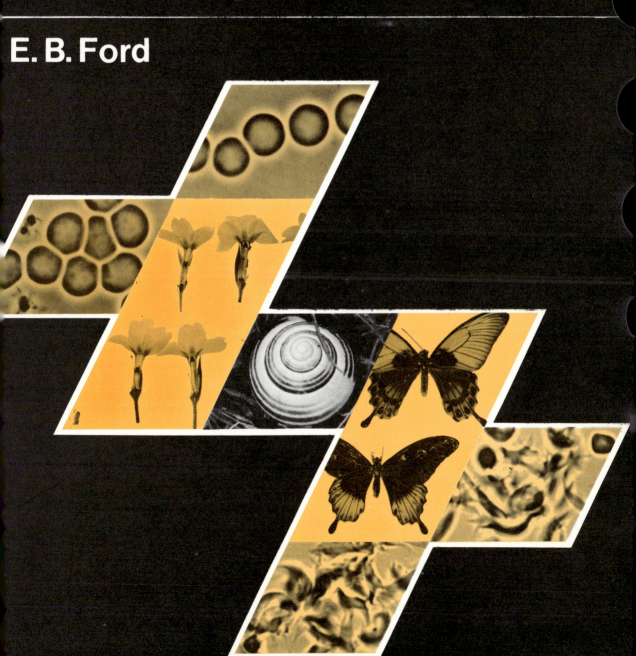

Oxford University Press, Ely House, London W.1

GLASGOW NEW YORK TORONTO MELBOURNE WELLINGTON
CAPE TOWN SALISBURY IBADAN NAIROBI DAR ES SALAAM LUSAKA ADDIS ABABA
BOMBAY CALCUTTA MADRAS KARACHI LAHORE DACCA
KUALA LUMPUR SINGAPORE HONG KONG TOKYO

ISBN 0 19 914143 6

Professor E. B. Ford, F.R.S., is a Fellow of All Souls College, Oxford, and was Professor of Ecological Genetics in the University of Oxford. He has written many books and research articles on genetics and evolution and has contributed two volumes to the New Naturalist Series (Collins).

Acknowledgements
Illustrations comprise an important part of this work and I have been most fortunate in having Mr. J. S. Haywood, whose skill as a photographer of biological material is well known, to collaborate with me. All the photographs were taken by him except where stated to the contrary. Others, and specimens loaned for illustration, are acknowledged on the relevant pages.

FILMSET AND PRINTED IN GREAT BRITAIN BY
BAS PRINTERS LIMITED, WALLOP, HAMPSHIRE

Darwin and his contemporaries demonstrated the reality of organic evolution while he and Wallace explained its mechanism by means of natural selection. Yet that process was not observed in detail nor studied experimentally for many years. Major Leonard Darwin, one of Charles Darwin's sons, told me how much his father would have valued such work; but he believed, wrongly as we now know, that even minor evolutionary changes take place too slowly for it to be practicable.

That point of view has been revolutionized by the science of ecological genetics. It has demonstrated that selection for advantageous qualities is far more powerful than Darwin and his immediate successors had thought, and consequently that evolutionary adaptations take place far more rapidly than they had conceived possible. Even in 1930 R. A. Fisher, a great authority on the subject, suggested that a one per cent selective advantage represents an approximate upper limit in natural conditions, and he was criticized for putting it so high. Today we know that 20 or 30 per cent, and

often much greater values, are common and usual. It is the purpose of this article to describe briefly, with relevant examples, how such information has been obtained, from a combination of ecology and laboratory genetics, and some of the conclusions to which it has led.

Before embarking on that inquiry, one general statement must be made. Selection, even when powerful, does not necessarily lead to *change*, quite the contrary. Mendelian segregation, and to a minute extent mutation (including chromosome reconstructions), inevitably produce variability which is checked only by selection for a constant form if this be an asset; as, for instance, in the correct development of the heart. But when we find that some small and apparently trivial quality, such as bristle number in the fly *Drosophila*, does not vary at random but is maintained at a definite frequency, we have clear evidence that though it may be of neutral survival value, the genes controlling the character, with their *multiple effects*, are not.

2

Population numbers and selection

Genetic variability is normally opposed by selection, though this can on occasion favour diversity (p. 9). That general concept was employed in one of the early attempts to observe and investigate evolution in progress. It was recognized that the changes, ecological and climatic, to which natural environments are subject, must give rise to numerical fluctuations in animal and plant populations. For unusually advantageous conditions lead to an increase in numbers, since the reproductive capacity of all organisms is much greater than is required to maintain their normal populations: they must have a 'value for safety'. When, conversely, an adverse situation supervenes, as it is bound to do after an exceptionally favourable one, the adult total may be much reduced.

An increase in population, in which some aspects of selection must be mitigated, will therefore result in increased variability; for at such a time forms will survive which would be destroyed in more rigorous conditions. Here is a possibility for rapid evolution to take place since genes can thus be tried out in fresh combinations, with some of which they may interact in new and occasionally favourable ways: an opportunity which would take an immense time to achieve in a numerically constant community. Furthermore, an advantage arising in that way will generally be at a premium when the numbers stabilize, and particularly so when they diminish again. Thus numerical increase with relative diversity prepares the way for numerical decline with relative uniformity, and the reverse.

These conclusions were originally tested in an isolated colony of the Marsh Fritillary butterfly, *Euphydryas aurinia* (Fig. 1). The individuals having been rare and invariable for eight years, underwent a great increase in numbers from 1920 to 1924 accompanied by extreme variability. Subsequently, the community became relatively constant both numerically and in appearance, but the form so produced differed from the old one. That is to say, an opportunity for rapid evolution had occurred and the colony had made use of it (Ford and Ford, 1930).

We should notice that comparable short-term opportunities for selective adjustments may arise from the succession of the seasons; for in winter, species with a series of broods throughout the year become reduced to relatively small numbers, and consequently their variability is much greater as selection is relaxed during numerical expansion in the spring. That situation has been demonstrated and analysed with special precision in *Drosophila pseudoobscura* (Birch, 1955; Dobzhansky, 1961).

Evolution of dominance

It is important to remember that each single gene has numerous effects. These all reappear together whenever it arises by recurrent mutation, which they cannot do when the apparently multiple action is in reality due to close linkage between separate loci. Evidence that the genes are usually, and probably always, responsible for a number of distinct characteristics comes, among other sources, from research on *Drosophila melanogaster*. Of the hundreds of mutants studied in that species, many controlling the most trivial features such as a slight change in eye-colour or wing-neuration, every one influences also the viability and fertility of the fly. We find, moreover, that though the genes remain permanently distinct, their effects can interact in a variety of ways, sometimes reinforcing and sometimes transforming one another.

In view of these two considerations, it is evident that the genetic characters of the organism combine to produce an internal environment or *gene-complex* (Ford 1931, pp. 38–9) within which every gene must operate. Thus if the genotype of a plant or animal be altered by segregation or mutation, the actions of alleles at other loci are in addition liable to be affected.

Bearing these facts in mind, R. A. Fisher pointed out in 1928 that selection, acting upon this general genetic variability, is capable of altering the effects

FIG. 1. *Euphydryas aurinia.* Stable forms, before 1920 (top left) and after 1925 (top right). Below are two of the variants which occurred during the period of rapid increase.

20 mm

of genes scattered throughout the chromosomes, and that in doing so it will tend to enhance the advantageous characters they produce, if any, and to diminish those that are harmful. For in each generation it is the better-adapted forms which will reproduce the more successfully. Particularly when we are concerned with an uncommon mutant, such adjustments will influence almost wholly the effects of the heterozygotes. This is because these generally comprise all the individuals in which even a single mutant gene is present at a given locus, in contrast with the homozygotes which must be relatively very rare since in them both mutant alleles must be present together. Because of these facts, we can change our terminology and say that selection will tend to make any useful effects of a gene *dominant* and, by progressively obliterating their action in single dose, to make disadvantageous ones *recessive*. Dominance and recessiveness, therefore, are properties of characters not of genes and they are the result of evolution.

It should therefore be evident that if we put a gene into a new genetic setting to which it has not been adjusted, the dominance and recessiveness of its effects will tend to be destroyed. For example, Kettlewell was able to hybridize the black form (*carbonaria*) of the British moth *Biston betularia* (Fig. 13) with a closely related North American species *Biston cognataria*. After three generations of backcrossing the segregating *carbonaria* with *cognataria*, dominance had broken down, giving a complete range of intermediates between the normal pale and black specimens (Fig. 14).

R. A. Fisher himself, working with poultry, had been the first to demonstrate dominance-modification experimentally. This was done for the first time in wild material by Ford, using the moth *Abraxas grossulariata*. Choosing respectively the more and the less extreme heterozygotes for breeding, the action of a gene for yellow, rather than the normal white, wing-colour became approximately dominant in one line and recessive in the other after only *four generations of selection*.

A cognate piece of work, with unusual implications, has been carried out on the moth *Panaxia dominula*, in which a gene with striking effects in the homozygote and detectable ones in the heterozygote (the *medionigra* form) occurs in an isolated marsh, Cothill, near Oxford. Selection experiments conducted for four generations in the laboratory made the heterozygous effects approach the dominant state in one line and the recessive in another.

Similar adjustments are now taking place in the wild, for the appearance of the heterozygotes has become more pronounced at Cothill during the last twenty years and less extreme (more recessive) at Sheepstead Hurst, where in 1954 *medionigra* had been introduced into another isolated population. This seems to be the first occasion in which experimental work in the laboratory has anticipated an evolutionary change taking place in nature.

Observing evolution in wild populations

When first attempting to observe evolution in wild populations and to study it experimentally, it seemed necessary to identify situations in which the process is likely to take place rapidly; for the generally high level of selection for advantageous qualities had not been discovered, as it subsequently was; an outcome indeed of that very project. Four such opportunities have proved to be available. (1) The first to be examined results from marked numerical fluctuations in plant and animal populations. The other three also need to be considered and briefly illustrated. They comprise (2) those occasions in which a species invades new territory; (3) instances in which multifactorial variability (i.e. variability controlled by a number of genes having similar and cumulative effects) can be studied either over a widely diversified area, or in isolated populations; and (4) all changes involving genetic polymorphism (p. 9).

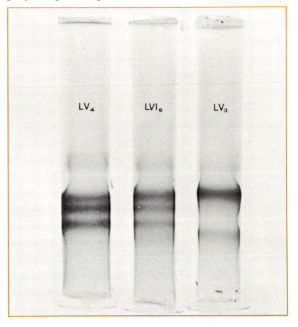

4

Adaptation on invasion of new territory

One striking example of evolutionary adaptation to a new territory is provided by the House Sparrow, *Passer domesticus*, which was introduced into North America in 1852 and has now become extremely common there. Adjustments to this novel environment have arisen due to genetic variation and selection, which has been analysed by Johnston and Selander (1964). They find that, among various local changes, the birds are now markedly dark on the coast of British Columbia where they arrived about 1900, and pale in the southwest of the U.S.A., California, and Arizona: they reached the Death Valley in 1904. While travelling further to the southeast, to the neighbourhood of Mexico City, the Sparrow has become much darker than normal and differs in other ways also, though it was not recorded there until 1933. These local distinctions do not merely represent averages, but characterize most of the individuals comprising the populations concerned.

The evolution of geographical races in birds had generally been regarded as a slow process: indeed the minimum time required was assessed by R. E. Moreau in 1930 as about 5000 years. The ineptitude of that conclusion when applied to *Passer domesticus* in colonizing North America is due to the fact that it was made without information on selection-pressures for advantageous qualities in nature. Yet we must not assume that the adaptations which prove necessary in a new habitat can always be recognized by a change in appearance. They may be cryptic, affecting physiological qualities only.

Many instances of that situation are known. They may be illustrated by a butterfly, the Essex Skipper, *Thymelicus lineola*. This European, and English, species was introduced into the New World at London, Ontario in about 1910. It feeds on grass, and has now spread to New Brunswick, through New England and down to Pennsylvania, becoming common enough in some places to constitute a minor agricultural pest. It has done so without any visible modifications. The same may

be said of the Canadian Musk, *Mimulus guttatus*, a plant which travelled in the reverse direction; for it is now widely distributed in Britain, where it was first reported in 1830. Yet we cannot doubt that these species have had to adapt themselves to their new and very distinct surroundings.

Methods are now available for detecting at least a part of such cryptic variation. Mutation produces a change in one or more of the amino acids in a protein; that controlled by the gene concerned. This may affect the charge on the protein, so altering its rate of movement in an electric field. If this migration is made to take place in a starch or an acrylamide gel, it can be observed by the special enzyme-staining methods that have been developed in recent years (Fig. 2). Some mutants controlling changes in the proteins of the organism, which are not overt characters, can thus be detected. We may anticipate that some of the apparently unadjusted colonists among plants and animals will be distinguishable by their proteins from the populations which gave rise to them in distant lands.

From what has already been said, we should expect such purely physiological variation to be influenced by fluctuations in numbers or by the conditions that give rise to them. That situation has already been discovered in the Field Vole *Microtus agrestis* by Semeonoff and Robertson (1968).

Multifactorial variation over a wide area

We can now turn to multifactorial variation occurring either over a wide and diverse area or in isolated communities; for this provides another of the situations in which evolution may proceed rapidly. It does so owing to the large amount of continuous variation under genetic control to which such qualities are subject and, as pointed out (e.g. Ford 1971, p. 46), to the small effect produced by segregation at each of the loci involved, which individually can do little to disturb the existing genetic balance. Instructive examples of this kind are provided by one of the commonest of British, and European, butterflies, the Meadow Brown, *Maniola jurtina* (Satyridae). It has proved to be tolerant of a very wide range of conditions; dry or damp habitats, open grassland on alkaline soils, acid meadows, along rides in woods, and in Atlantic and Continental climates. The caterpillar feeds on grass and there is but one generation in the year.

In order to study the adaptations of this insect

FIG. 2. Protein variation in the beetle *Coccinella septempunctata* showing relative mobilities of gene products in an acrylamide gel. The origin is at the top. The upper heavy band appears to be monomorphic and invariable. Below are three bands controlled by three alleles. Two of these bands (the upper very near the monomorphic one) appear in the left and middle gels. These are heterozygotes. The third (lowest in the right-hand gel) indicates a homozygote. (Photo G. Bell.)

we required some feature varying multifactorially, and we chose the occurrence of spots on the underside of the hindwings. These are placed near the margin, and may be absent or present in any number up to five (Fig. 3). Their frequencies are different in the two sexes (Fig. 4). At approximately 15°C, the heritability of spotting is rather high in the females, 0.63 ± 0.14 (that is to say, about 49 to 77 per cent of the variation is genetic, the remainder being environmental), and low in the males: 0.14 (with an error that cannot usefully be calculated) (McWhirter 1969). It increases with a rising temperature: at about 22°C it amounts to 0.78 ± 0.16 in the females and 0.47 ± 0.2 in the males. There is therefore considerable genetic diversity upon which selection can work.

It is possible that these small spots are of no importance to the individual, but that is not true of the genes controlling them, for these are powerfully influenced by selection. They affect, among other features, rate of development, and survival during the female aestivation which occurs in Italy.

The males are nearly always unimodal (that is to say, with a single maximum) at two spots, though the frequencies vary. Throughout the mainland of Britain, except Devon and Cornwall, the females are unimodal also but at 0 spots (the 'Southern English' type) (Fig. 4a). This is the most general condition. It characterizes the species throughout the greater part of Europe. The situation in the males, though in some respects particularly important, is the less striking, so that it will be appropriate to limit this account to the females. In west Devon (extending in some years even into Dorset) and east Cornwall (Fig. 4b), also in many other outlying habitats in Europe, these have a bimodal spot distribution, with a higher maximum at 0 and a lower at 2 spots.

The change from the Southern English to the East Cornish type occurs along a band running approximately north and south across Devon. Sometimes it involves a restricted area where the spot values are intermediate, but it may be too narrow for that: a few yards indeed, though the species is quite powerful on the wing and individuals are constantly crossing from the region of one stabilization to the other, between which there may be no apparent ecological or geographical distinctions. Thus there may be a sharp transition from the Southern English to the East Cornish stabilization along a line crossing a field. Evidently the difference between them is not a simple environ-

FIG. 3. *Maniola jurtina* females, natural size, showing the spots on the undersides of the hindwings. Lettered downwards in two columns: 0 to 4 spots. (c) and (d) show two spots differently arranged.

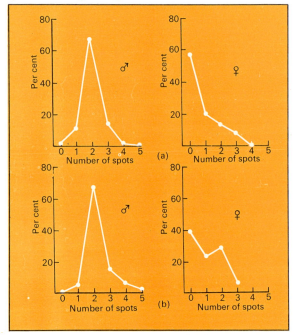

FIG. 4. Spot-frequencies of the butterfly *Maniola jurtina* (Satyridae). (a) Southern English type. (b) East Cornish type. (Reproduced with permission of Associated Book Publishers Ltd.)

6

FIG. 5. Itton Moor, Devon. In 1965 the boundary between the East Cornish and Southern English stabilizations of *Maniola jurtina* ran directly away from the observer and passed through the gap in the centre of the hedge. (Photo Dr E. R. Creed.)

(a)

Spots	0	1	2	3	4	5	Total	Spot average
Eastern area	48	34	26	7	1	–	116	0·96
Western area	49	21	32	17	1	1	121	1·20

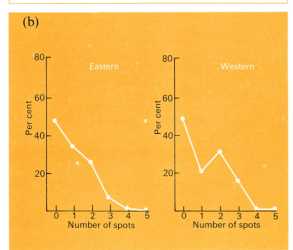

mental one, for they may not respond at all to great ecological differences. Thus in 1965 the Southern English type inhabited the eastern (right hand) half of the area at Itton Moor shown in Fig. 5, and the East Cornish the western half (Fig. 6).

Yet these forms respectively occupy the habitats illustrated in Fig. 7, where they are Southern English, and Fig. 8, where they are East Cornish. These places are obviously very different from that shown in Fig. 5, in which the dividing line between the two stabilizations runs across the middle of the picture away from the observer.

Indeed these two spotting-types represent two methods of adjustment, the interface between which may shift many miles in a generation. Wherever it may be, it is maintained by powerful selection, for a 'reverse cline' may develop there. That is to say, the characteristics of the two types may become more accentuated towards the line where they meet; indicating that the less well-adapted specimens which arise from interbreeding between them tend to be eliminated.

FIG. 6. Female spot-frequencies at Itton Moor, 1965 (Creed *et al.* 1970). (a) Table of data, (b) graph drawn from data.

7

FIG. 7. Near Swinbrook, Oxfordshire, where the *Maniola jurtina* stabilization is Southern English.

FIG. 8. Larrick, Cornwall, where the *Maniola jurtina* stabilization is East Cornish. (Photo W. H. Dowdeswell.)

Multifactorial variation in an isolated area

Evidence for such powerful selection may also be obtained from accurate adjustment to changing conditions which can take place where the species is strictly localized. One of several instances that have been studied is in the Isles of Scilly off the English Coast. It occurred on the Island of Tean where a small herd of cattle, long maintained there, was removed in the autumn of 1950. A profound alteration in the ecology ensued to which *Maniola jurtina* adjusted itself, as shown by the development of a new female spotting-type: it changed from a population bimodal at 0 and 2 spots to one unimodal at 2. This proved to be permanent, as the old one had previously been. It was possible to calculate the selection pressure necessary to achieve that change.

It amounted to 64 per cent against the non-spotted individuals in a single season.

Another aspect of evolution acting on continuous variation is illustrated by the work of Bradshaw and his colleagues who have demonstrated that it is possible for individual plants of certain species to evolve genetic tolerance to salts of heavy metals: lead, copper, zinc and others. Owing to the multifactorial nature of this quality it is likely that a few slightly tolerant plants may perhaps be found growing in ordinary places, and from them this adaptation might be built up. It enables specimens so endowed to establish themselves on the spoil-tips of old mines where metalliferous ores have been worked so that the ground is poisoned by them. For instance, McNeilly (1968) has studied the grass *Agrostis tenuis* on and immediately surrounding a small copper mine, Drws y Coed, in North Wales. Here the prevailing west wind is markedly canalized down the valley in which the mine is situated. Only copper-tolerant plants can grow on the mine tip, and up wind these are completely replaced by the normal non-tolerant type in one metre exactly at the edge of the mine. Down wind, copper tolerance fades out over 150 metres of uncontaminated ground. Here we have clear evidence for the effect of windblown pollen, and for powerful selection in favour of copper tolerance on the contaminated soil, and against it in normal conditions.

Adaptations involving genetic polymorphism

Variation controlled by major genes can also be used in the experimental study of evolution. We may ignore very small populations of 100 or so, in which chance can have relatively large effects. Otherwise, if two alleles are of equal survival-value one can replace the other only at an exceedingly slow rate; so slowly indeed that the number of *individuals* carrying a gene derived from a single mutant cannot differ much from the number of *generations* since its origin. Moreover, mutation is so rare that recurrences of it can do little to hasten the process.

If the less common of two alleles occurs in, say, one per cent or more of individuals, it cannot have attained that value fortuitously. It and its alternative must have reached, or be reaching, a state in which they are balanced by selective forces. This will favour the spread of either allele if it becomes too rare and oppose the spread of either if it becomes too common. That situation constitutes genetic polymorphism, which was defined by Ford (1940) as the occurrence together in the same locality of two or more discontinuous forms of a species in such proportions that the rarest of them cannot be maintained merely by recurrent mutation. It is one, moreover, in which coadapted genes, cooperating for the good of the individual, become, by selection, clustered together on the same chromosome. They then form a 'super-gene' the members of which are too close to be separated by crossing-over except as a rare event, when the incorrect combination can be eliminated by selection as can a disadvantageous mutant. It will be useful to give a few examples of selection acting upon polymorphism.

Cepaea nemoralis

Shells of the common snail *Cepaea nemoralis* may be dark brown or pinkish or else yellow (greenish when the animal is within). These colours are controlled genetically by multiple alleles of one gene, brown being dominant to pink and both dominant to yellow. To this coloured surface up to five bands may be added. They are generally

FIG. 10. *Cepaea nemoralis* on leaf litter. The two shells on the left are brown and the one on the right is yellow; the fourth is pink. All are unbanded.

FIG. 11. *Cepaea nemoralis*. Banded and unbanded yellow shells among long grass and mixed herbage.

FIG. 9. *Cepaea nemoralis*. Yellow and brown shells on short grass.

blackish. Absence of bands is dominant to banding of any kind and, when present, the number is controlled by modifying genes. The genes for colour and banding form a super-gene. They lie so close together on the chromosome that for long no cross-overs were detected between them. Then Cain *et al.* (1960) found two families in which about 2·25 per cent of interchanges were taking place. Such variation represents the kind from which selection can build up a super-gene by which appropriate adaptations may be kept together. This can be done by means of chromosome reconstructions, particularly inversions.

In many areas *C. nemoralis* is subject to predation, especially by thrushes which break the larger specimens on stones. Here the remains accumulate so that it is possible to discover whether the birds collect the snails at random. This they do not do, for they destroy relatively few of the least conspicuous types: yellow (greenish) upon grass (Fig. 9), brown upon leaf litter in woods (Fig. 10); banded shells upon a diversified background, as mixed herbage (Fig. 11), and unbanded in a relatively uniform environment. Cain and Sheppard (1950) found that, among the localities they were studying, the lowest percentage of yellow shells on a *green* background was 41 and the highest on a *brown* one was 17. Also that the lowest percentage of unbanded on a *uniform* background was 59 and the highest percentage on a *varied* one was 22. Here, then, we see natural selection at work.

Yet though the inappropriate colours and patterns are constantly being eliminated in nature, the populations do not become invariable. For the controlling genes also have physiological effects which help to maintain the polymorphism. Some of these will be disadvantageous and recessive, others advantageous and dominant; thus the heterozygotes will, for this and other reasons (Ford 1971, pp. 100–2), have the advantages only, so generating a genetic balance. There are, moreover, many areas, particularly on dry calcareous soils, where these physiological effects outweigh those produced by visual predation and there the average colour-pattern of the snails is unrelated to their backgrounds.

FIG. 12. Primrose flowers dissected to show the sexual organs. The style of each flower has been pulled aside to demonstrate more clearly the relation of anthers to stigma. Left, pin; centre, thrum; right, homostyle. (Photo J. S. Haywood. Reproduced with permission of Associated Book Publishers Ltd.)

Heterostyly in *Primula* spp.

These contending influences, physiological and morphological, are ever present where the latter are involved in polymorphism at all. Consider the primrose, *Primula vulgaris*, or the cowslip, *P. verus*. Here, as country children have always known, we have two forms of flowers borne on separate plants: the species are 'distylic'. There is the 'thrum' type with anthers at the top of the corolla tube and the stigma half way down, and the 'pin' in which these positions are reversed (Fig. 12). The two are controlled genetically; one set of genes affects the position of the male organs and another that of the female. These two are tied together into a super-gene giving a thrum type (dominant) and a pin type (recessive), so producing the alternative situations that are ordinarily needed. For as long as these flowers are distylic they can only be pollinated by insects and, as Darwin showed, the existence of the two conditions favours crossing between the unlike forms. Since these cannot occur on the same plant, the arrangement leads to outbreeding which must promote variation on which selection can act.

Moreoever, the genes have physiological effects also. Pin pollen is relatively ineffective (it does not often achieve fertilization) even when it does reach the high pin stigma; while thrum, whether selfed or crossed with another thrum, is almost completely sterile: Darwin's illegitimacy mechanism.

Yet rare cross-overs can occur within this super-

10

gene combining in the same flower high (thrum) anthers, with their physiology, and the long style and high stigma of pin (Fig. 12), with its physiology: such plants are no longer distylic but homostylic. This combination is self-fertile, since it is equivalent to a pin × thrum cross, so that it leads to inbreeding with consequent uniformity.

We may expect to find places in Britain, partly as a consequence of the 'enclosures' and of modern forestry, where it pays the primrose to adjust to a favourable environment by remaining relatively invariable. One is known near Sparkford, Somerset, another on the Chilterns; an outcome of evolutionary adaptation in modern times. A reprehensible practice is that of spraying land with insecticides. The destruction so caused may kill many of the insects that pollinate primroses in such places, so favouring the establishment of homostyles there, for these can fertilize themselves.

Melanism in moths and beetles

A further evolutionary effect of civilization must be mentioned here. Eighty or so species of moths have within the last hundred years become blackish in the industrial areas of Britain, sometimes spreading far outside them; and similarly in continental Europe and North America. We owe to the labours of Kettlewell (1961) a remarkably thorough study of one of these insects, the Peppered Moth, *Biston betularia*. This is normally whitish with minute black dots and pencillings. It is powerfully pro-

tected from bird predation, to which it is much subject, by its wonderful resemblance to lichen. On the other hand, the extreme melanic form, *carbonaria*, which is inky black (Fig. 13), is the better concealed on the uniform dark bark of trees in industrial areas where lichen cannot grow, poisoned chiefly by sulphur dioxide. The melanic form is a complete dominant to the pale one. The first specimen of it was recorded in Manchester in 1848. By 1895, 98 per cent of the individuals there were black, and this type is no commoner today because the gene controlling it has established a polymorphism due to its physiological effects. If we take the viability of the (black) heterozygotes as 100, then that of the black homozygotes and of the normal pale specimens proves respectively to be about 92 and 50.

Kettlewell (1965) has been able to demonstrate that the dominant black form *carbonaria* has evolved. This he has done by crossing it with the

FIG. 13. *Biston betularia*. Normal (pale) and black forms.

FIG. 14. (below) Breakdown of dominance (lettered horizontally, in 3 rows). The black form *carbonaria* of *Biston betularia* (b) is a complete dominant to the pale one (Fig. 13). The backcross therefore produces sharp segregation in equality of pale and black specimens. When crossed with the typical pale *B. cognataria* (a), a related North American species, for three generations, the *carbonaria* gene is placed in a new genetic setting largely composed of *cognataria* genes to which it is not adjusted. The *carbonaria* form is then no longer dominant, producing a range of intermediates ((c) to (f)) when it should give clear-cut segregation. Dominance is therefore not a property of the *carbonaria* gene but of that gene operating in its correctly selected gene-complex. (Material supplied by Dr H. B. D. Kettlewell, who carried out the research which it illustrates.)

11

closely related North American species *B. cognataria* (Fig. 14). The heterozygotes proved to be no longer dominant and completely black but variable and of an intermediate shade in the hybrid gene-complex to which they were not adjusted.

The lady beetle *Adalia bipunctata* is the only insect violently distasteful to birds in which black forms have spread in manufacturing regions (up to 90 per cent), and here the genes for melanism are known to have physiological effects. It is of much interest that both in this species and in *Biston betularia* they have become significantly rarer where smokeless zones have been established (Creed 1971; Clark and Sheppard 1966).

Mimicry in butterflies

Few situations throw so much light on polymorphism as butterfly mimicry, in which species palatable to predators copy distasteful ones that have acquired a striking and easily recognizable appearance which birds learn to avoid. Many such 'mimics' are polymorphic, resembling different 'models'. They increase in numbers as long as the deception favours them, but not when they become relatively so common that their colour-pattern on the whole suggests something edible instead of something to be avoided. That balance of advantage and disadvantage provides a clear basis for polymorphism; but hidden within it is another and physiological one; for the viability of the heterozygotes proves to be greater than that of the homozygotes.

The mimicry may affect colour-pattern only, when it will generally be controlled by a single major gene. This will have arisen suddenly by mutation, yet the resemblance for which it is responsible evolves gradually, by selection acting upon the gene-complex to improve any slight initial similarity to a poisonous form. We can see the reverse of that process if we cross mimics with their own species obtained outside the range of their particular models. A striking instance of the kind was obtained with the beautiful female mimicry by which the butterfly *Papilio dardanus* so well copies distasteful species in South Africa. For when mated with males from Madagascar, where *P. dardanus* is not mimetic at all, the accurate resemblances broke down in the offspring, though each form is controlled by a single gene. On backcrossing these to South African specimens, the deceptive appearance built up again. It is thus an inescapable conclusion that the mimicry is perfected by the genetic structure of the

South African population, so that selection can bring about its gradual improvement.

The situation becomes more complex but of great interest when mimicry involves a variety of features: colour-pattern of the wings, body-colour, wing-shape and habits. These distinct qualities are generally due to the action of individual genes built into a super-gene, so that they segregate together.

The mimicry discussed here is of the Batesian type, in which a relatively palatable form gains protection by copying a relatively distasteful and warningly-coloured species. There is also Müllerian mimicry in which a number of distasteful or poisonous species come to resemble one another, so that when young birds discover the inedibility of one, that lesson will apply to all. Müllerian mimicry seldom involves polymorphism since the object of this type of adaptation is to produce similarity.

Though Batesian mimicry occasionally affects both sexes, it is usually restricted to the females, never to the males. This is achieved by sex-controlled variation, not by the sex-linked type which, indeed, would not have that effect. The need for limiting Batesian mimicry to the female is probably determined by the stress of courtship. This is partly visual in butterflies, the female being stimulated by the colour-pattern of the male. Also that sex does not so urgently require protection: a single male can fertilize several females. Exposure during egg-laying is a particularly dangerous period for butterflies.

The evolutionary adjustment of genetic material is well illustrated in *Papilio memnon* from southeast Asia. It is normally tail-less in both sexes but one female form mimics a tailed species, *Atrophaneura coon*, and has acquired tails to resemble its model (Fig. 15). The gene producing this segregates with those for the correct colour-pattern since they have been brought together into the same super-gene. In the island of Palawan, however, *P. memnon* is exceptional since it always has tails (Fig. 16). These too are unifactorial, but the gene evoking them has not been brought into a super-gene with others for colour and pattern since in that place wing shape is not polymorphic and is unrelated to mimicry (Clark *et al.* 1968).

FIG. 16. *Papilio memnon*. Male (below) and female of the exceptional race from the Island of Palawan, where both sexes are always tailed. (Specimens supplied by Professor and Mrs C. A. Clarke.)

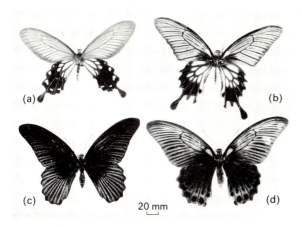

FIG. 15. *Papilio memnon*. (c) male, (d) male-like female, (b) female form possessing tails and the correct colour-pattern to mimic the poisonous species *Atrophaneura coon* (a). (Specimens supplied by Professor and Mrs C. A. Clarke.)

Cryptic polymorphism

The physiological component of polymorphism may be the only one operative when, of course, all the individuals look alike: their variation is cryptic. That situation is found in the polymorphic chromosome inversions that are widespread in flies of the genus *Drosophila*. Sixteen of these are known in the third chromosome of the North American species *D. pseudoobscura*, and in each inversion the order of a group of genes is reversed. Thus if we consider the normal arrangement to be ABCDEFGHIJK..., one inversion might produce the series ABCGFEDHIJK.... Another might be in a different region of the chromosome or the two could overlap, as would happen if the relation of the second to that just indicated were ABCDEFIHGJK.... Such inverted groups act as super-genes, for they retain their reversed order permanently. This is due to the fact that cross-overs cannot become established between the inverted and uninverted segments, since their corresponding loci no longer lie opposite one another (Fig. 17).

We are indebted to the great work of Dobzhansky (see his 1961 paper) and his colleagues for our knowledge of the evolutionary significance, as well as the genetic analysis of these inversions. He found that each has been built into a coadapted system selected to adjust the organism to particular environments. It is, moreover, normally in the heterozygous state that the inversions gain their superiority, becoming polymorphic thereby, since the disadvantageous effects which they will produce, in addition to their advantages, will tend to be recessive (p. 4).

The various inversion-types are given names, and are generally represented by a pair of upper-case letters. In many instances their frequencies undergo a seasonal cycle of changes. Thus at a locality in southern California, chromosomes carrying the 'Standard' (ST) inversion regularly decrease from 53 per cent in March to 28 per cent in June. Later in the summer, the proportion containing ST rises again until it reaches the level characteristic of early spring. Such a situation clearly indicates the operation of selection.

Dobzhansky has also encountered a recent instance of evolution affecting one of the inversions of *D. pseudoobscura*. Up to 1946, that labelled Pike's Peak (PP) was almost, though not quite, unknown in California; it had been detected in four chromosomes out of 20 000. Yet by 1954,

8 per cent of the chromosomes carried it through-out the vast area and in the immense ecological diversity represented by the greater part of that State. Its frequency has subsequently declined somewhat, though remaining far above its original value. It is still uncertain whether these striking changes are due to an alteration in the environment resulting from the residue of the pesticides so widely used in that region, or to genetic adjustments consequent upon the evolution of new super-genes in the gene-complex. Research to decide between these alternatives is being actively pursued at the present time.

Human blood groups

A further example of cryptic variation is provided by the human blood groups, of which one at least, the O,A,B series, is familiar to everyone owing to its importance in transfusion. The red cells may carry on their surfaces substances known as *antigens* each of which can interact with a cor-responding *antibody* in the plasma. When this occurs, the blood cells clump together or agglutinate (Fig. 18). They are then destroyed, giving rise to complex and dangerous symptoms in the process. A corresponding antigen and antibody cannot therefore normally coexist and ought not to be brought together: in this resides the danger of transfusion. The several members of each blood group series, of which there are about thirteen, are controlled by major genes and, as pointed out by E. B. Ford in 1942, they are polymorphic with all that this implies. Thus though they have no effect upon bodily structure, we can yet be sure that their diversity is maintained by a selective balance: partly by heterozygous advantage. From this it was predicted by Ford in 1945—and has now been proved—that the different blood groups can be associated with liability to specific diseases (e.g. the increased tendency to duodenal ulcers shown by individuals of blood group 0). They are also responsible for differential foetal mortality.

As might be expected (p. 9), the major genes controlling the blood groups are in several instances known to be collected into super-genes. It is probable that this is true also for some of those in which that arrangement has not yet been identified.

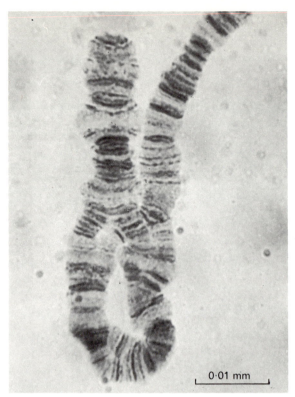

FIG. 17. Chromosomes pairing in an inversion heterozygote of *Drosophila pseudoobscura*. (Photo Professor Th. Dobzhansky.)

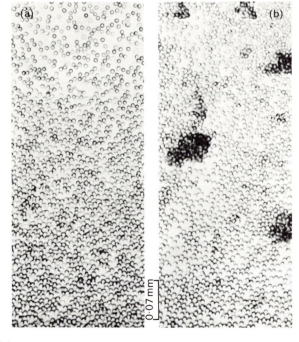

FIG. 18. (a) Normal human blood. (b) Human blood showing groups of agglutinated red corpuscles after a corresponding antigen and antibody have been brought together. (Photo Dr R. R. Race.)

14

Sickle-cell anaemia

The human blood groups seem to be very ancient. Some of them occur also in the Great Apes, and their controlling genes are so built into the gene-complex that it is difficult to shift their gene-frequencies even when these may differ greatly from one race to another, in each of which they will have distinct sets of genetic interactions. Thus a Hindu people, the Gypsies, retain their highly characteristic Indian O,A,B frequencies in Central Europe after spending hundreds of years there. Such genes are described as 'Palaeogenes' by McWhirter, in contrast with 'Neogenes' which, being responsible for relatively recent polymorphic conditions, are less firmly entrenched in the gene-complex. Thus there is a marked distinction between the blood groups and another situation connected with human blood: resistance to subtertian malaria (due to *Plasmodium falciparum*). This appears to be a relatively recent disease (Darlington, 1964). Consequently, being neogenic, the frequencies of the various polymorphisms concerned in checking its occurrence prove to be easily adjusted both in time and in space. One of them, the sickle-celled condition, has often been described as it was the first of its kind to be analysed on the basis of its clinical aspects. It does not, however, provide a particularly good example of polymorphism in general, as it does not appear to be controlled by a super-gene.

So far as we know, it is due to a single major gene concerned in modifying the chemistry of haemoglobin, of which an exceptional type 'S' completely replaces the ordinary 'adult' form 'A' in the homozygotes (except for traces of the juvenile phase, which has normally disappeared a year after birth). This affects the structure of the erythrocytes, making them curved and distorted (Fig. 19). It leads also to their rapid destruction in a way which gives rise to severe anaemia: four out of five such individuals die before the age of reproduction, and the remainder do not survive for long.

The heterozygotes, on the other hand, are quite healthy. Though about half their adult haemoglobin is of the abnormal 'S' type, the shape of their erythrocytes is unaffected when in circulation. Yet if a drop of their blood be sealed off from the air, the corpuscles assume the sickle shape but to a less pronounced degree. It will be noticed that one effect of the gene, the anaemia, is recessive while another, the formation of abnormal haemoglobin, is not.

In spite of the fact that the homozygotes suffer from a fatal disease, the heterozygotes are common in certain regions. They amount to 17 per cent of the population in parts of Greece and up to 45 per cent in some African tribes (Fig. 20). Thus they evidently have some advantage, the nature of which is now known. Their abnormal haemoglobin

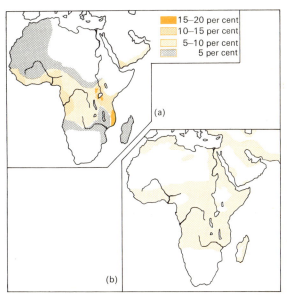

FIG. 19. Red blood cells from a patient who is heterozygous for the sickle-cell gene. (a) Normal shape in presence of oxygen, (b) distorted shape in absence of oxygen. (Phase-contrast micrographs Dr A. C. Allison.)

FIG. 20. (a) Distribution of sickle-cell gene in Africa. (b) Distribution of malaria (shaded areas) in Africa. (After de Beer).

confers marked immunity against malaria when due to *P. falciparum*, which is a parasite of the red blood corpuscles. Thus the sickle-cell condition occurs only where that disease is prevalent, and to an extent at which the disadvantage of the homozygotes is balanced by the advantage of the heterozygotes. The frequency of the gene rapidly declines in regions where measures have been taken to eliminate malaria. In this respect, the comparison with the palaeogenic condition of the O,A,B blood groups in the Gypsies is noteworthy and revealing.

We have here examples which illustrate the fact that the powerful natural selection so widely detected in plants and animals occurs also in Man. But throughout living organisms the direction of selection may alter with time, so that a particular quality may have but a small total advantage or disadvantage over a number of generations even when the selective forces operating upon it are great. Yet their high values make it possible for such features to adjust rapidly and in different ways during the periods when they are respectively harmful and beneficial. The same may be said for balanced polymorphism, the phases of which are normally maintained by selective forces in equilibrium. The power of selection against disadvantageous characters has been obvious since the time of Darwin. But its intensity in favour of advantageous ones could be recognized only when evolution was studied by observation and experiment, using the methods of ecological genetics.

FURTHER READING

General

CREED, E. R. (ed.) (1971). *Ecological genetics and evolution*. Blackwell Scientific Publications, Oxford.

FORD, E. B. (1971). *Ecological genetics*, 3rd edn. Chapman and Hall, London.

SHEPPARD, P. M. (1971). *Natural selection and heredity*, 3rd edn., repr. Hutchinson, London.

For reference

BIRCH, L. C. (1955). Selection in *Drosophila pseudoobscura* in relation to crowding. *Evolution* **9**, 389–99.

CAIN, A. J. *et al.* (1960). New data on the genetics of polymorphism in the snail *Cepaea nemoralis. Genetics* **45**, 393–411.

—and SHEPPARD, P. M. (1950). Selection in the polymorphic snail *Cepaea nemoralis. Heredity* **4**, 275–94. 275–94.

CLARKE, C. A. *et al.* (1968). The genetics of the mimetic butterfly *Papilio memnon. Phil. Trans. R. Soc.* B **254**, 37–89.

—and SHEPPARD, P. M. (1966). A local survey of the distribution of industrial melanic forms in the moth *Biston betularia. Proc. R. Soc.* B **165**, 424–39.

CREED, E. R. (1971). Industrial melanism in the two-spot Ladybird and smoke abatement. *Evolution* **25**, 290–3.

—*et al.* (1970). Evolutionary studies on *Maniola jurtina*. In *Essays in evolution and genetics* (ed. M. K. Hecht and W. C. Steer). Appleton-Century-Crofts, New York. pp. 263–87.

DARLINGTON, C. D. (1964). *Genetics and man*. Allen and Unwin, London. p. 263.

DOBZHANSKY, Th. (1961). On the dynamics of chromosomal polymorphism in *Drosophila. Symp. R. ent. Soc., Lond.* No. 1, 30–42.

FORD, E. B. (1931). *Mendelism and evolution*, 1st edn. Methuen, London. pp. 38–9.

—(1945). Polymorphism. *Biol. Rev.* **20**, 73–88.

FORD, H. D. and FORD, E. B. (1930). Fluctuation in numbers and its influence on variation in *Melitaea aurinia. Trans. R. ent. Soc. Lond.* **78**, 345–51.

JOHNSTON, R. F., and SELANDER, R. K. (1964). House Sparrows: rapid evolution of races in North America. *Science* **144**, 548–50.

KETTLEWELL, H. B. D. (1961). The phenomenon of industrial melanism in the Lepidoptera. *Ann. Rev. ent.* **6**, 245–62.

—(1965). Insect survival and selection for pattern. *Science* **148**, 1290–6.

McNEILLY, T. S. (1968). Evolution in closely adjacent plant populations. *Heredity* **23**, 99–108

McWHIRTER, K. G. (1969). Heritability of spot-number in Scillonian strains of the Meadow Brown Butterfly (*Maniola jurtina*). *Heredity* **22**, 314–18.

SEMEONOFF, R., and ROBERTSON, F. W. (1968). A biochemical and ecological study of plasma esterase polymorphism in natural populations of the Field Vole, *Microtus agrestis. Biochem. Genetics* **1**, 205–27.

See these other titles in the Oxford Biology Readers series:

1. *Some general biological principles illustrated by the evolution of man*. Sir Gavin de Beer.
3. *The mysterious origin of flowering plants*. K. R. Sporne.
10. *Studying the past by pollen analysis*. R. West.
11. *Homology: an unsolved problem*. Sir Gavin de Beer.
13. *The origin of life*. J. D. Bernal and A. Synge.
18. *The origin of chordates*. Q. Bone.
22. *Adaptation*. Sir Gavin de Beer.
68. *Evolution of flight and flightless birds*. Sir Gavin de Beer.

18

Oxford Biology Readers
Edited by J.J. Head and O.E. Lowenstein

The Origin of Chordates

Q. Bone

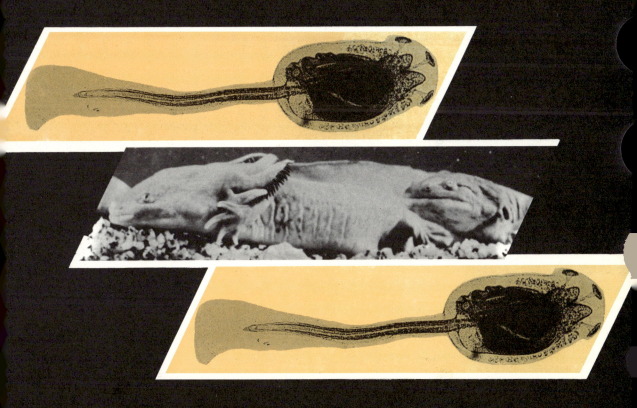

Oxford University Press, Ely House, London W.1

GLASGOW   NEW YORK   TORONTO   MELBOURNE   WELLINGTON
CAPE TOWN   SALISBURY   IBADAN   NAIROBI   DAR ES SALAAM   LUSAKA   ADDIS ABABA
BOMBAY   CALCUTTA   MADRAS   KARACHI   LAHORE   DACCA
KUALA LUMPUR   SINGAPORE   HONG KONG   TOKYO

ISBN 0 19 914118 5

Q. Bone is a zoologist at the Plymouth Laboratory,
Marine Biological Association of the U.K.

PHOTOSET AND PRINTED IN GREAT BRITAIN BY
BAS PRINTERS LIMITED, WALLOP, HAMPSHIRE

Dermal denticles of fish of two kinds are known from the lower Ordovician Glauconitic sands of Leningrad, and slightly later, small pieces of bone and bone-like material have been found in the Harding sandstones of Colorado, belonging to two different groups of fish (Fig. 2a). These first traces are fragmentary, but the kinds of fish to which they belong (Fig. 2c) are better known from later, more complete fossils (Fig. 2b). About 500 million years ago, then, bone had been evolved, and chordates were of considerable structural complexity. It is safe to assume that this implies a long period of chordate evolution prior to the Ordovician, and the group presumably arose at the base of the Cambrian, some 570 million years ago.

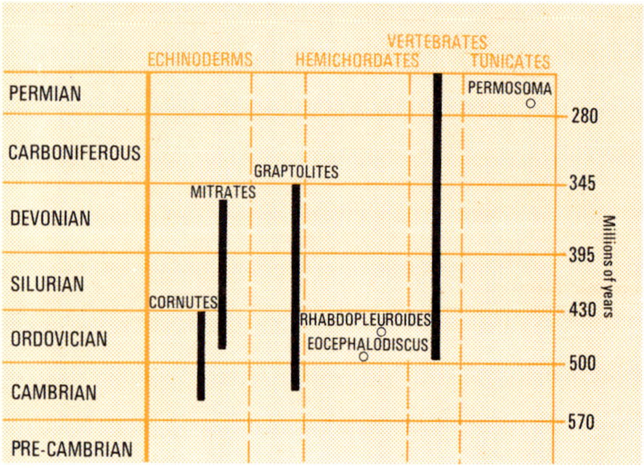

It is hardly surprising to find that there is still a good deal of room for argument about the ancestral group which gave rise to the chordates, and that, at one time or another, zoologists have proposed almost every invertebrate group as chordate ancestors. St. Hilaire's cockroach lying on its back (to make the nerve cord dorsal) was one of the first suggestions, but even in the last ten years cephalopods, annelids, nemertines, tunicates, and echinoderms have been suggested as chordate ancestors. Great ingenuity has been expended (and is required) in suggesting ways whereby the characteristic chordate features (shown in Figs. 3 and 4)—dorsal tubular nerve cord, notochord, and gill slits—could have arisen from structures in different invertebrates.

Ingenious though these attempts are (one of the most bizarre was Patten's (1912) closely argued case for arachnids as chordate ancestors), most of them are now seen to be based upon similarities due to analogy rather than homology, and there is fairly general agreement today that it is amongst the ancestors of two present day invertebrate groups—echinoderms and hemichordates—that we should seek to locate the origin of the chordates. (Analagous similarities are those resulting from the modifications of quite different structures to the same end, e.g. wings of bats and insects. Homologous similarities are due to the common origin of the similar features from an

FIG. 1. The appearance of different groups in the fossil record. (Partly after Cowie.)

2

ancestor of the two forms, e.g. the modifications of the pentadactyl limb in vertebrates.)

Modern schemes of classification differ in the way in which they express the relationships of hemichordates, tunicates, acraniates, and vertebrates; often all are accorded equal rank as subphyla of the phylum Chordata. Other schemes place the hemichordates in a phylum of their own, and retain only the acraniates and tunicates in the Chordata (Fig. 5). This is certainly rather confusing when we examine different books, and confusion is made worse because the same group of animals may be given different names, for example Tunicata or Urochordata; Acrania or Cephalochordata. But, of course, exactly how we associate large groups of like forms in these higher taxonomic units, or what we call them, merely reflects our view on the way in which the units are related, and different writers have different views. The formal scheme adopted here is shown in Fig. 5, but references to phyla and superphyla are henceforth omitted; groups are referred to by their (shortest) common names.

The 'orthodox' position, deriving chordates from echinoderms or hemichordates (Tarlo 1960, Bone 1960), will be adopted here, but it is important (and only fair) to make clear at the outset that other positions are held by present day zoologists who are conversant with the arguments which we shall consider, but are not convinced by them.

Fig. 6 shows several other schemes of chordate origins which have been suggested recently, ranging from Nursall's extreme view of separate origins of all major phyla from the protozoan level, to Jensen's revival of a theory of nemertine origin, and Gutmann's reconsideration of a segmented annelid ancestor. We may feel in this situation like Patten's despairing biologist, that 'Palaeontology is mute, Comparative Anatomy meaningless, and Embryology lies'! Still, we can review the arguments, and the reader must then

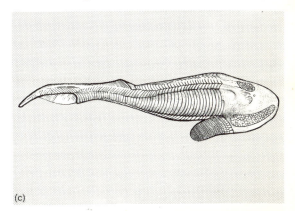

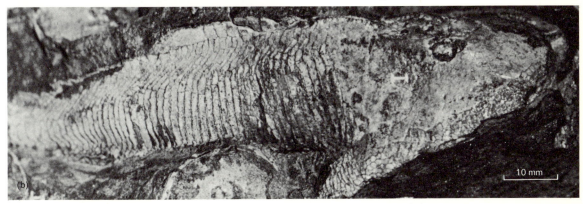

FIG. 2. (a) Portion of Ordovician rock from Canyon City, Colorado, showing fragmentary nature of the earliest known vertebrate fossil material. (b) Fossil of the primitive ostracoderm fish *Hemicyclaspis* (British Museum of Natural History) (Photos by J. J. Head). (c) Drawing of *Hemicyclaspis* fossil (after Stensiö).

3

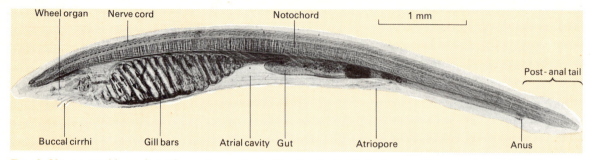

Wheel organ   Nerve cord   Notochord   1 mm

Buccal cirrhi   Gill bars   Atrial cavity Gut   Atriopore   Anus   Post-anal tail

FIG. 3. Young amphioxus just after metamorphosis showing the typical chordate features of dorsal tubular nerve cord (with black-pigmented light receptor cells); notochord; and gill bars and slits. The delicate gill apertures in adult amphioxus are protected by the metapleural folds and the pterygial muscle (see Fig. 4); in the larva, they are exposed (see Fig. 12). The arrows on the gill bars indicate the direction of ciliary beat.

FIG. 4. Transverse section of adult amphioxus showing atrium, dorsal nerve cord, and notochord.

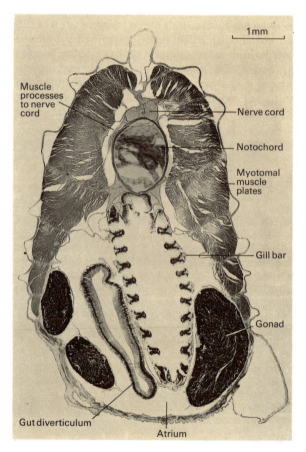

1mm

Muscle processes to nerve cord
Nerve cord
Notochord
Myotomal muscle plates
Gill bar
Gonad
Gut diverticulum
Atrium

make up his own mind which is the most convincing case. Most of the data we have to take into account are from the comparative anatomy and embryology of living forms, but some interesting work has been done in comparative biochemistry and there is some newly investigated fossil material.

Conditions under which the early chordates arose

The earliest chordate remains are from marine deposits, and the groups implicated in chordate origins are all marine, so that although it was at one time supposed that the vertebrates arose in fresh water, this is no longer accepted. Morphological arguments about the origin of the kidney and circulatory system which were held to suggest a freshwater origin, or suggestions about the origin of the chordate tail as a device to allow the animal to make headway up streams, have been rejected (see Robertson 1959). Geological evidence tells us something about the likely changes which took place in the marine environment in the pre-Cambrian and Cambrian, when the chordate line presumably arose. There is today in the earth's atmosphere about 21% of oxygen, and although there are variations due to lack of circulation and to oxygen depletion by animals, dissolved oxygen throughout the depth of the sea is more or less uniformly distributed and ample to allow animals to survive. Primevally, however, the atmosphere was a reducing one and only about 0·1% of oxygen was present. Oxygen levels in the early atmosphere gradually rose (through photosynthesis) until between 1000 and 600 million years ago the atmosphere contained about 1% of oxygen, enough to allow animals to occupy the surface layers of the sea. Further increase in atmospheric oxygen to the present level probably took place 400–450 million years ago as a result of the development of land floras.

In addition to the change in dissolved oxygen in the Cambrian sea, it seems that there were also

4

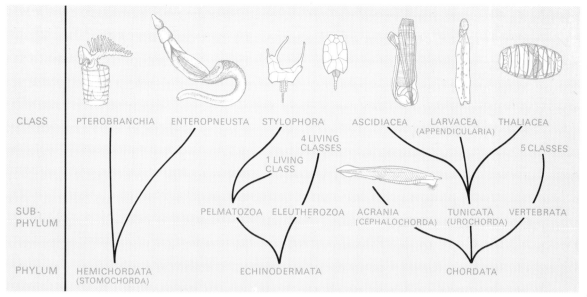

FIG. 5. Scheme of classification adopted in this Reader, alternative names for groups given in brackets. The systematic position of the Stylophora is uncertain; they have also been placed as a separate sub-phylum of the Chordata.

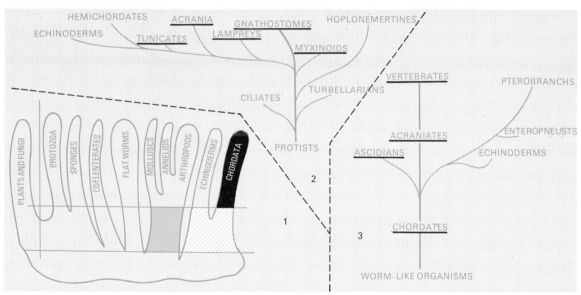

FIG. 6. Three recent schemes of the phylogeny of chordates. 1. The separate origin of all major phyla from a protozoan level of organization (Nursall). 2. Derivation of vertebrates, echinoderms, and hemichordates from a nemertine stage via a myxinoid stage (Jensen). 3. Origin of chordates, echinoderms, and hemichordates from a segmented annelid-type ancestor (Gutmann).

ionic changes, in particular a decrease in magnesium in the early Cambrian, held to favour the use of hard carbonate material for skeletal material. Calcium was abundant but as in most present day marine waters, the concentration of phosphorus seems to have been low. These tentative conclu-sions about the environment in which the chordates arose suggest two things. First, it is likely that the surface waters were colonized by planktonic forms feeding directly or indirectly upon the surface phytoplankton; secondly, it was only later (when the oxygen content of the atmosphere increased), that the shallow seabed around the coasts was made available for colonization by a great variety of sessile or semi-sessile forms. These animals could form protective skeletons of various kinds, calcareous, phosphatic, or even chitinous. Phosphatic skeletons may well have been adopted as a device for storing scarce phosphorus; there are good reasons to suppose that bone, and the more primitive material resembling bone but without cells (aspidin), arose in this way as calcium and phosphate reserves. If initially planktonic forms gave rise secondarily to sessile forms of different kinds on the seabed, as seems plausible, it is obvious that only the latter are likely to appear in the fossil record.

The whole of chordate structure, in particular bilateral symmetry and the notochord and myotomal segmented muscle blocks controlled by the dorsal nervous system, implies that the first chordates were active motile animals. At first sight it would seem that we should seek chordate ancestors amongst planktonic forms, and that the sessile or semi-sessile fossils of the Cambrian, the echinoderms, hemichordates of various sorts, graptolites, and so on are to be excluded from chordate ancestry. Several zoologists would take this view, and since palaeontology is indeed likely to be mute about the early planktonic forms, speculation comes to a halt. But there is a way out of this impasse, made clear by Garstang, whose views have had much influence, owing to their basis of profound morphological knowledge lucidly expressed (see Hardy 1954). He showed that it was possible to retain sessile animals (with motile distributory larvae) as chordate ancestors, by invoking the principle of neoteny.

Neoteny

If modifications occur in the developmental processes and the life cycle so that larvae become sexually mature, neoteny has occurred. A useful discussion of neoteny, and of other developmental changes, including a consideration of the role of neoteny in chordate origins, is given by de Beer (1940).

Living amphibia show various degrees of neoteny, and in the well-known case of the tiger salamander, *Ambystoma* (Fig. 7), the larva (the axolotl) may become sexually mature and reproduce without undergoing metamorphosis. Under

FIG. 7. Adult tiger salamander (*Ambystoma*) on right and its larva, the axolotl. Note external gills of axolotl, and dorsal fin.

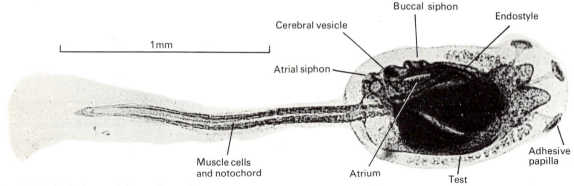

FIG. 8. Tadpole larva of the sessile tunicate *Amaroucium* showing chordate features of notochord and dorsal hollow cerebral vesicle (with eye spot). The internal arrangement of such a larva is seen in Fig. 9.

6

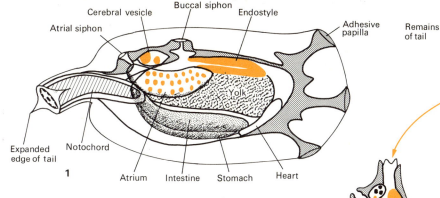

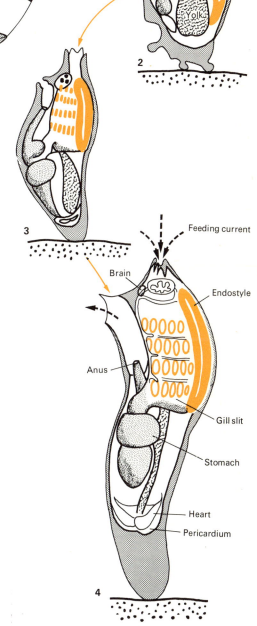

FIG. 9. Stages in the metamorphosis of the tunicate tadpole larva of *Amaroucium*. 1. Newly hatched tadpole larva (posterior part of tail omitted) showing structure. Compare with Fig. 8. The tadpole larva does not feed, and the buccal and atrial siphons are not functional. 2. Larva after attachment to substrate by the adhesive papillae. The tail has been almost completely resorbed, and the rotation of the pharynx has begun. 3 and 4. Further stages in the rotation of the pharynx leading to the young adult (cf. Fig. 5, Ascidiacea). As in amphioxus (Fig. 4) the gill slits open into an atrium, leading to the atrial opening (after Scott).

different environmental conditions, axolotls metamorphose into the adult, which then reproduces. Given the appropriate environment (as in the alkaline lakes of southern Canada) adult tiger salamanders occur only rarely, and it is a small step from such a situation to one where the adult stage might be discarded entirely, leaving us with a new form showing none of the characteristics of the *adult*, such as colour pattern and semi-terrestrial existence.

There is no question that neoteny *can* occur, nor that *if* it occurred during the origin of the chordate line, here we have a possible mechanism whereby a sessile adult with a motile larva could have given rise to the free-swimming early chordate.

Modern sessile tunicates like the solitary *Ciona* or the colonial *Botryllus* are as adults highly specialized sessile animals which show little which is chordate-like in their morphology apart from the pharyngeal filtering mechanism. The coelom is reduced or absent, there is no notochord, and the brain is not like that of a chordate. The relatively simple structure of such an animal is seen in Fig. 9.

However, as Kowalevsky showed in 1867, they possess motile larvae which have a strikingly chordate-like organization, so much so that they

7

are often loosely termed tadpole larvae. There is (Fig. 8) a dorsal tubular nervous system, with special sense organs at the anterior end, a notochord, and (unsegmented) muscle fibres for operating the propulsive tail. We will consider these larvae in detail later on, but they are obviously of great interest, and many zoologists have been impressed by the possibility of deriving the vertebrate line by the neotenous transformation of tadpole larvae belonging to a primitive sessile tunicate. This is indeed an attractively simple idea, but others, notably Carter, have found it rather naive and have not been convinced that the concept of neoteny is valuable or necessary in phylogenetic speculation. They admit that neoteny *can* occur, but that it *has* occurred seems to them doubtful, and they unhesitatingly cut the ground from under the feet of the supporters of neoteny with a flourish of Ockam's razor.

If the idea of neoteny as a means by which a group is enabled to escape from adult sessile specializations and begin a new mode of life in the larval habitat is to be seriously considered, we have to think carefully how it could have occurred, and why; it is (as Carter rightly emphasized) simply not enough to point to the resemblance of the adults of one group to the larvae of another and assert that the former arose neotenously from the latter. We need to consider the possible mechanism for such a change, and ask whether the transformation is likely to have been sudden, or very gradual over many generations? If gradual, what selective advantages can it have conferred on the larvae which progressively became closer to the new type of sexually-reproducing neotenous adult form?

Neoteny in chordate ancestry
The larvae of the living groups implicated in chordate ancestry (the echinoderms, hemichordates, tunicates, and acraniates) are morphologically very different from one another, but they fall into two distinct types depending upon the duration of larval life.

The ascidian (tunicate) tadpole which has a free swimming life of a few hours only, and which does not feed as a larva, is one type. In contrast to this, the larvae of echinoderms, acraniates, and enteropneust hemichordates spend many weeks or even months feeding and growing in the surface layers before they undergo metamorphosis to an adult on the seabed. Berrill (1955) has clearly shown

that the structural and behavioural pecularities of the ascidian tadpole larva fit it to choose a site for metamorphosis that will suit the needs of the adult ascidian. It is in fact a highly specialized stage in the life-history, designed for one role only, short-range site selection for the new adult. Once it settles, the tail is lost, the pharynx rotates, the nervous system is remodelled, and the new adult begins to feed (Fig. 9). Interestingly enough, where exact site selection is of much less importance (when the adults are found in a rather uniform habitat, as are the species which live buried in sand flats), development is direct and the tadpole larva is secondarily suppressed.

The ciliated larvae of echinoderms, hemichordates, and acraniates have a very different larval life. They drift and feed for long periods in the

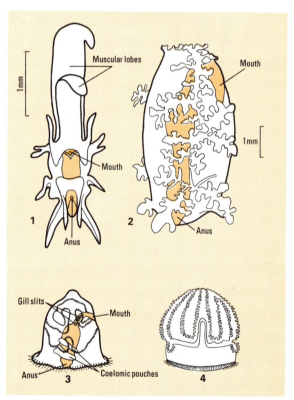

FIG. 10. Echinoderm and hemichordate larvae specialized for maintaining their position in the surface layers. 1. Bipinnaria of the starfish *Luidia* with large muscular lobes. 2. Auricularia of a Bermudan starfish with complex tentaculated ciliary bands (both after Garstang). 3. Tornaria larva of hemichordate with enlarged ciliary bands, and with gill slits in the larval stage (after Ritter and Davis). 4. Tornaria larva with complex tentaculated ciliary bands (after Stiasny).

plankton and show several kinds of adaptations to allow them to remain in the surface layers. For example, the larger tornaria larvae of enteropneusts (Fig. 10) have very complicated and enlarged ciliary bands, and the larvae of the starfish *Luidia* possess muscular lobes which assist in swimming. Acraniate larvae, though retaining their surface ciliation, are capable of muscular movement from a very early stage, and in some species at least, exhibit diurnal vertical migrations requiring muscular effort.

When they eventually metamorphose, the onus of site selection for the new adult weighs less heavily on these long-lived larvae than it does upon the ascidian tadpole larva, for the adult forms are only semi-sessile. If the young amphioxus does not care for the site its larva chose for

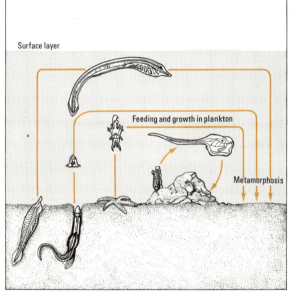

Surface layer

Feeding and growth in plankton

Metamorphosis

FIG. 11. Diagram to show the two different larval types, one (the ascidian tunicate tadpole larva) having a short free-swimming existence and metamorphosing close to the adult site; the other (hemichordate tornaria, acraniate amphioxus larva, and echinoderm auricularia) living for a long period in the plankton, and metamorphosing far from the parent site.

metamorphosis, it can move to a more favourable patch of sand (adult amphioxus are very choosy about the substrate in which they live); but the young ascidian is much more dependent upon a correct decision by its larva. Adult *Ciona* (ascidian) are apparently capable of very slow movement over the rocks to which they are attached but this is too slow and uncommon a process to

enable them to change the attachment site chosen by their larvae. Fig. 11 shows diagrammatically this dichotomy of larval types in the different groups.

Evidently each type of larva is specialized in a different direction; which is more likely to have undergone neoteny to give rise to a new adult form? Since it seems that the neotenic production of a new form would have been a gradual process, the larval phase being prolonged so that gonads appeared and became functional over a long series of generations by gradual steps, it would seem that the echinoderm–hemichordate–acraniate types of larvae are much more suitable than is the ascidian tadpole type.

We may suppose the lengthening of larval life and the consequent gradual onset of neoteny to be a response to the availability of the phytoplankton on which the larva was feeding in the surface layers. As the adult habitat on the seabed became more crowded, more and more of the life history took place at the surface.

In other words, the neotenic transformation giving rise to the chordate line (and ourselves) was more or less an accidental by-product of selection pressure on the larva to exploit the surface phytoplankton of the Cambrian! This view is strengthened by the remarkable fact that we can actually see what seems to be a similar kind of gradual neoteny of a long-lived larval form taking place at the present time.

Acraniate larvae (e.g. amphioxus) usually live for some months in the plankton, achieving a length of 4–5 mm and some 14–15 primary gill openings. At the beginning of the century however, three species of acraniate larvae were found in plankton collections which were up to 9 mm long, and had as many as 27 gill openings (Fig. 12). Because gonad rudiments were found in some of these specimens, it was at first supposed that they were adult animals, and they were put into a new genus, *Amphioxides*. Later, others were found, which could be assigned to known adult acraniate species, and it is now clear that these giant amphioxides larvae are acraniate larvae which do not metamorphose at the usual stage but go on to spend a much longer planktonic life; in some species at least, gonad rudiments appear during this larval phase. We are here very close to the pattern of gradual neoteny of a long-lived larva which we inferred for the origin of the chordates, and whilst this does not mean that acraniates gave

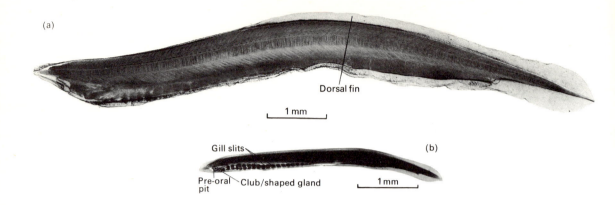

(a)

Dorsal fin

1 mm

Gill slits

(b)

Pre-oral pit   Club/shaped gland   1 mm

FIG. 12. 'Amphioxides' larva, to same scale as normal acraniate larva (below), showing enlarged fin. (a) Larva of 'Amphioxides' pelagicus (note pre-oral pit, pigment cells in cord, and notochord). (b) Late stage larva of the Amoy amphioxus (*Branchiostoma belcheri*) showing 19 gill slits. Note pre-oral pit, club-shaped gland, and absence of atrium (cf. Fig. 3).

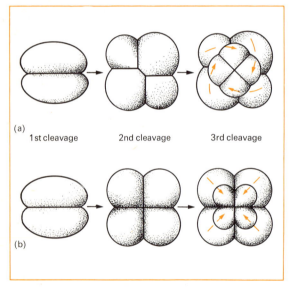

(a)
1st cleavage   2nd cleavage   3rd cleavage

(b)

FIG. 13. The two major types of cell division in the development of the embryo. (a) Spiral cleavage, characteristic of annelids and other invertebrates. (b) Radial cleavage, characteristic of echinoderms, hemichordates, and chordates.

The adult ancestral forms

Acraniates (as we might expect) are absent from the fossil record. The earliest tunicate (*Permosoma*) is known from the Permian, but there are several different types of echinoderm and hemichordate in the Cambrian and Ordovician (Fig. 1). The evidence is that there was a large radiation of hemichordates and possibly allied forms such as graptolites at the base of the Ordovician; and also that there were echinoderms showing some chordate features, which were quite unlike the starfish and sea-urchins familiar today. There is no question but that living echinoderms and hemichordates, very unlike though they are as adults, share certain basic developmental features which can only have been inherited from a common ancestor, and all of these features except the last are also found in acraniates:

1. Cleavage of the zygote is radial (unlike the spiral cleavage of invertebrates) and the anus forms near the site of the blastopore (Fig. 13).
2. The coelom is tripartite and arises by outpouching from the endoderm (enterocoely).
3. The anterior coelom develops an external pore on the left side of the larva, and a dorsal pulsating vesicle arises in association with the anterior coelom.

rise to vertebrates, it is certainly encouraging to find actual larvae alive today which seem to be recapitulating the course of events in Cambrian seas.

So far, we have considered the problem of chordate origins in a general way suggesting that neoteny allows semi-sessile or sessile forms to be considered as ancestors, and noting the kind of way in which the neotenous transformation is likely to have occurred. We now have to review the evidence for the nature of the *adult* animal that produced the larva which underwent neoteny.

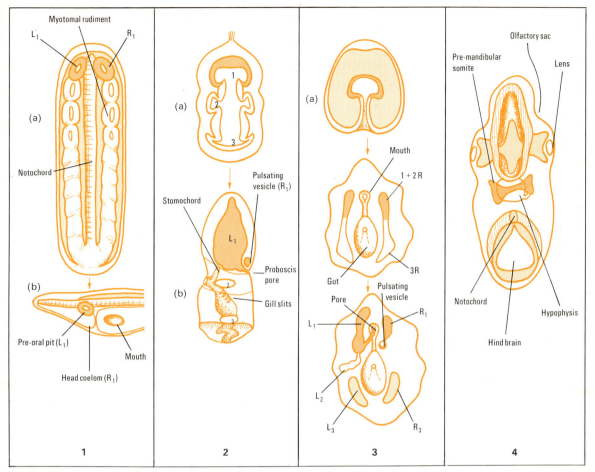

FIG. 14. Homology of the coelom in chordates, hemichordates, and echinoderms. Paired pouches arising from the gut undergo a characteristic asymmetric development as ontogeny proceeds. The right and left first coelomic pouches (R1 and L1) develop in a remarkably similar way in each group producing a proboscis pore which connects the left first coelomic pouch with the exterior. The pre-mandibular somite in the gnathostome vertebrate (4) is homologous with the first coelomic pouch but the proboscis pore is transient and may only be virtual in some classes. 1. Acraniates. (a) Horizontal section of early stage. (b) Side view of early larva (after Conklin). 2. Hemichordates. (a) Early stage (after Bateson). (b) Left side view of late tornaria larva of *Saccoglossus* (after Knight-Jones). 3. Echinoderms. Schematic stages in asteroid development (after various authors). 4. Vertebrates. Section through the head of a young elasmobranch (after Goodrich). To different scales.

4. The ciliated tornaria (hemichordate) and auricularia (echinoderm) larvae are very similar (Fig. 17).

The way in which the coelom develops, its asymmetrical anterior opening and associated pulsating vesicle in the three groups (Fig. 14) provide what is perhaps the one sure foothold in these shifting sands of phylogenetic speculation, for it is inconceivable that such peculiarities could have arisen independently. What is more, the opening of the premandibular coelom occasionally found in craniates is in its position and relations

exactly the same as in these three groups, though a dorsal pulsating vesicle is no longer found.

In tunicates, these relationships of the coelom are absent, because the coelom itself is reduced or absent in modern forms, but the embryological development of ascidians is so similar to that of acraniates (making allowances for the reduced cell numbers in ascidians) that it is probable that ancestral tunicates possessed similar coelomic arrangements.

On the other hand, tunicates share with acraniates and hemichordates a similar pharyngeal

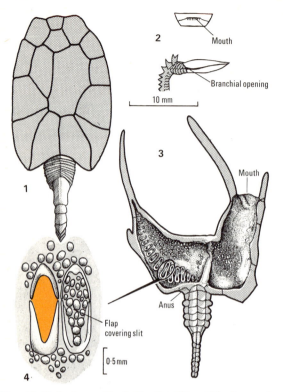

2   Mouth

Branchial opening

10 mm

1

3

Mouth

Anus

Flap
covering slit

0·5 mm

4

FIG. 15. Stylophora, fossil chordate-like echinoderms. 1. Mitrate type in dorsal view. 2. The same in anterior and side view. 3. A cornute type showing row of 'gill' slits and position of mouth and anus. 4. Enlarged view of 'gill' slits, showing one slit covered with flap valve. (After Jefferies.)

arrangement, with pre-oral ciliary organ, endostyle, and (a striking detail) apparently non-adaptive asymmetry of ciliary beat of the gills. The lateral gill bar cilia in the three groups beat in an anti-clockwise direction around the gill opening, when viewed from the external aspect, on either side of the body, as shown by the arrows on Fig. 3. It is simplest to suppose that this asymmetry was inherited from a common ancestral form (Fig. 18).

There are then, clear links between the echinoderm–hemichordate–acraniate and vertebrate in coelomic structure, and great similarities between tunicates, acraniates, hemichordates, and vertebrates in pharyngeal arrangement and nervous structure. The tunicates do not have to be disqualified from the first group, since they have evidently much reduced the coelom, but can we add the echinoderms to the second group—or do the endostyle, peripharyngeal bands, and

gill slits of the second group suggest that echinoderms are off the chordate line?

One of the most interesting recent fossil investigations has been the remarkable reconstruction of a group of echinoderms from the Cambrian to Devonian eras by Jefferies (1968). Jefferies places these fossils (Fig. 15) in a new sub-phylum, the Calcichordata, allied to the Chordata. He considers that it is from adult animals of this group that the chordates arose. They are small fossils preserved as hollow moulds with most of the original calcite skeleton dissolved away, which allows the most minute details of the original animals to be observed by the technique of latex casting (Fig. 16). These details are certainly, in some respects, chordate-like. There are for example, series of branchial slits, covered by flaps and valves; there are traces which allow the reconstruction of the course of the gut and what seem to be the limits of an atrium (see Figs. 4 and 9); and there are other grooves and pores which can be interpreted as blood vessels, nerves, and so forth. Jefferies' interpretation of these fossils, elegant and complex as it is, deserves more detailed consideration than it has yet received, but it still must be admitted that most zoologists, and in particular, echinoderm specialists, have not been fully convinced by his views, though the *facts* revealed are not in dispute. For example, the complexity and size of the central nervous system suggested by Jefferies resemble those of a relatively advanced craniate such as an ostracoderm (Fig. 2b), rather than any simpler form, and moreover, his view implies that the chordate brain and cranial nerves were unsegmented from their origin in phylogeny. In other words, he does not accept Goodrich's theory of head segmentation and the segmental nature of the cranial nerves.

There is no dispute, however, that here are echinoderm-like fossils (whatever phylum we may choose to place them in) which are apparently adapted for ciliary feeding using a series of branchial openings in an essentially chordate manner, and which seem to show the same sort of pharyngeal asymmetry found in tunicates and hemichordates. This new evidence certainly links echinoderms, with the second group of forms mentioned above and distinguished by their pharyngeal arrangements.

There are other pieces of evidence which link the various groups we have been considering. All employ acetylcholine as a neuromuscular transmitter, as do the vertebrates. Some hemichordates and acraniates smell strongly of iodoform, and in hemichordates this is found to be due to their production of the biologically peculiar 2,6-dibromophenol. It would be most interesting if this substance occurs also in acraniates.

In both acraniates and tunicates, the endostyle binds iodine, and utilizes it to synthesize thyroxine, as it does in the ammocoete larva of the lamprey (Agnatha), where the endostyle becomes transformed into the thyroid gland of the adult at metamorphosis. In hemichordates, iodine is bound by the mucus glands of the collar, but thyroxine is not formed.

It was at one time supposed that phosphagens (compounds with high-energy phosphate bonds used in energy transfer, for example in rephosphorylating adenosine diphosphate to adenosine triphosphate (ATP)) provided useful phylogenetic data. All vertebrates, and acraniates, were found to utilize phosphocreatine (PC) whilst in many invertebrates the phosphagen is phosphoarginine (PA).

Naturally enough, when both PC and PA were found in hemichordates and echinoderms, and only PA in tunicates, this seemed to remove the tunicates from consideration as chordate ancestors; unfortunately later work on other species of tunicates demonstrated PC, and it was found that PC is present in a number of different annelids. There are reasons why PC is more efficient than PA, and the use of creatine in the phosphagen is probably opportunistic. Curiously enough, PC occurs in both the hemichordate *Saccoglossus* and the tunicate *Styela*, but a key enzyme in its synthesis is absent, and it seems that these animals obtain PC from the environment instead of synthesizing it as do echinoderms and vertebrates. Perhaps further work on phosphagen kinases will prove more helpful phylogenetically than the distribution of the phosphagens themselves.

Some recent histological work has also provided new links between the different groups. In both echinoderms and acraniates, it has been found by electron microscopy that motor nerve fibres do not run from the central nervous system to the muscle fibres; rather, the reverse occurs, and the ventral roots of amphioxus are actually composed of bundles of thin extensions from the myotomal muscle fibres (Fig. 4).

Similarly in echinoderms, the ribbon 'axons' passing to the tube feet of starfish are now known to be extensions of muscle fibres towards the radial nerve cord. This is a peculiar arrangement (found also in nematodes, one of the few groups which has not yet been suggested as chordate ancestors) and it would be very interesting to know if it is also found in hemichordates or tunicates. In adult tunicates, motor nerves seem to end on muscle fibres in the normal vertebrate manner, but we do not yet know if this is the case in the tunicate tadpole larva before the whole nervous system undergoes remodelling at metamorphosis, nor do we know anything about the details of the neuro-muscular junction in hemichordates. In amphioxus, an even more surprising finding (by the same worker, Flood (1970)) has been that the notochord contains muscle fibres, and physiological investigation shows that it alters its stiffness under the control of the nervous system. This looks like a trial in the development of a sinusoidal swimming pattern that has not been adopted by the vertebrates, nor by tunicate tadpoles, and again, places the acraniates in an isolated position.

There are, then, different lines of evidence all linking in one way or another, the echinoderms,

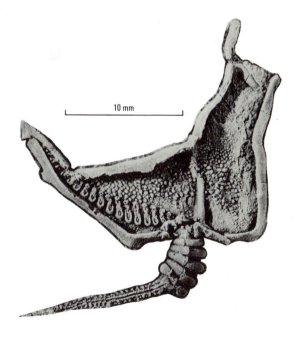

10 mm

Fig. 16. Latex cast of *Cothurnocystis* from Upper ordovician, Scotland. (From Jefferies.)

13

hemichordates, tunicates, and acraniates. What is the most probable way in which these groups are related, and related to the line which gave rise to the vertebrates?

Most recent workers adopt the position taken by Garstang in 1928, who derived the tunicates, acraniates, and vertebrates from the echinoderm–hemichordate stock, supposing the motile chordate-like larva to have been evolved by this stock before the other groups arose from it. Sessile tunicates, and from them by further neoteny the various types of pelagic tunicate, retained metamorphosis and both adult and larva became specialized away from the general chordate line; the larva by the reduction of larval life, the adult by adoption of a completely sessile habit. Acraniates are perhaps best regarded as derived by neoteny from the original chordate-like larva, the metamorphosis of which is seen today as a secondary phenomenon related to the secondary adoption of a semi-sessile habit. Acraniate metamorphosis is the least drastic of any in the groups we have been considering, and the adults retain in the central nervous system patterns of giant cells concerned with the regulation of swimming appropriate for a more free-swimming existence than they show today.

There are difficulties in this view of the relationship between acraniates and tunicates as independently derived from the ancestral echinoderm–hemichordate stock. Both groups show remarkable similarities in their embryonic development, and a good case can be made for explaining the asymmetry of the amphioxus larva in terms of ascidian metamorphosis.

However, the embryological features that the two groups share are just those which lead to the motile chordate-like larva, those also which we should expect would be retained by both groups, although they employ this similar larval organization for different ends. The asymmetry of acraniate larvae has also been explained as adaptive, and secondary, resulting from the way in which the larva feeds; and in any case, the total absence of any trace of asymmetry in the development of the pharynx of modern chordates suggests that the chordates were derived from symmetrical larval forms.

Echinoderms and hemichordates share today a very similar type of ciliated larva (Fig. 10), quite unlike the tadpole type of larva from which the chordate line is supposed to have arisen by neotony.

Once again, it was Garstang who provided a most ingenious way out of this difficulty and his *auricularia* theory has been accepted by many zoologists, fantastic though it seems at first sight.

Auricularia theory

Garstang supposed that the notochord and gill slits first arose in the *adult* form ancestral to echinoderms and hemichordates, but that the dorsal tubular nerve cord arose in the *larva*, by the approximation of dorsal ciliated bands (Fig. 17). If the notochord and tail muscles appeared in the larval stage, this would cause the flattening of the larva laterally, so that the ciliated bands would become apposed and finally roll in and fuse. This bold hypothesis explains in a neat way several features of the chordate nervous system, such as the dorsal position of the sensory columns and the existence of the neurenteric canal joining the posterior end of the nerve cord and the gut, found in vertebrate ontogeny; and what is more, the origin of the nerve tube in *phylogeny* by the inrolling and fusion of ciliated bands is consistent with its origin in vertebrate *ontogeny*.

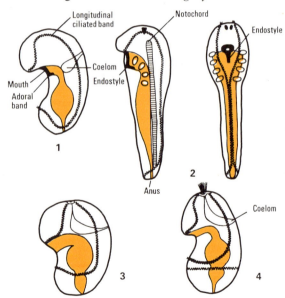

FIG. 17. Garstang's auricularia theory. 1 and 2 show stages in the formation of a chordate from an echinoderm auricularia larva involving rolling together and fusion of ciliary bands, the development of gill slits and of a notochord. Note that the anus and the posterior ends of ciliary bands are linked. 3. Echinoderm auricularia larva. 4. Hemichordate tornaria larva, compared with 3 to show similarity of structure. (After Garstang and Stiasny.)

The fusion of the edges of the neural plate to form the dorsal neural tube of the vertebrate is still incomplete in enteropneust hemichordates such as *Balanoglossus* and *Ptychodera*, where the posterior part of the nerve cord is still open dorsally; the giant cells in the cord which seem to be linked with control of movement are found in the fused region. It can be argued, therefore, that the inrolled dorsal nervous system arose in the adult hemichordate as an adaptation to protect this crucial motor system, and that *all* the three chordate characters (gill slits, notochord, and dorsal tubular nerve cord) originally appeared in the adult stage, and were then pushed back in ontogeny to give the tadpole type of larva which underwent neoteny to give rise to the chordate line.

Together with the appearance of the dorsal nerve cord and the notochord and muscles (as adaptations to assist the larva to remain in the surface layers), the gill slits of the original adult also appeared in the larva. It is interesting that gill slits are found today in some enteropneust

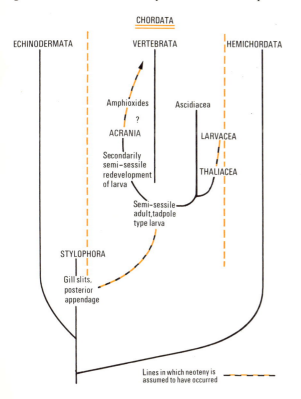

FIG. 18. Suggested relationships of groups implicated in chordate origins.

tornaria larvae before metamorphosis begins (Figs. 10 and 14). If we accept Garstang's auricularia theory, we are able to derive the chordate line directly from echinoderms or hemichordates despite the fact that there is no trace today in larva or adult of either the notochord or segmented muscle blocks. After all, whether we are convinced by Garstang's idea or not, these features of the chordate must have been evolved at some stage.

Jefferies' detailed studies of the fossil echinoderms which he regards as giving rise to the chordate line led him to suppose that it was from *adult* animals of this type that the chordates arose (by progressive modification in the direction of thinning and reduction of the exoskeleton, actually observed in the group), so that his view does not involve neoteny. If we feel, as many zoologists do, that these forms are as *adults* already too specialized to have given rise directly to the chordate line, we are certainly able to ask whether it is possible that their *larvae* could have done so via a neotenous transformation of the kind which we have already considered. His demonstration of adult filter-feeding echinoderms with a movable posterior appendage (but apparently, no tube feet) provides an adult form with just the chordate characters required for the adult form whose larva could be transformed to the ancestral chordate.

If certain of these adult characters, such as gill slits and, perhaps, the initial development of the movable posterior appendage, appeared in the larva before metamorphosis, we are near to the tadpole type of larva and would have a plausible chordate ancestor.

Rather similar arguments would also apply, of course, to hemichordates, and we know from the fossil record that these were both abundant and diverse at or just after the presumed time of origin of the chordates. At present, it is probably best to admit that we are not yet in a position to decide between hemichordates and echinoderms as chordate ancestors (perhaps we never will be), nor to decide the relationship between these two groups. If we must make a choice, it could be as in Fig. 18. In the space of this Reader, it is not possible to do other than state one case for chordate origins, picking facts to support it, and perhaps neglecting others which could be used to support a quite different case. I have tried to indicate that many zoologists would disagree with the views suggested here, though I trust as many would be found to be in general agreement with them.

Zoologists have long found the problem of the origin of the chordates a fascinating one, and very many papers have been written about it. Only a few are mentioned below, but they will show the range of disagreement and give a wider view of the problem.

FURTHER READING

General

BARRINGTON, E. J. W. (1965). *The biology of Hemichordata and Protochordata.* Oliver and Boyd, Edinburgh.

DE BEER, G. R. (1940). *Embryos and ancestors.* Clarendon Press, Oxford.

HYMAN, L. H. (1955; 1959). *The invertebrates. Vol. 4, Echinoderms. Vol. 5, Smaller coelomate groups (Hemichordata).* McGraw-Hill, New York.

KERKUT, G. A. (1959). *The Invertebrata, a manual for the use of students*, 3rd edn. Cambridge University Press.

YOUNG, J. Z. (1962). *The life of vertebrates*, 2nd edn. Clarendon Press, Oxford.

For reference

BERRILL, N. J. (1955). *The origin of vertebrates.* Clarendon Press, Oxford.

BONE, Q. (1960). The origin of chordates. *J. Linn. Soc.* Zool. **44**, 252–69.

FLOOD, P. R. (1970). The connection between spinal cord and notochord in amphioxus (*Branchiostoma lanceolatum*). *Z. Zellforsch. mikrosk. Anat.* **103**, 115–28.

GARSTANG, W. (1928). The morphology of the Tunicata and its bearing on the phylogeny of the Chordata. *Q. Jl microsc. Sci.* **72**, 51–187.

HARDY, A. C. (1954). Escape from specialization. In *Evolution as a process* (eds. J. Huxley, A. C. Hardy, and E. B. Ford). Allen and Unwin, London.

JEFFERIES, R. P. S. (1968). The sub-phylum Calcichordata (Jefferies 1967). Primitive fossil chordates with echinoderm affinities. *Bull. Br. Mus. nat. Hist.* Geol. **16**, 243–339.

PATTEN, W. (1912). *The evolution of the vertebrates and their kin.* Churchill, London.

ROBERTSON, J. D. (1959). The origin of vertebrates—marine or freshwater? *Advmt Sci., Lond.* **61**, 516–20.

TARLO, L. B. (1960). The invertebrate origins of the vertebrates. *Int. geol. Congr.* **21**, 113–23.

See these other titles in the Oxford Biology Readers series:

3. *The mysterious origin of flowering plants.* K. R. Sporne.
11. *Homology: an unsolved problem.* Sir Gavin de Beer.
13. *The origin of life.* J. D. Bernal and A. Synge.
53. *The uniqueness of the echinoderms.* D. Nichols.
54. *Buoyancy in marine animals.* E. J. Denton.

Metamorphosis

J. R. Tata

Oxford University Press, Ely House, London W.1

GLASGOW   NEW YORK   TORONTO   MELBOURNE   WELLINGTON

CAPE TOWN   SALISBURY   IBADAN   NAIROBI   DAR ES SALAAM   LUSAKA   ADDIS ABABA

BOMBAY   CALCUTTA   MADRAS   KARACHI   LAHORE   DACCA

KUALA LUMPUR   SINGAPORE   HONG KONG   TOKYO

ISBN 0 19 914142 8

J. R. Tata is a senior member of the scientific staff of the Medical Research Council at the National Institute for Medical Research, Mill Hill. His published work includes several reviews and papers on the action of growth and developmental hormones and on the regulation of protein synthesis in animal cells. He is the author of *The Thyroid Hormone* (with R. Pitt-Rivers) (Pergamon 1959) and *The Chemistry of Thyroid Diseases* (with R. Pitt-Rivers) (Charles C. Thomas 1960).

Figs 1(b), 5 and 8 drawn by Derek Whiteley.

FILMSET AND PRINTED IN GREAT BRITAIN BY

BAS PRINTERS LIMITED, WALLOP, HAMPSHIRE

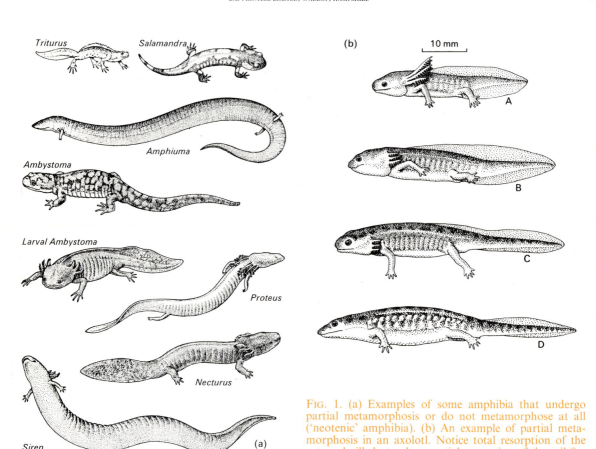

FIG. 1. (a) Examples of some amphibia that undergo partial metamorphosis or do not metamorphose at all ('neotenic' amphibia). (b) An example of partial metamorphosis in an axolotl. Notice total resorption of the external gills but only a partial resorption of the tail fin.

Introduction

Most of us as children have wondered how tadpoles turn into frogs or caterpillars into butterflies. This dramatic transformation, known as metamorphosis, has been studied by biologists for over two centuries, but it is only during the last fifty years or so that we have begun to unravel its biological nature. Progress in understanding the physiological and biochemical regulation underlying metamorphosis has been so rapid in the last thirty years that some of our present-day notions of developmental biology have been acquired from its study.

What makes metamorphosis so attractive is that it allows us to look at autonomous embryos in which all the fundamental mechanisms are analogous to those in which embryonic development cannot always be easily studied.

Metamorphosis, which according to its Greek derivation simply means a change in form or structure, can also be described as late but abrupt embryonic maturation found in some insects and amphibia. There are however two very important qualifications to this definition. First, unlike the early stages of differentiation, metamorphosis is obligatorily under hormonal control. Second, the morphological and functional changes that take place in every tissue during metamorphosis occur in anticipation of a change in environment; metamorphosis is not an adaptation to changes that have already occurred, it prepares the larva for a future change in environment, diet, etc. The magnitude and importance of this preparation is obvious when one considers that the tadpole while still in water has to be equipped for a terrestrial life, breathe air and alter its digestive tract to suit a carnivorous diet.

Metamorphosis does not occur in all amphibian and insect groups, nor does it occur to the same extent. Thus, for example, the process is a major developmental event in amphibia such as anurans (frogs and toads), it may occur in a partial or limited way in some urodeles (newts, axolotls and salamanders) or may never occur in some 'neotenic' species such as *Necturus* (mud-puppy) or *Proteus*. (Neoteny is the condition in which a larval form attains sexual maturity.) Fig. 1(a) shows some of the amphibia in which metamorphosis may be partial or absent. In the Insecta, some primitive apterygotes show continuous growth with no distinct stages. There are various reasons for incomplete metamorphosis or non-metamorphosis. In the 'neotenic' amphibia (e.g. *Necturus*), the animal may throughout life produce the thyroid hormones which normally trigger off amphibian metamorphosis (see below); failure to metamorphose into a terrestrial form lies in the loss of hormonal response in the target cells. Such amphibia, which are extremely rare, go through sexual maturation and breed without reaching an adult stage (neoteny). In the axolotl or *Ambystoma*, Fig. 1 (a, b), metamorphosis may or may not occur and if it does it is only partial, which means that the larva may lose its external gills but not its tail. In the axolotl the 'defect' seems to be in the failure of the pituitary to produce thyrotrophic (thyroid stimulating) hormone (TSH) which in turn initiates the formation and secretion of thyroid hormones. When one considers that neotenic amphibia (*Necturus* and *Proteus*) have survived for over 250 million years, it is difficult to see the advantage of partial or total metamorphosis, other than that it usually permits one species to exploit two different environments. In insects partial metamorphosis is not known, and metamorphosis often proceeds in discrete stages rather than being a continuous process as in amphibia. In this account we shall consider metamorphosis only in those amphibia and insects where there is a major transformation from the larval to the adult form or function in almost every cell type.

Hormonal control of metamorphosis

In 1911, Gudernatsch, a German biologist working in Naples, was studying the effects of different dietary conditions on tadpoles, when he quite accidentally discovered that feeding extracts of mammalian thyroid glands caused them to turn very precociously and abruptly into frogs. The thyroid gland principles responsible for inducing amphibian metamorphosis are the hormones L-thyroxine and tri-iodo-L-thyronine (Fig. 2). Later work showed that surgical removal of the inactive thyroid gland of the tadpole prevented it from ever turning into a frog and that the availability of the tadpole thyroid hormones at the right time was indispensable for metamorphosis to occur. In numerous experiments the tadpole stage of a frog has been prolonged for several months by removing the pituitary or thyroid gland. During this period the tadpole may continue to grow in size, but ceases to do so and turns into a normal frog once the missing pituitary or thyroid hormones are replaced. Similarly in a wide variety of insects, the metamorphosis of larvae or pupae into the adult

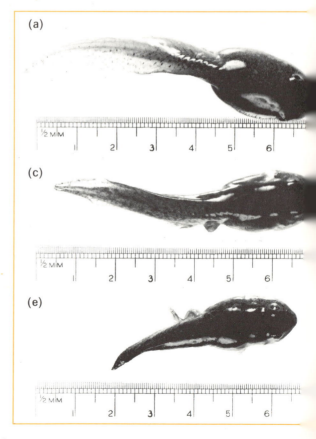

(a)

L-Thyroxine

3,3,'5 − Tri-iodothyronine

(b)

α − Ecdysone

$CH_3-CH_2-C-CH-CH_2-CH_2-C=CH-CH_2-CH_2-C=CH$

Juvenile Hormone
(Methyl 10 − epoxy −7 − ethyl − 3, 11 − dimethyl − 2,6 − tridecadienoate)

FIG. 2. Chemical formulae of the principal hormones directly controlling (a) amphibian, and (b) insect metamorphosis.

forms obligatorily requires the steroid hormone ecdysone (Fig. 2). It is common practice among insect physiologists or endocrinologists to arrest the transition of insect pupae into adult forms by removing a large part of the brain; this prevents the thoracic glands from secreting ecdysone. A debrained pupa can survive for long periods of time in some sort of 'suspended animation'; this can be interrupted by the injection of ecdysone which leads to the appearance of an almost normal-looking adult insect despite the absence of a large part of the brain.

The induction of precocious metamorphosis by the administration of hormones from the outside has become a well established practice for studying sequentially the cellular and biochemical changes that occur during late embryonic development. This is illustrated in Fig. 3 for the bullfrog tadpole which normally has a long premetamorphic life of two or even three years. In addition to the most obvious morphological changes there are others as important that cannot be seen in the photographs in Fig. 3. For example, if the same tadpoles were to be dissected one could readily see the total loss of gills, about 80 to 90 per cent resorption of the intestinal tract, the rapid appearance of lungs and also more subtle changes in the shape of the head that lead to a 'repositioning' of the eyes. Similarly Fig. 4 shows the very dramatic external morphological changes that accompany the transformation of the American silkmoth *Cecropia* from the larval to the pupal and then the adult forms.

(a)

(c)

(e)

4

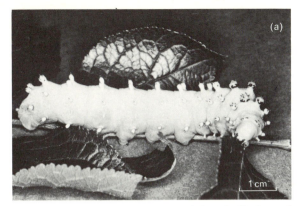

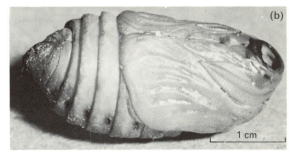

FIG. 3. Dorsal (a), (c), (e), and ventral (b), (d), (f) views of a tadpole of the American bullfrog *Rana catesbeiana*, induced to metamorphose artificially by injection of the hormone tri-iodothyronine. (a), (b) Day of injection. (c), (d) Four days after injection. (e), (f) Nine days after injection. Notice the regression of the tail, the sequential emergence of hind and front limbs and the pigmentation. Other changes not obvious in the photographs are regression of gills and the gut, formation of lungs and change in the shape of the head. Normally a bullfrog tadpole of the size indicated may take 1 to 2 years to metamorphose.

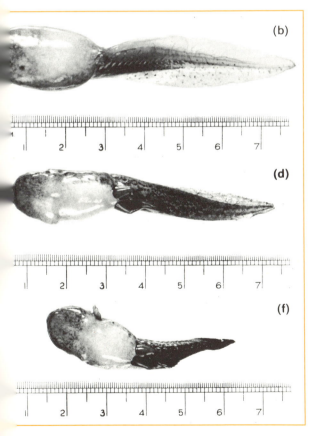

FIG. 4. The larval (a), pupal (b), and adult (c) stages of the American silkmoth *Cecropia*. Much of the brain in this pupa has been removed to prevent spontaneous metamorphosis that would otherwise result from 'brain hormone' activating the synthesis and secretion of ecdysone. Such a pupa can be kept for long periods without any change until ecdysone is injected, when a virtually adult form will be produced. Notice the wing bud (and antenna) in the pupa. (*Cecropia* photographs kindly provided by G. R. Wyatt.)

Neurohormones

How does the amphibian tadpole or the insect larva or pupa know when to release the hormone which triggers off metamorphosis? Neuroendocrinologists have established in the last 20 years that the activities of virtually all endocrine glands (glands producing and secreting hormones, such as the gonads, adrenal, thyroid) are controlled by a series of hormones produced in the brain and known as neurohormones. The job of the neurohormones is to put the organism in tune with the environment. Fig. 5 illustrates in a simplified way how such a 'tuning in' operates in insects and amphibia. It is interesting to note the analogy between the processes in the two groups of animals. The initial trigger for metamorphosis may be a stimulus such as a sudden change in temperature or length of daylight which then indirectly leads to the production of 'brain' hormones which in turn act on the pituitary in the

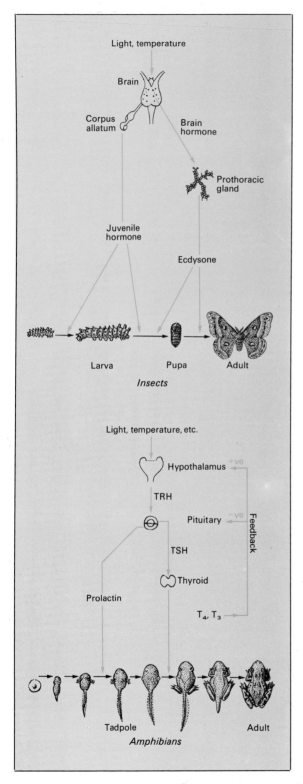

Light, temperature

Brain

Corpus
allatum

Brain
hormone

Prothoracic
gland

Juvenile
hormone

Ecdysone

Larva

Pupa

Adult

Insects

Light, temperature, etc.

Hypothalamus   +ve

TRH

Pituitary   −ve   Feedback

TSH

Thyroid

Prolactin

T₄, T₃

Tadpole

Adult

Amphibians

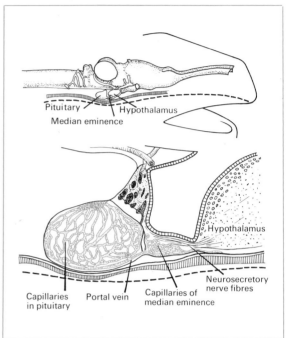

Pituitary   Hypothalamus
Median eminence

Hypothalamus

Capillaries   Portal vein   Capillaries of   Neurosecretory
in pituitary   median eminence   nerve fibres

FIG. 6. Schematic view of the connexion between the hypothalamus and pituitary of the frog via the median eminence. Before metamorphosis, the latter region of the brain is poorly developed in the tadpole and its maturation actually takes place as metamorphosis progresses.

case of tadpoles or the corpora allata in insects. Fig. 6 shows in greater detail the connexions in the frog brain between the two principal hormone-producing centres of the brain, the hypothalamus and the pituitary. The hypothalamus receives the external stimuli via the appropriate nerves and begins to secrete the hormone TRH, a simple tripeptide, which reaches the pituitary via the median eminence. The thyrotropic cells of the pituitary now begin to form TSH, a large polypeptide, which is drained into the portal vein and then to the thyroid gland via general circulation. In premetamorphic tadpoles the median eminence is poorly developed and it is actually during metamorphosis that its nervous tissue thickens and become prominent. This process forms part of a complex 'positive feedback' system in the metamorphosing tadpole which we shall discuss later (Fig. 7).

FIG. 5. Scheme showing the analogy between insects and amphibia in the hormonal transmission of external stimuli leading to metamorphosis. TSH = thyroid stimulating hormone; TRH = TSH releasing hormone; T₄ and T₃ = thyroxine and tri-iodothyronine.

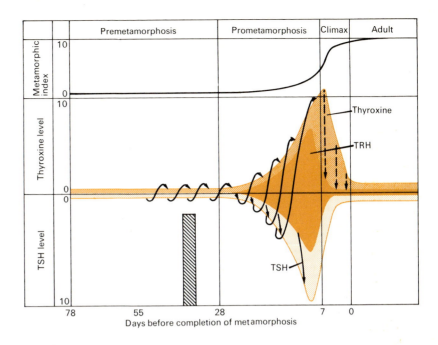

FIG. 7. Scheme proposed by Etkin to explain the onset of an abrupt metamorphic climax in the tadpole. There is an explosive release of thyroxine at the beginning of metamorphic climax because the positive feedback via neurosecretory activity at prometamorphosis is a self-accelerating process. Solid arrows indicate positive feedback, broken arrows negative feedback. The direct negative feedback of thyroid hormones via the hypothalamus and pituitary, indicated in Fig. 5, is not shown here. The metamorphic index is derived from the ratio of hind leg/tail length. The vertical hatched bar indicates the stage at which the hypothalamus of the developing tadpole acquires sensitivity to thyroxine. The levels of circulating TSH and thyroxine are expressed on an arbitrary scale of 0–10 and the levels of TRH were not actually measured.

Hormones which retard metamorphosis

Another important feature shown in Fig. 5 is the action of hormones that actually retard metamorphosis. This concept stems from the discovery by Sir Vincent Wigglesworth in Cambridge of 'juvenile hormone' in insects. In most insects juvenile hormones have been shown to belong to a class of relatively simple chemicals known as 'terpenes'. They are produced in a loose collection of cells called the corpora allata once these are stimulated by a hormone, as yet poorly defined, produced by nervous tissue and known as 'brain hormone'. A variety of juvenilizing hormones have now been discovered in insects, and they seem to have a dual action. A juvenile hormone can induce a moulting on its own and also delay or prevent the separate moult induced by the metamorphic hormone ecdysone. During normal development, the balance between the relative amounts of juvenile hormone and ecdysone determines the rate and manner in which the larval and pupal moults

proceed, leading to the emergence of the adult. In amphibia a principle whose function is similar to that of insect juvenilizing hormone has recently been discovered. The hormone prolactin, a polypeptide hormone formed by the pituitary (and which in mammals controls lactation), promotes growth of tadpoles and retards or inhibits the onset of both natural or artificially-induced metamorphosis. While amphibian prolactin or a juvenilizing hormone has not yet been convincingly identified, it seems that the tadpole pituitary does produce and release a substance with the properties of mammalian prolactin.

There is much interest in the possible use of insect juvenile hormones in controlling insect populations by preventing metamorphosis. Unlike agents such as DDT, the hormone would be species specific, would have no toxic effects on other forms of life and would therefore not upset ecological balance to the same extent as do wide-spectrum toxic substances. It is also interesting in the context of ecological balance to note that several species of plants have now been identified which contain substantial amounts of ecdysone and substances which have a powerful juvenile hormone action on a variety of insects. The biological significance of this is difficult to assess in ecological terms; but the formation of substances with juvenile hormone activity would have great survival value for plants, which could thereby prevent the development and breeding of predacious insects.

Natural timing of metamorphosis

What determines the timing and rate of metamorphosis in nature? The initial trigger would be a change in the environment, but how is metamorphosis accelerated so rapidly? Etkin has proposed a novel explanation for the sudden onset of maturational changes (often called 'metamorphic climax') in amphibia. In most situations a feedback mechanism operates by which the amount of circulating thyroid hormone regulates the activity of the neurosecretory tissues of the pituitary and hypothalamus. Etkin has divided the period before the tadpole enters into metamorphic climax into two stages: an initial stage of premetamorphosis and an intermediate stage of prometamorphosis. As we have already noted, the pituitary and the hypothalamus produce TSH (thyroid stimulating hormone) and TRH (TSH releasing hormone) respectively. In prometamorphosis, regulation is thought to occur by positive feedback in which a small amount of thyroid hormone itself activates the hypothalamus. This would lead to a self-accelerating system which would result in a rapid build-up of the level of circulating hormone just before metamorphic climax. At this point an inhibitory or negative feedback system supersedes the positive one and results in a return to low levels of thyroid hormone (Fig. 7). Negative feedback is the classical concept of how the activities of all endocrine organs in adult animals are regulated. It means that formation of neurohormones by the hypothalamus and pituitary is inhibited by the hormones of their target glands. A similar situation of positive feedback leading to self-acceleration just before metamorphosis occurs has not been discovered in insects.

Local effects of metamorphic hormones

A well established fact in metamorphosis in both insects and amphibia is that the metamorphic hormones act directly on every tissue of the pupa or larva. The effect of the hormone is local and the response of each of the different cell types to it is quite different. By *local* one means that the hormone itself has to be continuously present in or near the cells it acts on. This is in contrast to indirect or *systemic* effects which at one time were believed to occur. For example, it was previously thought that the regression of the tadpole tail might be caused indirectly by the invasion of the tissue by macrophages (scavenging cells) and not by the direct action of thyroid hormone on some cells of the tail tissue. Some examples are illustrated in Fig. 8. This feature forms the basis of much work in which the interaction between the hormone and the target cells can be studied *in vitro* in organ cultures, and has been applied to tadpole tail regression and the development of wing scales in *Cecropia* moths.

FIG. 8. Some examples illustrating the principle of direct and local, as opposed to systemic, effects of hormones in inducing metamorphosis. (a) Thyroxine (T_4) dissolved in a fatty base was applied directly to one side of a tadpole tail fin and the base alone (C = control) on the other; after a few days only the side to which the hormone was applied underwent resorption. (b) Tadpole eyes contain the larval visual pigment porphyropsin (P), whereas the adult frog has the terrestrial visual pigment, rhodopsin (R); the eye to which thyroxine was applied contained rhodopsin after a few days, whereas the control eye receiving only the base continued to have porphyropsin. (c) Just before metamorphosis an insect larva is ligated below the prothoracic glands which synthesize the hormone ecdysone (see Fig. 5); after a few days the top half of the larva turns into an adult while the bottom half still retains its larval form and composition.

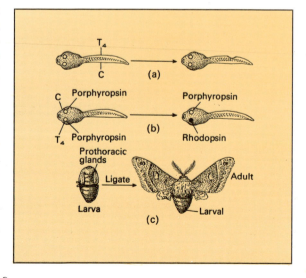

Biochemical effects of hormones during metamorphosis

Behind the obvious morphological changes in metamorphosis, such as insect wing eruption or tadpole tail resorption, are some very fundamental biochemical changes. These involve induction of the synthesis of new proteins or the preferential synthesis of proteins, some of which are summarized in Fig. 9. Virtually no pupal or larval cell escapes the impact of thyroid hormones or ecdysone. Some changes are of great biochemical and physiological importance, such as the formation of urea cycle enzymes in tadpole liver, or the laying down of a cuticle by epithelial cells of insects. These two metamorphic changes are of immense survival value for the developing organism. With regard to the first, ammonia which is rapidly lost by diffusion in aquatic animals is extremely toxic to terrestrial forms of life if not converted to a non-toxic substance like urea before its elimination from the body. Fig. 10(a) shows how the nitrogenous waste excreted changes from ammonia to urea during the metamorphosis of a tadpole into a frog. The biochemical basis of this shift from *ammonotelism* to *ureotelism*, knowledge of which stems from the classical work of P. P. Cohen of the University of Wisconsin, is the induction of formation of the enzymes of the urea cycle (Fig. 10(c)). It is interesting to note that the same pattern of induction of urea cycle enzymes in the liver has been noted during foetal development in man and other mammals.

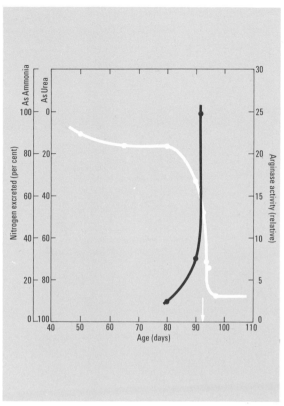

FIG. 10. (a) Change in excretion of nitrogenous waste from ammonia to urea (white line) during the metamorphosis of a tadpole into a frog. The onset of metamorphosis is denoted by the vertical arrow. Arginase (black line) is a key enzyme in the production of urea.

Thyroid hormone, amphibians		Ecdysone, insects	
Tissue affected	Changes induced	Tissue affected	Changes induced
Liver	Maturation; induction of urea cycle enzymes; serum albumin	Fat body	Resorption and reorganization
Tail, gut	Resorption; synthesis of hydrolases (cathepsin, nucleases, collagenase)	Salivary gland	Regression; accumulation of hydrolases; chromosomal puffing
Skin	Hardening and pigmentation; collagen deposition; induction of Na–K ATPase	Epithelial cells	Cuticle and pigment formation; induction of DOPA decarboxylase, polyphenol oxidase, cocoonase
Limb buds, bone	Cell division, growth	Wing	Cell division, scale and pigment formation
Eye	Enzymes for conversion of vitamin A to rhodopsin		

FIG. 9. Some morphological and biochemical changes characteristic of metamorphosis induced by thyroid hormone in amphibian larvae and by ecdysone in insect larvae.

9

The survival value of metamorphic changes taking place in the insect skin during metamorphosis, known as sclerotization, is also quite obvious. The hardening or cuticle formation protects the organism against dehydration, and pigmentation offers camouflage in a new environment. The basic biochemistry of pigmentation in insect epithelial cells during metamorphosis is now well established and is illustrated in Fig. 10(d). The survival value of some other metamorphic changes such as the changes in amphibian visual pigments are less easy to understand.

The changes listed in Fig. 9 can be subdivided into two main categories: those involving maturation and growth and those involving tissue resorption and cell death. These will be considered separately, but it is well to realize that a common mechanism underlies the apparently opposite phenomena of cell growth and death.

How are these biochemical changes brought about? To have a precise answer one must know the identity and characteristics of the cellular component with which the hormone initially interacts and which in turn sets off a chain of events leading to the final manifestations of metamorphosis. We do not yet know what such components are, but we do know the sequential biochemical responses of the cells undergoing a prematurely-induced metamorphic change. Since the appearance of new functions is almost always a consequence of the induction of specific proteins, most of the biochemical responses studied have been those connected with the regulation of protein synthesis in animals.

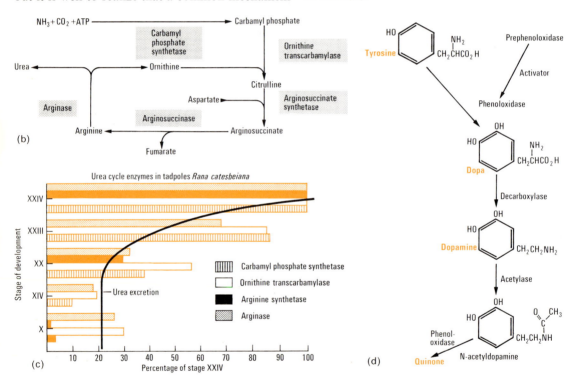

FIG. 10. (b) A simplified scheme showing the main substrates, products and enzymes of the urea–ornithine cycle in amphibia. This accounts for the fixation of ammonia and carbon dioxide and leads to the elimination of nitrogen as urea. (Note that in (c) arginine synthetase indicates the sum of the activities of the enzymes arginosuccinate synthetase and arginosuccinase.) (c) Urea excretion in a metamorphosing tadpole is accompanied by the formation of four urea cycle enzymes. This experiment of P. P. Cohen utilized bullfrog tadpoles which take a very variable time (2–3 years) to metamorphose. The stage numbering system on the horizontal axis is based on the ratio of hind limb length/tail length: stage X corresponds to an early larval stage, stage XX to metamorphic climax and stages XXIII and XXIV to froglets. (d) The major biochemical steps leading to the insect cuticle as it is formed during metamorphosis. This particular scheme illustrates the metabolism of the amino acid tyrosine in the tanning of the blowfly puparum (based on the work of Karlson and Sekeris). DOPA = dihydroxyphenylalanine and it is the quinone derived from Dopamine which contributes to the coloration of the insect cuticle.

Functional maturation of tissues during metamorphosis

We will take two examples: (a) the acquisition of urea cycle enzymes, mitochondrial proliferation (as indicated by accumulation of cytochrome oxidase) and the capacity of the liver of the metamorphosing bullfrog tadpole to make serum albumin (Fig. 11); and (b) the changes in wing epidermal cells of the silkmoth *Cecropia*, leading to the acquisition of scales (Fig. 12). In the first system the changes were induced by injecting tri-iodothyronine into intact bullfrog tadpoles. In the second, wing buds were removed from *Cecropia* pupae which had been injected with ecdysone; these were then incubated.

In both systems, the responses can be temporally divided into three phases. First, after a lag period, there occurs an activation of nuclear function which then leads to an enhanced rate of synthesis of all types of RNA (messenger RNA, transfer RNA, ribosomal RNA). The events occurring during the lag period are still a mystery. It is thought, though not yet definitively proven, that some of the additional RNA made by the nucleus is a new messenger RNA coding for the new proteins to be synthesized. What is certain is that the nucleus is also stimulated into making a large amount of ribosomal RNA and into rapidly transferring newly-made ribosomes into the cytoplasm. The importance of this is shown by injecting the classical inhibitor of RNA synthesis, actinomycin D, soon after the tadpole or insect larva has received the metamorphic hormone; this abolishes all metamorphic changes.

The second phase involves an apparent lull before

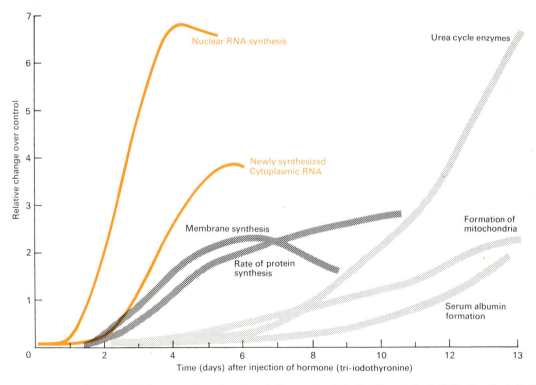

FIG. 11. Sequence of biochemical events that take place in liver cells of the bullfrog tadpole following the injection of tri-iodothyronine at zero time to induce metamorphosis artificially. After an initial lag period of one day, the nucleus is stimulated to make all the major types of RNA at an accelerated rate; this is followed by an increased rate of transfer to the cytoplasm of newly made ribosomes and messenger RNA (as polyribosomes). The next phase consists of a marked change in the pattern of distribution of ribosomes on the endoplasmic reticulum (see Fig. 14), and the rate at which these membranes are assembled is increased. This structural change is accompanied by an activation of the ribosomes in their capacity to synthesize proteins. About two days later one notices the onset of biochemical changes that characterize the metamorphic transformation in hepatic cells, namely production of enzymes of the urea cycle, increased rate of formation of mitochondria, and the production and secretion of plasma albumin.

11

the new proteins appear. During this period more and larger polyribosomes accumulate in the cytoplasm. Polyribosomes are the complex of ribosomes and messenger RNA which, together with transfer RNA and various soluble factors, form the unit of protein synthesis. It seems that not only is more protein-synthesizing machinery being assembled but that it has to be topologically distributed in a certain manner before the metamorphic changes can occur. The topological changes largely involve the deployment of new ribosomes on membranes of the endoplasmic reticulum, initially around the cell nucleus. Biochemically these structural changes are reflected in an enhanced rate of accumulation of both membrane protein and of phospholipid associated with membrane-bound ribosomes.

Only after these biochemical and structural changes in the cytoplasmic protein synthetic apparatus have taken place will the cell enter the third phase of responses. This involves the appearance of proteins found in adult frogs or moths and the gradual acquisition of new functions. The sequential events of the three different phases are summarized in Fig. 13. I need hardly emphasize that this is an over-simplified story based only on the regulation of protein synthesis. Other important biochemical changes such as conformational changes in proteins, readjustment of permeability barriers and alteration of levels of metabolites must undoubtedly also play a part.

Not much attention has been paid so far to the close association between ultrastructural changes and the qualitative or quantitative modification of

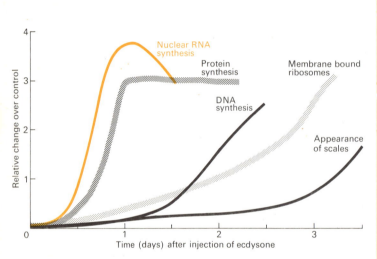 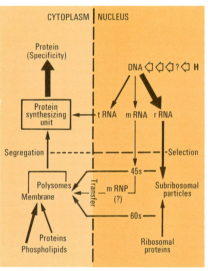

FIG. 12. Sequence of events leading to the formation of wing scales in wing bud tissues of the American silkmoth *Cecropia*, induced to metamorphose artificially by administration of ecdysone. The rates of synthesis of RNA, protein and DNA were measured in whole tissue and not in subcellular fractions as for tadpole liver in Fig. 11. The rate of proliferation of membranes to which ribosomes were bound, and the appearance of scales, was estimated from electron microscopic evidence. (Data of Wyatt.)

FIG. 13. Scheme summarizing the major sequential events associated with hormone-induced development as in metamorphosis. According to this model the ultimate nature of the developmental change would reside in the types of proteins synthesized; and this would be regulated by a series of steps following hormone administration. Initially the hormone (H) would interact with 'receptor' whose nature is not known and which in turn would trigger off some change (open arrows) leading to the

activation of nuclear RNA synthesis. All types of RNA are made at an accelerated rate but the increase is most marked for ribosomal RNA (the width of the solid arrows reflects the magnitude of response). By a process of selection, whose nature is not yet understood, the right species of RNA molecule is selectively transferred into the cytoplasm and combined there with ribosomal proteins or proteins associated with messenger RNA, to form polyribosomal precursor particles. There is some mechanism in the cytoplasm which co-ordinates the assembly of membranes from membrane proteins and phospholipids. A large proportion of newly assembled polysomes may then be bound to newly formed membranes of the endoplasmic reticulum resulting in *functional segregation*, which means that populations of ribosomes formed at different stages of development manufacture different classes of proteins; rRNA, mRNA and tRNA are ribosomal, messenger and transfer RNAs respectively, and mRNP is messenger ribonucleoprotein particle.

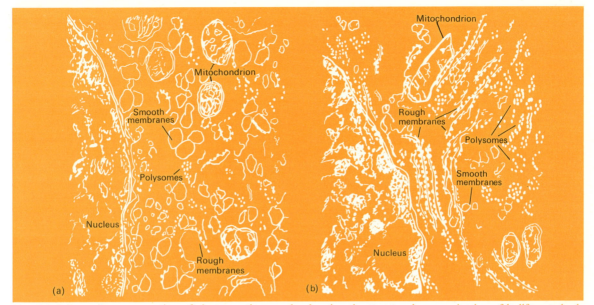

FIG. 14. Schematic representation of electron micrographs showing the structural reorganization of bullfrog tadpole liver cells during thyroid hormone-induced metamorphosis. (a) Liver cell in premetamorphic tadpole showing small groups of polysomes mainly around simple rounded membrane vesicles. (b) Similar cell in a tadpole six days after induction of metamorphosis with tri-iodothyronine, showing dense ribosomal accumulation near the nucleus, mainly associated with double walled membranes. At the same time the nucleus is enlarged and heavily filled with particles, presumably precursors of ribosomes. Also, the mitochondria within the dense ribosomal clusters around the nucleus have a different appearance (elongated with few cristae as against rounded and rich in cristae) from those seen in premetamorphic tadpoles. This is the stage of development at which one observes the first appearance of new or additional proteins characteristic of metamorphosis. Magnification × 25 000.

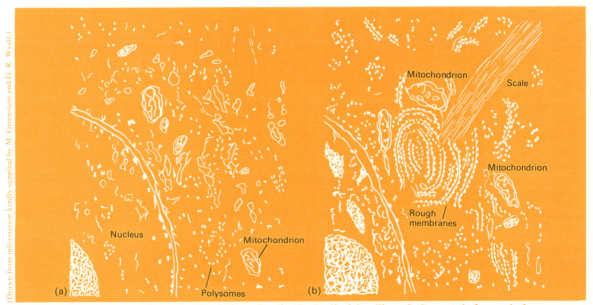

(Drawn from information kindly supplied by M. Greenstein and G. R. Wyatt.)

FIG. 15. Similar representation as in Fig. 14 but of a wing-bud cell of the silkmoth *Cecropia* before and after treatment with ecdysone. (a) Cell from a pupa in diapause showing little rough endoplasmic reticulum. (b) Cell from an animal five days after treatment with ecdysone, showing the dense region of rough endoplasmic reticulum (membrane attached to ribosomes) and a scale growing in its vicinity. Magnification × 20 000.

13

protein synthesis. This applies not only to metamorphosis but to a wide variety of non-metamorphic developing tissues and to adaptational responses. It is possible that the advantage of co-ordinated proliferation of ribosomes and membranes to which they are bound during rapid development is to bring about a topological segregation of different classes of ribosomes carrying out the synthesis of different classes of proteins. It would therefore be interesting to see if the mitochondria appearing in the newly formed perinuclear populations of membrane-bound ribosomes in the induced tadpole hepatocyte (liver cell) (Fig. 14) have a different composition from mitochondria in premetamorphic tadpoles. It is known that many of the enzymes of the urea cycle are located in mitochondria. More visual evidence of such a topological segregation of specialized ribosomes is perhaps provided by wing bud cells of the metamorphosing *Cecropia*, in which the scale appears to grow from the dense membrane-bound ribosomal units around the nucleus (Fig. 15).

FIG. 16. Puffing of a specific chromosomal region (I-17-B) of one of the polytenic (giant) chromosomes of *Chironomus* larval salivary glands induced by the injection of ecdysone. During normal or artificially hormone-induced metamorphosis the salivary gland, which consists of only about 50 cells, undergoes regression after a series of puffs has occurred in the chromosomes. The puffs are thought to represent areas in which RNA is synthesized on genes that were dormant before the hormone was administered. (a) and (b) Chromosomes from uninjected control. (c) and (d) Chromosomes from ecdysone-injected larvae. (Drawn from the work of Clever.)

Tissue regression and cell death during metamorphosis

The loss of the tadpole tail and the regression of the larval salivary gland in many insects are classical examples of tissue regression accompanying growth and maturation during metamorphosis. During this period the larva stops feeding and a new digestive tract more suitable for a new type of diet replaces the larval gut. The major source of nutrients for new metamorphic structures, for example amphibian limbs and lungs, or insect wings and cuticle, is that derived from the breakdown of tissues that will not persist during adult life. Many investigators have asked how it is that the same hormone that stimulates growth and maturation also provokes regression, and whether its activity is fundamentally similar in both cases.

Activities of lysosomal enzymes

From earlier studies it was believed that the hormone would somehow activate lysosomal enzymes, such as cathepsins (protein degrading enzymes) nucleases, phosphatases, etc. in cells programmed for death. Lysosomes are intracellular sac-like bodies which are present in all animal cells and contain all the major hydrolytic enzymes (enzymes whose function is to degrade nucleic acids, proteins, phosphate esters, etc.). Normally the enzymes are present in inactive forms and have to be converted to their active forms when they are released to digest the contents of the cell, or foreign substances such as bacteria, drugs, etc. Despite the widely held idea of activation of pre-

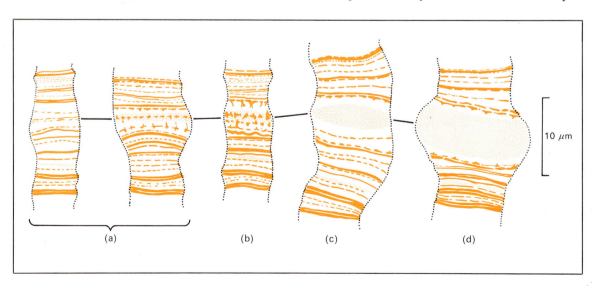

10 μm

(a)   (b)   (c)   (d)

existing lysosomal enzymes, work in the last decade suggests that the same fundamental mechanisms involving the regulation of protein synthesis do indeed provoke growth and maturation or regression and lysis, depending on the way different types of cells have been programmed at earlier stages of differentiation. Both the involution of the insect salivary gland and the resorption of the tadpole tail are now classical examples of the dependence of cell death on synthesis of new proteins.

Gene activation: chromosomal puffs

Activation of dormant genes by ecdysone to produce RNA is an important event anticipating the regression of insect salivary glands. In many insect larvae, especially in *Chironomus*, the salivary gland is composed of a few cells (about 50) which contain as few as four giant or polytenic chromosomes. Work from the laboratories of Beerman and Clever in Germany showed that exposure of these cells to ecdysone soon results in activation of dormant genes which are now transcribed into RNA. This activation occurs in a well defined sequential pattern and it can be followed microscopically as chromosomal puffs (Fig. 16). The exact nature of the puff is not yet established. However, it is agreed that in them the DNA (in association with the basic proteins known as histones) is present in a diffused or unwound state, making this region of DNA readily available as a template for making RNA. By injecting radioactive precursors of RNA

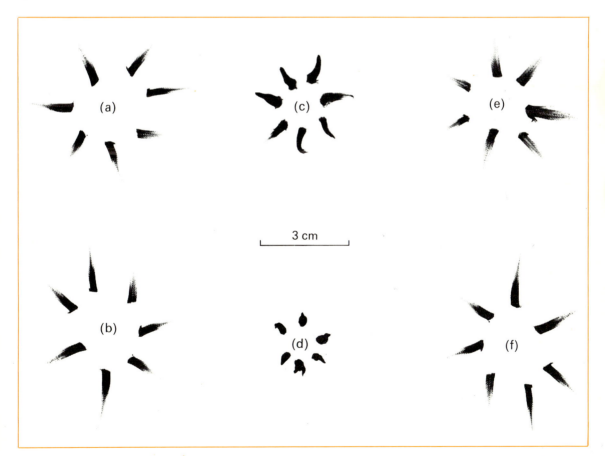

3 cm

(a) (c) (e)

(b) (d) (f)

into the animal it can be shown that RNA is made on these puffs. Although there is no proof yet that part of this RNA is new messenger RNA, we do know that such RNA synthesis is an important event leading to regression, since blocking RNA synthesis with actinomycin D prevents the eventual involution of salivary glands. The study of chromosomal puffs in insect salivary glands is a major tool for studying gene activation.

Similarly, both RNA and protein synthesis are important processes for tadpole tail resorption. In recent years it has been possible to study regression in organ culture in which tri-iodothyronine is added to isolated tails. Fig. 17 shows that inhibition of RNA synthesis by actinomycin D protected the tissue against hormone-induced regression, and the same results were obtained when drugs inhibiting protein synthesis (cycloheximide and puromycin) were added to the culture medium. When the nature of the lytic enzymes made in the tadpole tail undergoing regression was carefully examined, it was found that newly formed cathepsin and deoxyribonuclease had different properties both from the enzymes present before the hormone was added and from those extracted from non-regressing tissues. If this is generally applicable, it seems that a new class of enzymes may be induced in regressing tissues to accomplish the function of cellular death, which is an integral part of embryonic development.

FURTHER READING

General

ETKIN, W. A., and GILBERT, L. I. (eds.) (1968). *Metamorphosis*. Appleton-Century-Crofts, New York.

TATA, J. R. (1968). Hormonal regulation of growth and protein synthesis. *Nature, Lond.* **219**, 331–7.

For reference

COHEN, P. P. (1970). Biochemical differentiation during amphibian metamorphosis. *Science* **168**, 533–44.

ETKIN, W. A. (1964). Metamorphosis. *Physiology of the Amphibia* (ed. J. A. Moore). 427–68. Academic Press, New York.

FRIEDEN, E., and JUST, J. J. (1970). Hormonal responses in amphibian metamorphosis. *Biochemical Actions of Hormones*, Vol. I (ed. G. Litwack). 2–52. Academic Press, New York.

KARLSON, P., and SEKERIS, C. E. (1966). Ecdysone, an insect steroid hormone, and its mode of action. *Recent Progr. in Hormone Res.* **22**, 473–93.

ROLLER, H., and DAHM, K. H. (1968). The chemistry and biology of Juvenile Hormone. *Recent Progr. in Hormone Res.* **24**, 651–78.

SCHNEIDERMAN, H. A., and GILBERT, L. I. (1964). Control of growth and development in insects. *Science* **143**, 325–33.

TATA, J. R. (1971). Hormonal regulation of metamorphosis. *Symposia of the Society for Experimental Biology*. 26, 163–81.

WYATT, G. R. (1971). Insect hormones. *Biochemical actions of hormones*, Vol. II (ed. G. Litwack). 386-490 Academic Press, New York.

See these titles in the Oxford Biology Readers series:

16. *The nucleolus.* E. G. Jordan.
25. *Gene expression during cell differentiation.* J. B. Gurdon.
51. *Development of pattern and form in animals.* L. Wolpert.
70. *Insect hormones.* V. B. Wigglesworth.
75. *Transcription of DNA.* A. A. Travers.
79. *The cellular effects of hormones.* P. J. Randle and R. M. Denton.

Oxford Biology Readers
Edited by J. J. Head and O. E. Lowenstein

Primates and their Adaptations

J. R. Napier

Oxford University Press, Ely House, London W.1

GLASGOW   NEW YORK   TORONTO   MELBOURNE   WELLINGTON
CAPE TOWN   SALISBURY   IBADAN   NAIROBI   DAR ES SALAAM   LUSAKA   ADDIS ABABA
BOMBAY   CALCUTTA   MADRAS   KARACHI   LAHORE   DACCA
KUALA LUMPUR   SINGAPORE   HONG KONG   TOKYO

PHOTOSET AND PRINTED IN GREAT BRITAIN BY
BAS PRINTERS LIMITED, WALLOP, HAMPSHIRE

Dr. John Napier is the Director of the Unit of Primate Biology at Queen Elizabeth College in the University of London. He is the author with his wife of *A handbook of living primates* (Academic Press 1967). He has also written *The origins of man* (Bodley Head 1967) and *The roots of mankind* (Allen and Unwin 1971).

Figs. 10, 17, and 18 drawn by Ken Orton Williams.

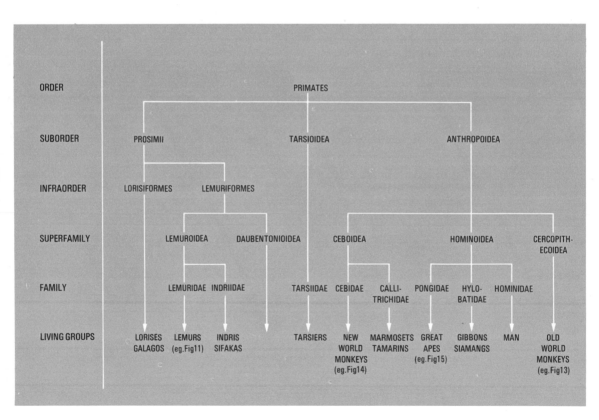

FIG. 1. A simplified classification of the Primate Order based on Simpson (1945). Note that the names of taxa are in left hand column. Tree-shrews have been omitted in view of their uncertain status.

2

Introduction

The mechanism of adaptation is natural selection acting on the genetic variability of a population. Adaptation produces change, and change occurring with time is evolution; adaptations therefore constitute evidence for evolution. Evolution, adaptation, and natural selection are *processes* or the results of processes but the material on which these processes operate is particulate matter, the genes. All adaptations start as genetic changes which, unless they are instantaneously lethal, are ultimately subjected to the test of fitness. Fitness means reproductive success. The test-bench of all genetic novelties is the environment and if they are to play any further role in evolution they must pass an extensive series of trials in much the same way as a new aero engine.

Structural adaptations are the physical basis of the *habit* or behaviour of an animal. The environment in which an animal lives is its *habitat*. Habit and habitat are causally related because what an animal does is wholly dependent on where it lives.

The adaptations of living primates are the hallmarks of their success; they are the certificates that they have acquired during their long apprenticeship. This article examines some of the most important adaptations, particularly those that have played a prominent part in the emergence of the quintessential primate, man.

The arboreal way of life

Primates owe their distinctiveness to the fact that, historically, they are arboreal animals; they not only live in trees but are specifically adapted to do so. Many animals utilize trees: insects, birds, snakes, lizards, and mammals. But, like the Felidae (cats) and Sciuridae (squirrels) among the mammals, many are facultative arborealists who spend the critical moments of their lives on the ground or in the air; for them trees are merely refectories, dormitories, or safe-deposits. Primates are professional tree-livers, though some of them have secondarily adopted a ground-living way of life. Arboreal adaptations, far from restricting the potential of primates to colonize other environments, have actually made such an ecological shift possible. Paradoxically, the most successful primate species are those that have renounced life in trees in favour of life on the ground. Man and the cercopithecines (baboons of Africa and the macaques of Asia (Figs. 12 and 13)) are obvious examples. Collectively, a more successful group of animals could not be envisaged; between them they are the dominant form of mammalian life throughout the Old World. Human and cercopithecine populations have been in competition for thousands of years. Man has the upper hand at present and will probably retain it by the basic procedure of eliminating his competitors.

Before discussing some of the critical adaptations of primates, it might be as well to know a little bit about the taxonomy and geographical distribution of the Order.

Classification of primates

A very simplified classification of primates is shown in Fig. 1. At subordinal level three major groups separate out. This indicates a very wide disparity of adaptive characters and argues a very ancient separation of the prosimians (lemurs and lorises), the tarsiers, and the anthropoids (monkeys, apes, and man). Fig. 2 shows that this event dates back to the Eocene epoch some 50–60 million years ago. Nevertheless prosimians, tarsiers, and anthropoids share a sufficient number of characters of common inheritance to make it certain that all three stocks at one time shared a common ancestor.

The classification in Fig. 1 is by no means universally accepted. It is based on Simpson (1945) but differs in one particular, the position of the tarsiers. Following Romer (1966) they are here placed in a separate suborder Tarsioidea, indicating their intermediate position between the prosimians on the one hand and monkeys and apes on the other.

The Anthropoidea, the division comprising the monkeys, apes, and man is subdivided into three families: the Ceboidea (the New World monkeys), the Cercopithecoidea (the Old World monkeys), and the Hominoidea (the greater and lesser apes and man himself). New World monkeys are often referred to as platyrrhines (meaning broad-nosed) and Old World monkeys as catarrhines (narrow-nosed). The separation of these groups is very ancient and took place sometime during the Eocene when the land-bridges between North America and Asia and between Greenland and Western Europe were still warm enough to permit the passage of tropical and subtropical forest animals. In spite of their remote common ancestry, New and Old World monkeys are so similar in their general appearance that anyone unfamiliar with primates visiting a zoo where they are exhibited in adjacent cages could be forgiven for not being able to tell them apart. New and Old World monkeys exhibit *parallel evolution*.

	Duration	Millions of years B.P.
Pleistocene	about 2 million years	2
Pliocene	about 10 million years	12
Miocene	about 14 million years	26
Oligocene	about 11 million years	37
Eocene	about 18 million years	55
Palaeocene	about 10 million years	65

Total duration approximately 65 million years

FIG. 2. Geological time scale during the Cenozoic.

Zoogeography of primates

Fig. 3 shows the distribution of primates. While the Old World can boast of prosimians, monkeys, and apes, the New World only has monkeys, though these display a variety of colour, shape, and size unmatched by the Old World group.

Among Old World monkeys there are two major divisions which are reflected in the classification as subfamilies. Their separation is not primarily geographical but ecological: the Colobinae, the leaf-eating monkeys of Africa and Asia (e.g. colobus monkeys and langurs), are essentially arboreal while the Cercopithecinae (e.g. baboons, mandrills, and macaques) are largely ground-dwellers. Some members of the Cercopithecinae may, in the not so distant past, have reverted to tree-living again and we find such arboreal cercopithecines as the African guenons and the mangabeys.

The great apes comprise the family Pongidae which is found in both Africa and Asia. Chimpanzees and gorillas are Africans and orang-utans are Asians, or more correctly Indonesians for they only occur today on the islands of Borneo and Sumatra. The lesser apes, the gibbons and the siamang, are

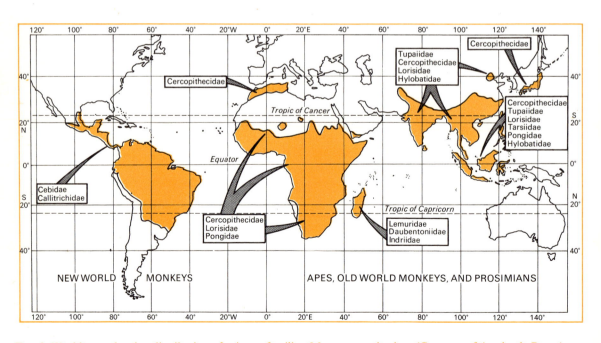

FIG. 3. World map showing distribution of primate families. Mercator projection. (Courtesy of Academic Press.)

4

more widely distributed in the tropical and monsoon forests of the south-east Asian mainland as well as the islands of Sumatra and Borneo.

Probably the best known of all prosimians are the lemurs (Infraorder—Lemuriformes); these exceptionally elegant animals have a remarkably limited geographical range and are only to be found on the island of Madagascar where they have been steadily extinguished at the hands of man for the last 2–3000 years. The second major infraorder are the lorises; this group includes the galagos and pottos of Africa and the slow and slender lorises of the Far East. The tarsiers, here regarded as a separate suborder of the primates, are found in Malaysia, Indonesia, and the Philippines. They occupy a special place in primatological thinking for several reasons: their ancient history, their rarity in captivity, and the possible role that their ancestors have played in the evolution of higher primates.

FIG. 4. Diagram illustrating the concept of the *Échelle des êtres*. For explanation see text.

'Échelle des êtres'

Turning from the zoogeography of the present to the fossil history of the past we must consider a concept that has bedevilled (and illuminated) evolutionary thinking for almost 200 years, the *échelle des êtres*, originally a philosophical conceit of the eighteenth century. Naturalists imagined a ladder or staircase as the perfect model to demonstrate the orderliness of the *Systema naturae*. On it mammals could be arranged, step by step, in the order of their special creation. At the bottom was the archetypal mammal and at the top man himself. The ladder represented the *discontinuity* of life which conformed to the contemporary belief in the immutability of species. Like many bright ideas, the *échelle des êtres* backlashed. Progressive naturalists like Geoffroy and Lamarck, who were convinced that the various grades of animal life were vertically related and that one form changed into another, were quick to use the ladder as a device illustrating continuity of life rather than discontinuity. In their hands the staircase became an escalator.

The *échelle des êtres* of the Order *Primates* is illustrated in Fig. 4. Points A–G represent a series

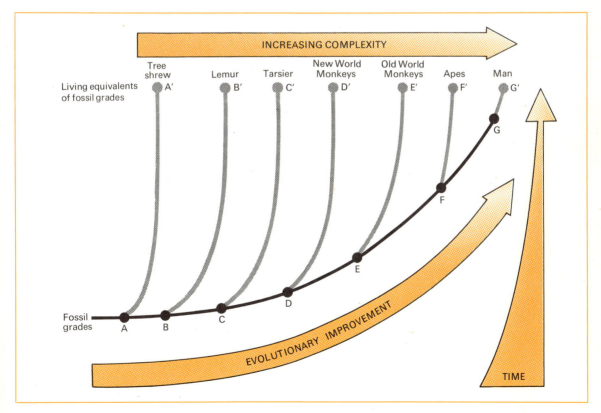

5

of ascending grades in the fossil record, while A′–G′ represent their modern descendants. The sequence A′–G′ comprises a staircase from the lowest to the highest, and between one step and another there is a *discontinuity*. Before accusing me of promoting the very philosophy that I have been arguing against, please backtrack along the shaded lines and there you will see, in an evolutionary context, that the gradation from A to G is continuous— a flowing process of improvement with time—and this is reflected in the fossil record. The early naturalists saw the system of nature as a series of discrete and independent grades because they lacked access to the fossil record. However it should be said that when the evidence of geology was presented to them they were too brainwashed by fundamentalism to do more than cry 'catastrophism', meaning that life forms were created anew in the wake of floods and earthquakes which were supposed to have eliminated previously-existing forms.

Primates are unique among the living mammals in displaying the 'staircase' phenomenon. In other orders the evolutionary sequence shows the successive replacement of archaic forms by new and improved ones. Take for example the horses. There is only one genus of horse today; there is no *Eohippus*, *Merychippus*, or *Pliohippus* to provide the milestones on the road to *Equus*. As each 'new' horse evolved, the 'old' horses went out of business; not so with the primates. As each new grade of primates evolved the old grades, which had established themselves in their own particular environmental niches, continued to flourish.

Arboreal life, the chosen way of primates offers a multitude of ecological opportunities. Primates can be either nocturnal or diurnal; they can be insect-eaters, leaf-eaters, grass-eaters, or fruiteaters; they can dwell in the shrubs of the forest-floor, in the primary, secondary, or tertiary strata of the forest canopy, or the tropical savanna, montane moorland, or temperate forest. The possible ecological permutations are vast.

Behavioural plasticity of primates

The *échelle des êtres* of present-day primates is a reflection of their behavioural plasticity, or adaptability. Adaptability, which is not the same phenomenon as adaptation, is the potential of an individual to change horses in mid-stream. The anatomy of primates is very generalized. They retain many of their basic, all-purpose, mammalian characters, eschewing the extreme specializations of many mammalian groups. The real glory of primates is their individualism. They have emancipated themselves from the shackles of a life of reflex and conditioned responses characteristic of the behaviour of non-primates. They are thinkers. Their responses are unpredictable and vary from individual to individual. The source and spring of this adaptability is, of course, the brain.

Although they are primitive and generalized mammals in terms of their skeletal and muscular systems, this is not true of their neurological mechanisms. The brain of higher primates with its highly developed associative functions, its aptitude for receiving, analysing, and synthesizing incoming sensory impulses and converting them into appropriate motor responses, provides the basis of their behavioural plasticity and durability.

Primate evolution

Fossil evidence suggests that the earliest primates were small, forest-living, insectivorous mammals not much bigger than a rat. The living tree-shrew is often quoted as the model for this grade of primate evolution, though there is controversy as to whether tree-shrews should be considered primates at all. Having rather short legs and, therefore, a low centre of gravity, tree-shrews can travel on quite slender branches by gripping with their claws. Like squirrels however they spend a lot of time foraging in low shrubs and on the ground. The typical adaptations, including grasping extremities, only developed when primates adopted the tree habitat as *a whole way of life*. We do not know what prompted the earliest primates to plump for the trees but this sort of 'opportunism', as Simpson (1949) has called it, is one of the outstanding characteristics of mammalian evolution. Forests were a bit of a novelty in the early Cenozoic; angiosperms did not really get into their evolutionary stride until the environment underwent a dramatic change at the end of the Cretaceous. Forests of flowering shrubs and trees attracted insects, and in their wake came birds and mammals, and this was the environmental opportunity that the earliest primates seized. The paradox that man who is so unquestionably a primate and a ground-based one at that should owe his very being to an arboreal heritage has already been pointed out. The reason why in fact there is no paradox will emerge as the story develops.

In order to exploit to the full the food resources of the forest canopy an animal must possess grasping extremities (as opposed to simple clawed extre-

FIG. 5. Hands and feet of primates. Hands: (a) tree-shrew, (b) potto, (c) tarsier, and (d) macaque. Feet: (e) tarsier, (f) potto, (g) gorilla, and (h) baboon. Note claws on tree-shrew hand (a) and simple toilet claw on feet of potto (f) and double toilet claws on foot of tarsier (e). (Courtesy of Academic Press.)

mities), be able to hold its body erect, and have highly mobile limbs. It needs specialized locomotor mechanisms for crossing gaps between the crowns of trees and a good visual mechanism for the accurate estimation of distance. Finally, to benefit from its climbing skill and mobility, an animal must be able to distinguish edible fruits and leaves from inedible ones by shape and colour, to pluck foods, and to feed itself manually.

Translated into terms of anatomy, physiology and behaviour the adaptations of an arboreally adapted primate are as follows.

1. Primates have grasping extremities

The thumb and big toe of primates are separated from the remaining digits and show varying degrees of independent movement. Complete opposability of the thumb is present *only* in Old World monkeys, apes, and man, but incomplete opposability ('pseudo-opposability') is found in New World monkeys and prosimians. Opposability of the big toe, which evolved earlier than that of the thumb, is 'pseudo' in all species except man whose big toe is not opposable at all. The opposable thumb ('pseudo' or 'complete') provides the basis of prehensility, though, in the immortal phrase, some thumbs are more opposable than others. The relative size and length of the thumb varies in different species in relation to particular locomotor requirements. Among the lorises—the potto for instance—the thumb is a vast digit which provides 50 per cent of the power of the hand (Fig. 5(b)). Pottos and other lorises rely on their powerful grasp to give them security as they climb clumsily, inexorably, and with unbelievable slowness from branch to branch. In some apes such as the orang-utan and the chimpanzee the thumb is small and stunted. Apes depend largely on their long, strong fingers to support their body weight as they hang from an overhead branch to feed. Man possesses quite the longest thumb among the Old World primates, and this is the basis of his exceptional manual dexterity. There are two thumb-less primate genera, one from the Old World and one from the New. The South American spider monkeys have at least the compensation of a prehensile tail but the colobus monkeys of Africa are simply thumb-less. Although this adaptation is related to the particular locomotor needs of the animals its precise significance is not understood.

The big toes of primates are almost as variable (interspecifically) as their thumbs. The potto's big toe is identical in size with its thumb and has a

similar function. Although the orang-utan has nearly lost its big toe, no primate has lost it altogether.

2. Primates have nails—not claws—on their digits

As a corollary to the prehensility of their hands and feet, primate fingers and toes bear nails. When we compare the anatomy of the nail with that of the claw (Fig. 6), it is easy to understand the functional significance of this adaptation. The mammalian claw is a horny spur which protrudes from the ends of the digits. When retractile (as in cats) it is slender and needle-sharp, and when non-retractile (as in dogs) it is blunt and stout. A claw is not compatible with prehensility because in order to grasp and manipulate an object the opposing surfaces must have a rich sensory nerve supply. This allows the quality of the object to be appreciated by the brain, and the outgoing impulses to the muscles of the hands to be adjusted accordingly. For example eyes provide limited information about the ripeness of fruit: more information comes from feeling and pressing it. Mammalian claws are insensitive and useless for obtaining tactile information. During the evolution of the nail from the claw there was a gradual shift from the end of the digit to its dorsal surface, exposing the soft pad of the finger-tip which could then function as an exploratory, tactile device. The function of nails is little more than to provide a rigid backing for the digital pads.

All primates except tree-shrews, marmosets, and tamarins have nails, and Le Gros Clark (1971) showed that marmoset and tamarin claws are intermediate in their anatomy between claws and nails. Tamarins are small, about the size of grey squirrels and their smallness accounts for the persistence of claws. The marmoset hand is not fully prehensile and the thumb is not opposable. The big toe of marmosets and tamarins, which is pseudo-opposable, bears a nail although the rest of the toes are clawed.

All other monkeys, lemurs, and apes have nails whose shape varies with the shape of the finger- and toe-tips. Man has exceptionally broad finger-tips and consequently very broad, flat nails. In contrast, the finger-tips of New World monkeys are very narrow and their nails are so curved that they are semicircular in cross-section. A claw with a very specialized function has been retained on the second toe of prosimians (Fig. 5(f)); it is called a toilet claw and is used in grooming the fur. Tarsiers have two toilet claws, one on the second and one on the third toe.

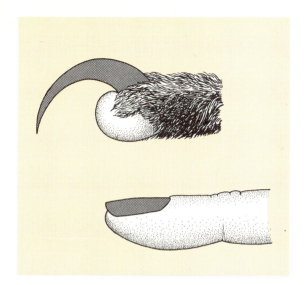

3. Primates have eyes that face forwards

A whole series of interconnected adaptations involving the skull and the brain have occurred during primate evolution resulting in a visual mechanism that produces sharp, well-focused stereoscopic images and colour vision. These adaptations facilitate advanced feeding and manipulative behaviour.

The bony orbits of most unspecialized mammals are directed laterally and there is no basis for stereoscopic vision (Fig. 7(a)). Tree-shrews have this arrangement, the long axes of their orbital cavities subtending an angle of 140°. Frontality of the orbits has become more advanced in lemurs and lorises (70°) and is still more developed in monkeys and apes (30°). The axis of the *eyeball* differs from the axis of the *orbit*, being more inturned, and in monkeys and apes the divergence between the eyeballs is virtually zero. As a concomitant of the forward shift of the eyes, certain changes occur in the bones of the face. A postorbital bar, a rim of bone, develops to close the lateral margin of the orbit in prosimians; in anthropoids the exclusiveness of the eyeball and its musculature has proceeded even further by the interposition of a plate of bone which completely closes off the outer side of the orbital cavity. Tarsiers are intermediate in this respect having an incomplete postorbital closure.

Frontality of the orbits in primates is associated with certain other changes that facilitate an advance in visual perception:

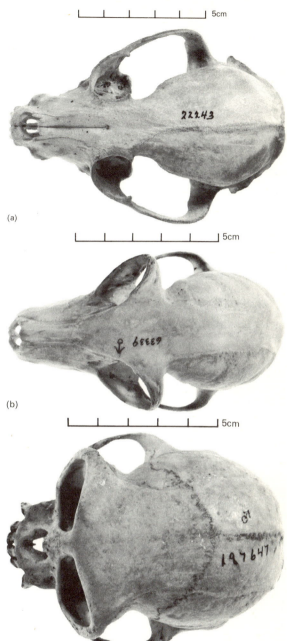

(a)

(b)

(c)

Fig. 7. Skulls of (a) raccoon, (b) lemur, and (c) Old World monkey, seen from above. Note particularly: in (a), absence of postorbital bar, sideways-facing orbits, long snout, and small pear-shaped neurocranium; in (b), postorbital bar, partial frontality of orbits, persistence of muzzle, and globular shape of skull; in (c), complete postorbital closure, complete frontality of orbits, absence of snout, non-projecting (orthognathous) face, and expanded neurocranium.

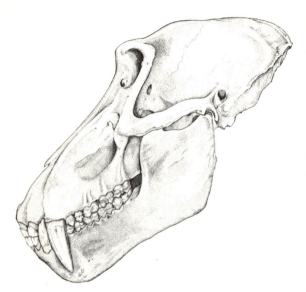

FIG. 8. Skull of adult male baboon. The muzzle is bent so that the eyes have uninterrupted vision over the dental muzzle. Note large sabre-like canines in upper jaw.

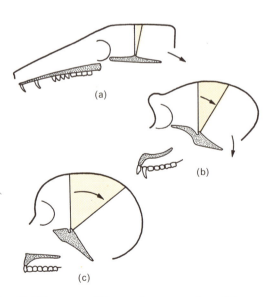

Flexure of the muzzle. The presence of a long snout in an animal like a dog inhibits stereoscopic vision by acting as a barrier between the eyes. Associated with the rotation of the eyes towards the front of the head, there is a bending of the snout relative to the plane of the eyes. A baboon has a long dog-like snout (Fig. 8) but because it is strongly flexed it does not limit the overlap of the visual fields; a baboon looks down its nose and over its muzzle. (The snout here is not *olfactory* but *dental*, an adaptation to accommodate a particular pattern of teeth. Baboons have large molars and exceptionally large canines which are used for offence, defence, and as potent social signals.) Most primates do not have a projecting snout but their facial skeletons are flexed to various degrees, acutely so in man (Fig. 9).

Development of cones in the retina. All primates, except the nocturnal prosimians (potto, loris, galago, etc.), possess two kinds of photoreceptor in the retina, rods and cones. Rods have little resolving power but respond to light of very low intensity. Cones provide the basis for fine visual discrimination and for colour vision. At the centre of the retina there is an area, made up wholly of cones, called the fovea; this is the area of maximum acuity in human and anthropoid eyes. Only one

FIG. 9. Skull of (a) man and (b) gorilla. Note particularly the large, rounded braincase of man, the absence of heavy facial buttresses and skull crests, the presence of a chin, and the extreme orthognathy (flatness) of the face.

FIG. 10. Diagram to illustrate how expansion of the cerebral cortex affects the position of the foramen magnum and consequently the balance of the skull on the vertebral column. (a) Tree-shrew, (b), chimpanzee, and (c) man. Skulls sectioned in sagittal plane. (After Dr. J. Biegert.)

Royal Free Hospital School of Medicine.

genus of monkeys (*Aotus*, the South American night monkey) lacks cones, though it possesses a fovea, meaning that it has secondarily adopted the nocturnal habit.

Elaboration of the occipital region of the brain. Changes in the skull and retina concerned with improved vision are reflected in the progressive enlargement of the part of the brain which receives and interprets visual impulses from the retina. Expansion of the visual cortex has been accompanied by a broadening of the back of the primate skull and displacement of the occipital condyles, which balance the skull on the atlas vertebra, from the rear of the skull to the undersurface (Fig. 10). This change has led in turn to a reorientation of the head relative to the body, an important adaptation in primates for whom sitting, standing, and walking are behavioural patterns of great significance.

Other brain adaptations associated with the refinement of visual sense are modifications in the optic chiasma and the lateral geniculate bodies, the relay stations for optic impulses passing from the retina to the visual area of the cortex.

Fig. 11. A prosimian, *Microcebus*—the mouse lemur. One of the smallest of all primates. Note prehensile hands and feet and nocturnal adaptations of eyes and ears. (Courtesy of Dr. Gilbert Manley.)

4. Primates have a diminished olfactory sense

Olfactory and visual systems supply the brain with similar sorts of information, and usually one dominates the other. It seems as if there is no room in the skull for both systems to be equally well developed. Non-primate mammals, on the whole, rely more on a sense of smell than on vision; amongst primates the reverse is true and this is expressed in the reduction in length of the snout and the loss of the 'wet' nose of other mammals. There are also changes in the brain involving reduction of the olfactory bulbs and the size of the area of cortical representation. The nasal chamber of 'olfactory' mammals is capacious and packed tightly with much-curled, wafer-thin, scroll bones or turbinals while the nose of 'visual' animals is small with scroll bones of much simpler construction.

Amongst primates, the prosimians retain much of a generalized mammalian smelling ability; their social behaviour is still based largely on olfactory communication by means of scent marking. Special skin glands are rubbed on branches or transferred to the tail which is then wafted in the air in the manner of a censer. Scent glands and urine are used in marking territories. Higher primates (monkeys and apes) are less dependent on olfaction but the sense of smell is not lost, it merely becomes less important than vision. Consequently higher pri-

Fig. 12. A young baboon. This photograph reveals a very great deal about primate adaptations. Note the involvement of the eyes, brain, hands, and mouth in 'hand-to-mouth' feeding. (Courtesy of Dr. C. K. Brain.)

mates place a high reliance on visual signals—on such functions as facial expression, gesture, and colouration, and physical peculiarities like the heavy mantle on the shoulders of the males of *Papio hamadryas*, the sacred baboon.

The 'wet' nose of mammals (the rhinarium) indicates that the snout is not only a smelling and feeding organ but an organ of touch. All prosimians possess 'wet' noses devoid of hair and their upper lips, which are firmly tethered to the underlying gum margin, are split by a septum. All anthropoids on the other hand have 'dry' noses which are more or less hairy and are linked with freely mobile upper lips which contribute to the great range of facial expressions by which monkeys, apes, and man communicate their emotions to one another. The passage from a prosimian to an anthropoid grade is characterized among other things by a radical change in communication systems.

5. Primates possess the ability to sit, stand, and walk upright

One of the principal adaptations to arboreal life was the acquirement of the upright posture. All primates with the exception of the tree-shrew, whose claim for primate status is somewhat dubious anyway, can either sit upright, stand upright, walk upright—or do all three.

The skeletal adaptations for uprightness are to be found (a) on the base of the skull; (b) in the curvatures of the vertebral column; (c) in the form of the pelvis; and (d) in the relative proportions of the limbs.

Skull morphology. The forwardly displaced position of the occipital condyles, and their articulation with the atlas was described on p. 11.

FIG. 13. An Old World monkey, *Macaca nemestrina*—the pig-tailed macaque. (Courtesy of Doris Sorby.)

FIG. 14. A New World monkey, *Callicebus moloch*—the dusky titi. (Courtesy of Gisela Epple.)

FIG. 15 (left). An ape, *Gorilla gorilla*—the lowland gorilla. The photograph is of a young male about 22 years old. (Courtesy of Doris Sorby.)

FIG. 16. Bipedal posture assumed by an orang-utan ('Jimmy') at Chester Zoo. (Courtesy of Doris Sorby.)

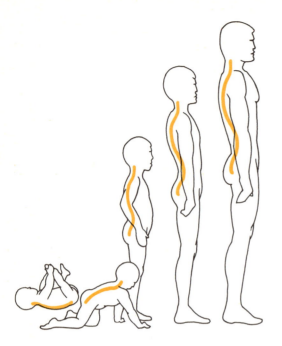

FIG. 17. Lumbar curvature. Changes in curvature of the human vertebral column from infancy to adult life. (After Harrison and Montagna 1969.)

The pelvis. The primate pelvis bears all the hallmarks of adaptation towards the upright position. The pelvic girdle of mammals is slung from the vertebral column in the manner of a cable-car from an overhead track; the symphysis joining the right and left halves of the girdle is extensive, involving not only the pubis but part of the ischium, and it operates to provide support for the abdominal viscera. In the primates the symphysis is reduced to a narrow conjunction that involves only part of the pubic bones (Fig. 18). The ischia, which are

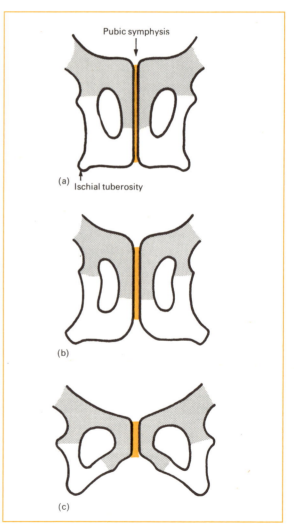

FIG. 18. Primate pelves. The figure shows the pubic symphysis in (a) quadrupedal mammal, (b) Old World quadrupedal monkey, and (c) man. Note the reduction in depth of the symphysis from (a) to (c) and the resulting flaring of the ischial tuberosities.

The vertebral column. The curvature of the vertebral column in most quadrupedal mammals is a single span, convex dorsally, extending from the skull to the pelvis. At the functional mid-point of this continuous pitch lies the anticlinal vertebra which represents the hinge of vertebral movement during locomotion. (Visualize, for example, the croquet-hoop posture of a greyhound at full gallop.) In higher primates, such as the monkeys, the pitch of the vertebral column is modified in the lumbar region where the curve is somewhat flattened. In the apes this flattening is accentuated and in some (e.g. the gorilla) the curve tends to be reversed to a slight concavity. In man the lumbar concavity, an important adaptation for upright bipedal walking, is well marked. At birth the human infant's spine shows a continuous convexity from skull to sacrum. As the infant learns to support the weight of its head a concavity appears in the neck region, and as it begins to sit up, and later to stand up, the lumbar concavity makes its appearance (Fig. 17). These ontogenetic changes in the human infant reflect the evolutionary changes of the Order. As the late Professor Wood Jones used to say 'The human child sits up before it stands; the human stock sat up before it stood.'

free of involvement in the symphysis, flare out laterally to form the sides of a basin which, when sealed off below by a muscular diaphragm, effectively supports the weight of the superincumbent abdominal viscera. At the same time the attachment of the iliac bones to the sacrum (sacro-iliac joint) is modified to facilitate weight transference from the vertical pillar of the spine through the pelvis to the hip bones.

As the body became increasingly vertical during the phylogeny of the primates, the method by which the viscera were suspended within the abdominal and thoracic cavities underwent mechanical alteration. In a dedicatedly quadrupedal animal the heart, the stomach, the liver, and intestines are slung from the dorsal wall of their respective cavities; in the primates, with their evolutionary trend towards the upright posture, the suspensory ligaments tend to be attached to the superior wall.

Locomotion. There are three principal categories of primate locomotion: (1) vertical clinging and leaping; (2) quadrupedalism; and (3) brachiation.

Vertical clinging and leaping is seen only in the Suborders Prosimii and Tarsioidea and is characterized by long hindlimbs and short forelimbs. Their habitats are the boles of trees and the larger, vertically orientated, primary branches. The resting posture is usually vertical and progression through the forest is by leaping from tree to tree propelled by enormously powerful thrusts from the hind legs (Fig. 11).

Quadrupedal primates essentially move on four limbs, but unlike most quadrupedal mammals their limbs have considerable freedom of movement at the shoulder and hip, so that they can climb, swing, and leap. When in a position of rest they are sitting (Fig. 13) and they bear on their rumps the *ischial callosities*, hard, insensitive skin pads which protect the animal from prolonged contact with the rough surfaces of the branches of trees. The great apes who build nests and sleep on their sides do not have these devices, nor does man; but lesser apes, like the gibbon, which are physically and behaviourally rather closer to monkeys, retain them.

Brachiation constitutes the most highly specialized form of arboreal locomotion found amongst the primates. Brachiators swing freely from overhead branches using the hands as suspensory hooks. Unlike vertical clingers and quadrupeds, whose adaptations to verticality are largely limited to the body above the hips, brachiators are vertical in respect of the whole body including the hips. It is

not surprising that anthropologists like Sir Arthur Keith saw in the arboreal posture of gibbons all the prerequisites for human uprightness. Keith's 'brachiating theory' of human evolution was widely accepted in the 1920s and 1930s but is no longer tenable for many reasons, not the least of which is that vertical uprightness derived from *suspension* by the arms is biomechanically a very different proposition from vertical uprightness derived from *support* by the legs.

In all these types of arboreal locomotion, verticality of the body is a salient feature. With this heritage pattern it is not necessary to be unduly puzzled by the seemingly aberrant direction taken by man's locomotion. Man has simply followed the evolutionary trend established some 50–60 million years ago by his Eocene forbears, many of whom were vertical clingers and leapers. Other primates have crossed the ecological threshold and taken up life on the ground—baboons, for instance, and the gorilla, though neither of these genera has adopted bipedalism as an habitual way of life. Clearly there were other evolutionary pressures operating on hominids that prompted the selection of this habit, and they are likely to have been primarily concerned with the use of the hands. Upright posture bestows the inestimable advantage of freeing the hands from their locomotor responsibilities so that they can be deployed in other directions; in turn the emancipation of the hands undoubtedly pushed along the evolution of the upright posture and also led to the emergence of the third of man's outstanding characters—his large inventive brain.

The adaptations discussed involving the hands, the skull, the brain, the posture, and the locomotion of primates are by no means definitive. There are other systems that we could have considered such as the teeth, the digestive system, the reproductive organs, and the endocrine system, all of which have features unique to primates. We could have considered their social behaviour and the relationship this bears to present-day ecology; or the evidence of the blood proteins which is currently modifying our thinking about primate phylogeny. This article has been designed simply as an introduction to a fascinating and important group of mammals. Non-human primates are a sort of mirror into which man may gaze and see not an image of himself but a reflection of all the faces of all the creatures that, over the millennia, have contributed their little bit to the incredible transformation of arboreal monkey into terrestrial man.

FURTHER READING

BUETTNER-JANUSCH, J. (1966). *Origins of man.* John Wiley, New York.

CLARK, W. E. LE GROS (1971). *The antecedents of man.* Edinburgh University Press.

HARRISON, R. J. and MONTAGNA, W. (1969). *Man.* Appleton–Century–Crofts, New York.

JOLLY, Alison (1967). *Lemur behaviour.* Chicago University Press.

NAPIER, J. R. (1971). *The roots of mankind.* Allen and Unwin, London.

—— and NAPIER, P. H. (1967). *A handbook of living primates.* Academic Press, London and New York.

NAPIER, P. H. (1971). *Monkeys and apes.* Hamlyn, London.

PILBEAM, D. R. (1970). *The evolution of man.* Thames and Hudson, London.

REYNOLDS, V. (1967). *The apes.* E. P. Dutton, New York.

ROMER, A. S. (1966). *Vertebrate paleontology*, 3rd edn. Chicago University Press.

SIMPSON, G. G. (1949). *The meaning of evolution.* Yale University Press, New Haven.

WOOD JONES, F. (1898). *Arboreal man.* Edward Arnold, London.

See these other titles in the Oxford Biology Readers series:
 1. *Some general biological principles illustrated by the evolution of man.* Sir Gavin de Beer.
11. *Homology: an unsolved problem.* Sir Gavin de Beer.
22. *Adaptation.* Sir Gavin de Beer.
32. *Fossil man.* M. Day.

28

Oxford Biology Readers
Edited by J. J. Head and O. E. Lowenstein

The Fossil History of Man

M. H. Day

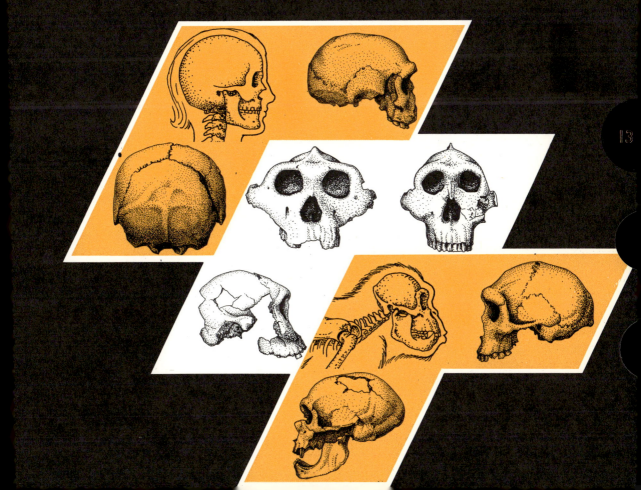

Oxford University Press, Ely House, London W.1

GLASGOW   NEW YORK   TORONTO   MELBOURNE   WELLINGTON
CAPE TOWN   SALISBURY   IBADAN   NAIROBI   DAR ES SALAAM   LUSAKA   ADDIS ABABA
BOMBAY   CALCUTTA   MADRAS   KARACHI   LAHORE   DACCA
KUALA LUMPUR   SINGAPORE   HONG KONG   TOKYO

ISBN 0 19 914128 2

Michael H. Day is Professor of Human Anatomy at St. Thomas's Hospital Medical School, University of London. He is the author of a number of books and scientific papers on palaeoanthropology, including *Guide to fossil man* (Cassell 1965) and *Fossil man* (Paul Hamlyn 1969).

Drawings by Audrey Besterman.

PHOTOSET AND PRINTED IN GREAT BRITAIN BY
BAS PRINTERS LIMITED, WALLOP, HAMPSHIRE

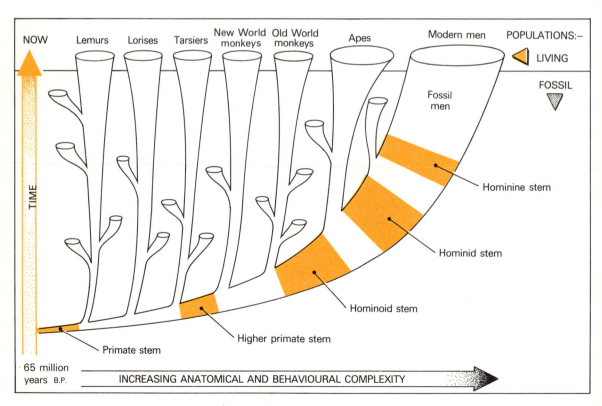

Introduction

The story of human evolution can begin with the statement that man is a primate; he belongs to the Order Primates, one of the major subdivisions of the Class Mammalia. The Order also includes all the living and fossil lemurs, lorises, monkeys, and great apes. Assignments of this kind are based on anatomical similarity and lead to the construction of a scheme of classification that puts some order (admittedly artificial) into our understanding of the animal kingdom. The only biologically real category is the species, a group of actually or potentially interbreeding natural populations that produces fertile offspring like itself; though even this concept has. been weakened by the discovery of species intergrades.

Darwin's proof that species were not individually and supernaturally created, but evolved through time from former species, opened the way for a sensible interpretation of the fossil record that did not have to involve successive biblical 'floods'. The similarities between fossil and living forms permits the idea of successive evolving populations that grade into each other and produce a trend, an evolutionary process known as *anagenesis*. On occasion, however, a split may occur which branches a population into two parts, each going its own way; this is evolution by *cladogenesis*. Both processes acting together produce a radiation of species that will exploit the available environmental opportunities in a variety of ways (Fig. 1). The primates exhibit such an adaptive radiation beginning 65–70 million years ago in the Palaeocene period. Many of the lines of this radiation have survived to the present day and the living primates can be arranged roughly in order of increasing anatomical and behavioural complexity. Although modern primates are the end products of their own evolutionary lines, they represent some of the major stages through which the order has passed during its evolutionary history. No *modern* ape or monkey is in any way ancestral to man; nonetheless man has shared a common ancestry with all primates for at least 65 million years, and with the great apes for 10–12 million years (Fig. 2).

Fig. 1. The Order Primates has existed for over 60 million years. During this time evolution in the order has taken place by successive change (*anagenesis*) and by splitting (*cladogenesis*).

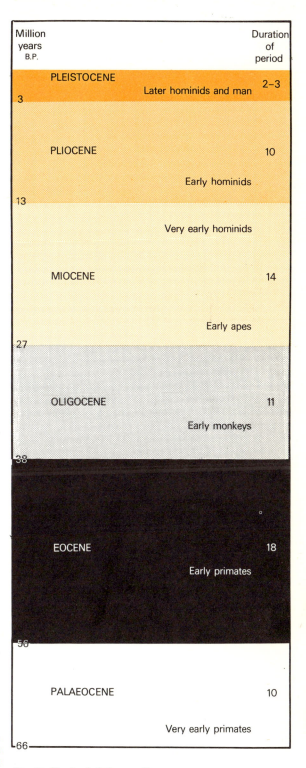

Fig. 2. Geological time-scale.

3

The hominids

In order to group a number of species into a genus, or a number of genera into a subfamily, the taxonomist must define the various groupings, or taxons, in terms of their anatomical features. The identification of anatomical features taxonomically relevant to primates has been the pursuit of anatomists and anthropologists for many years. It is important to look for individually significant anatomical features, but more useful are significant functional complexes such as a manipulative hand, a propulsive foot, or a tooth and jaw arrangement that is adapted for a particular diet. Sometimes it is possible to add these complexes together to form, in Le Gros Clark's phrase, the 'total morphological

pattern' of the species, and to base taxonomic decisions on these wider premises. Recently multivariate statistical methods have been used to quantify the similarities and differences between individual specimens and comparative populations. This enables probabilities to be calculated in the assignment of doubtful material to various functional and possibly taxonomic groups.

But what makes a hominid, or man-like creature, distinct from a pongid or ape-like form? The anatomical features of prime importance are cranial, masticatory, and locomotor; that is, related to the form of the skull and brain, the teeth and jaws, and the posture, gait, and manipulative ability. All these are reflected in the form of the skeleton (Figs. 3, 4, and 5). In brief, hominids are relatively large-brained, short-faced primates who have hands capable of manipulative skills and lower limbs which produce an upright stance and a bipedal gait. This combination is easily seen in modern man but is less apparent in some of his early ancestors. Fossil material is rarely complete enough to allow the identification of all of these features in one specimen or even one group of specimens; and the range of normal anatomical variation and sexual variation in populations adds to the problem; but a great

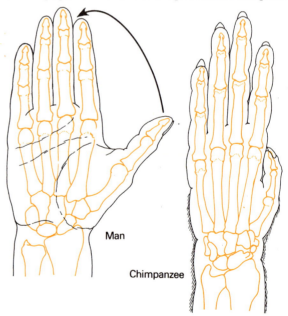

FIG. 3. The skulls of man and gorilla compared to show the relative sizes of the brain and face. In man the brain is large, the jaws and teeth small, and the face straight (*orthognathous*); in the gorilla the brain is small, the jaws and teeth large, and the face sloping (*prognathous*).

FIG. 4. The hands of man and chimpanzee compared. The chimpanzee hand is used as a suspensory hook in the trees and as a knuckle-prop on the ground. The human hand is not used in locomotion but can manipulate small objects between fingers and thumb.

4

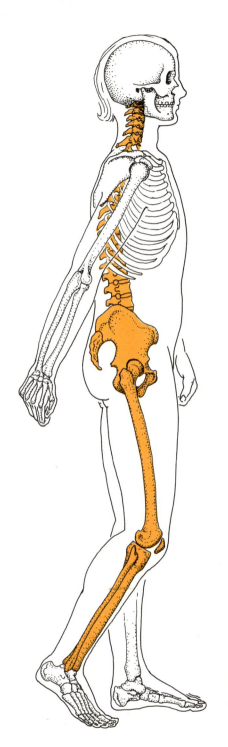

deal of information can be obtained from fossil bones and teeth when all their features are considered together. Decisions on similarity must be made on combinations of features. Isolated features may be misleading since they may be part of an evolutionary mosaic. *Mosaic evolution* means that different features of an organism evolve at different rates, so that primitive and advanced features may be found together in one individual. Mosaicism produces intermediate or linking forms between clear cut stages. A celebrated example is *Archaeopteryx*, intermediate between reptiles and birds. Thus, an undoubted fossil ape from the viewpoint of the limbs may have an advanced dentition; such a transitional form would pose a difficult taxonomic problem.

Hominid evolution has produced three principal levels of morphological organization linked by intermediate stages. These are the australopithecines (literally 'southern-apes', in fact 'ape-men'), the pithecanthropines (literally 'ape-men', in fact 'true men'), and the sapients (literally 'wise-men', in fact 'modern men'). The intermediate stages between these major groups and the relationships that exist between them all are the subject of much research interest at the present time.

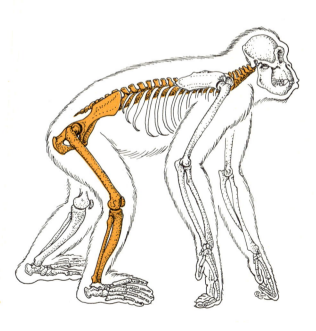

FIG. 5. The posture and gait of man and chimpanzee compared. The form of the vertebral column, pelvis, and hind limbs in the two forms reflects the differing types of locomotion that they employ.

The earliest hominids

During the Miocene and early Pliocene periods, 30–10 million years before the present (B.P.) the primate order contained a wide radiation of early apes from many parts of the world including Kenya, India, China, and parts of Europe. These fossil apes are known as dryopithecines (*Dryopithecus*) and are distinguished from fossil monkeys by a number of features including a special arrangement of the cusps and grooves of their molar teeth. The Indian representatives of the dryopithecines come from the Siwalik Hills of the Punjab while the African group (some of which are known as the genus '*Proconsul*') come from Kenya. Both of these assemblages have produced examples that are not dryopithecine, being more advanced in their dental and mandibular characters. The Indian specimens (*Ramapithecus*) were recovered from a 12–14-million-year-old deposit in 1932 by Lewis who then considered them to be hominid rather than pongid (Fig. 6). Few people accepted this view until it was revived by Simons, who showed in 1961 that the small low-crowned molars coupled with the apparently small canines and incisors had shortened the face of this primate and produced a rounded dental arcade typical of later hominids.

In 1962 Leakey described a new fossil primate from Miocene deposits at Fort Ternan, Kenya, which he called *Kenyapithecus wickeri*. Again the molars are low-cusped; they and the canines and premolars are small; and on these grounds *Kenyapithecus wickeri* has been regarded as an early hominid. Later Simons suggested that both these fossils should be known as *Ramapithecus*, because of their similar morphology, a suggestion rejected by Leakey. It must be admitted that the total quantity of early hominid material is very small and is entirely confined to the teeth and jaws—we know nothing of the cranium or the locomotor system of either *Ramapithecus* or *Kenyapithecus*, if indeed they are distinct. Their attribution to the Hominidae should be regarded cautiously until more specimens are available.

Following these scanty remains of the earliest hominids there is a very large gap in the fossil record from about 10 to about 5·5 million years B.P. when the oldest known member of the next major group occurs. This is a fragment of australopithecine jaw from Lothagam, Kenya. What the intermediate forms between *Ramapithecus* and *Australopithecus* were like we do not know, but this is an area for continuing exploration and research.

The ape-men of the Lower Pleistocene

In 1925 Professor Raymond Dart announced to a sceptical scientific world that he had recognized the infant skull of a fossil primate from Taung in South Africa as a representative of a new species and genus of hominid which he called *Australopithecus africanus* (Fig. 7). Subsequent finds have proved him to be right. His case was based on the rounded appearance of the skull, the size of the fossilized brain that it contained, some features of the teeth, and the roundness of the dental arcade. In all these

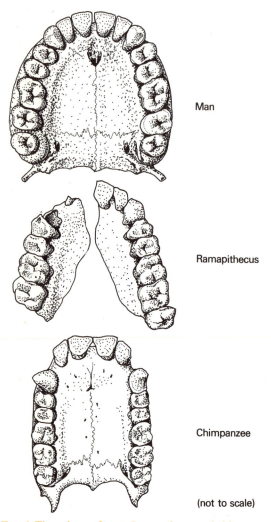

Man

Ramapithecus

Chimpanzee

(not to scale)

FIG. 6. The palates of man, *Ramapithecus* and chimpanzee, compared. The ape tooth-row is rectangular and the human rounded. The reconstructed tooth-row of *Ramapithecus* (an early hominid) is human rather than ape-like.

features the Taung infant differed from an infant modern ape. In addition he was able to point, indirectly, to a locomotor feature—the position of the foramen magnum at the base of the skull was such that the vertebral column must have been attached below the head as in a biped, and not at the back as in a quadruped. It was not until 1936 that there was confirmation from another site called Sterkfontein (Fig. 8). Here Robert Broom recovered a fine adult skull, of similar character, followed by jaws and teeth in quantity and some postcranial bones including an almost complete pelvis. Later finds from Makapansgat added more evidence of the anatomy of the skull, teeth, and pelvis of these ape-men. The picture that has been built up has suggested that *Australopithecus* was a small, lightly built hominid adapted for upright stance and bipedal gait, living in a relatively dry area of open country (Fig. 9). The dentition suggests an omnivorous diet, probably containing some meat.

Further discoveries at two other sites, Kromdraai and Swartkrans not far from Sterkfontein, produced the remains of another form of australopithecine that was much larger and more powerful. At the time these were thought to belong to another genus called *Paranthropus*. (The classification of the australopithecines is still a matter of confusion and acute controversy.)

The larger form is very different from the smaller in a number of respects, for example its molar and premolar teeth are massive grinders while its anterior dentition (incisors and canines) are comparatively small. The few postcranial bones found indicate that the larger form was capable of upright stance and a bipedal gait, though it may not have been particularly efficient at walking. Its more massive dentition may mean that its diet may have differed from that of the smaller ape-man, perhaps coping better with coarse vegetation, roots, stems, and seeds, than with meat; but this is still a matter of dispute.

New finds in East Africa from Olduvai Gorge in Tanzania and from northern Kenya have added a lot of new information. Since 1959 Dr. L. S. B. Leakey, with his wife, family, and a team of workers, has uncovered a remarkable series of hominid fossils. Most recently Mr. Richard Leakey has found more ape-man fossils in the area to the east of Lake Rudolf, Kenya (Fig. 10). From Olduvai there came a magnificent large australopithecine skull known as *Zinjanthropus* (later

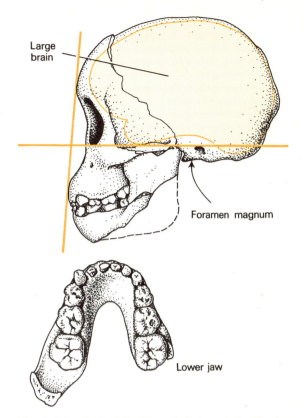

FIG. 7. The skull of the Taung infant australopithecine; its hominid features include a large brain, a rounded jaw, a straight face, and an attachment for the vertebral column that lies under the skull.

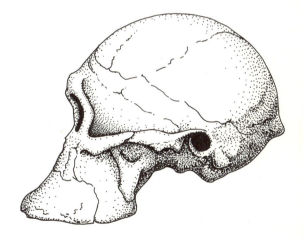

FIG. 8. The skull of an adult smaller australopithecine (Sterkfontein 5). It has a rounded cranium, a small browridge, and a 'dished' face. The vertebral column was attached beneath the skull.

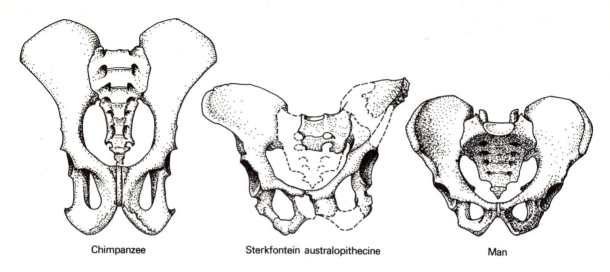

Chimpanzee   Sterkfontein australopithecine   Man

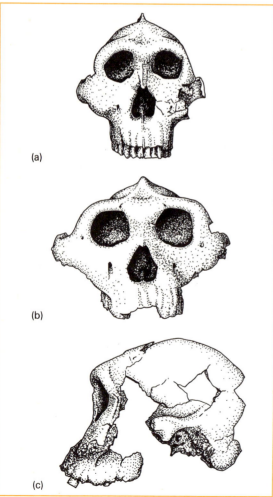

(a)

(b)

(c)

FIG. 9. The pelves of man, chimpanzee, and *Australopithecus* compared. The australopithecine pelvis has clear similarities to that of upright bipedal man.

FIG. 10. The skulls of two large australopithecines from East Africa. (a) *Australopithecus boisei* from Olduvai Gorge. (b) Similar skull from East Rudolf, Kenya. (c) New half skull from East Rudolf that may have belonged to a female. The Olduvai specimen has massive molar and premolar teeth.

FIG. 11. The foot skeleton of *Homo habilis* from Olduvai Gorge. It is small, arched, and clearly bipedal in character. There is no divergence of the great toe so it is a propulsive rather than a grasping foot.

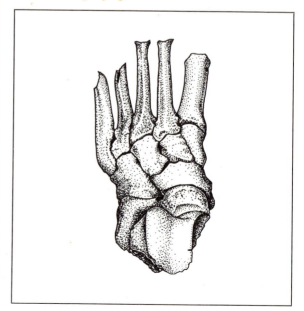

called *Australopithecus boisei*) and the remains of a number of smaller hominids, not unlike the smaller australopithecines in some ways but more advanced in others. From East Rudolf also large and small australopithecine skulls have been found and some interesting postcranial bones including parts of both the arm and the leg. While much of this new material is as yet unanalysed it has served to pinpoint the main controversy concerning these ape-men from Africa. (So far no unequivocal evidence exists of australopithecines outside the African continent.)

This controversy is whether both the large and the small variety are representative of one, two, or even three species belonging to one or even two genera. Each of the alternatives has its adherents and each its difficulties. If all these creatures belonged to one species, all the major differences between them probably related to sex—sexual dimorphism. The form of locomotion and the diet must have been the same in both, despite the differences in limb form and tooth form that some workers emphasize. On the other hand, if these fossils belonged to two species, then the large and small forms did not interbreed and we must expect differences such as differences in locomotion and diet. If both species lived in the same area at the same time (that is, were *sympatric*) there must have been some clear differences in their relationships to their environment, otherwise they would have competed with each other—the principle of competitive exclusion. The three-species idea is almost the same as this except that the two larger forms from East and South Africa are said to be specifically distinct. They may have been geographically isolated, and so have undergone genetic isolation and speciation. A less extreme form of this approach suggests that the large types from East and South Africa differed at only race or variety level, which would not have excluded interbreeding.

A small but important minority of anthropologists maintains that the differences between the larger and the smaller forms are great enough to merit generic separation. The larger forms are all referred to genus *Paranthropus* and the smaller to *Australopithecus*.

Whatever the outcome of this debate, perhaps the more pertinent speculations centre around the relationship of the australopithecines to the human evolutionary line. There is general agreement that the specialized larger form must be regarded as an

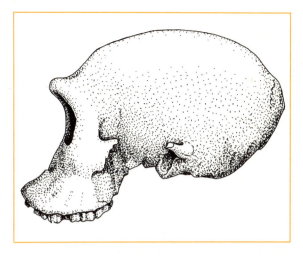

FIG. 12. This is the skull of Olduvai Hominid 24 (H.24) recently recovered from one of the lowest levels at Olduvai Gorge. It was badly crushed and has needed a great deal of conservation and reconstruction. (*Homo habilis*.)

extinct offshoot that took no further part in human evolution, but there is less agreement about the smaller form. Dr. Leakey and some of his colleagues entirely exclude the smaller ape-man from the human line on the grounds that it is preceded in time by fossils from Olduvai Gorge that they have already attributed to the genus *Homo*, as *Homo habilis*. This new species of man is said to be in advance of *Australopithecus africanus* in a number of respects including brain size, dental features, locomotor ability, and manual skill; the last two characteristics are shown by a remarkable, and clearly bipedal, foot skeleton and a hand structure that would have permitted tool-making (Fig. 11). Not all anthropologists agree with this attribution but regard *Homo habilis* as another australopithecine, perhaps an advanced one, which may be a linking form between the australopithecine stage of human evolution and the pithecanthropine stage which follows (Fig. 12).

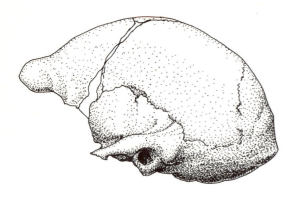

FIG. 13. A skull of *Homo erectus* from Java. It is low-vaulted and has heavy brow-ridges and a stout occipital ridge.

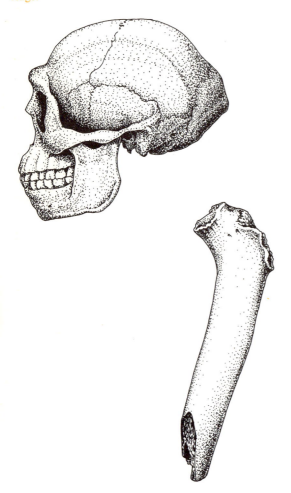

The men of the Middle Pleistocene

Unlike the ape-men there is no doubt that the representatives of the pithecanthropine group are true men and fully merit acceptance into the genus as *Homo erectus*. This form of man has been found at a wide range of sites in the Far East, Asia, North and East Africa, and possibly Europe, which have been dated to the Middle Pleistocene period around half a million years B.P. The first discovered remains came from Trinil, Java, in 1890 and consist of a skull cap of primitive form and a femur of quite modern appearance (Fig. 13). Subsequent finds in Java (Sangiran), at Choukoutien, near Peking in China, and also at Olduvai Gorge have confirmed the appearance of the original skull from Java, though not of the femur. The skull of *Homo erectus* differs from the modern human skull in being low-vaulted and having low capacity (900–1100 cm³: modern man has a mean cranial capacity of 1400 cm³). It has a beetling brow-ridge, a stout occipital ridge at the back, and small mastoid processes behind the ear openings (Fig. 14). The jaws are known from Java, Choukoutien, and North Africa; in general they are large and chinless with robust teeth that show several primitive features, such as a collar of enamel around the molars and some wrinklings of the biting surface of the tooth. One famous jaw from Heidelberg in Europe may belong to this group, but its dating is not absolutely certain.

Until recently very little limb material was known that could be attributed to *Homo erectus* other than the conflicting femora from Java and Choukoutien; however, a recent find from Olduvai Gorge has added a new femur and half a pelvis. The femur is very like the femora from Choukoutien. It was formerly widely believed, on the basis of the Trinil femur, that the limb bones of *Homo erectus* were very modern and that subsequent human evolutionary changes must have been confined to the skull, brain, jaws, and teeth. The new Olduvai femur and pelvis have cast doubt on this view, and we may have to think again about the details of this early man's posture and gait (Fig. 15).

The cultural evidence of *Homo erectus* is widely spread from China, North Africa, and East Africa, consisting of both flake-tools and hand-axes of the

FIG. 14. The skull and femur of *Homo erectus* from Choukoutien, China. The skull is low-vaulted, and has a heavy brow-ridge and a stout occipital crest. The jaw is robust and chinless. The femur is flattened and distinctive in a number of ways.

10

Acheulean culture. There is good evidence of fire-making from the Choukoutien site and of co-operative hunting, since the deposit contains the bones of large animals needing more than one man to capture them.

So by 500 000 years B.P. man was successful in that he had spread into a large area of the Old World; he had soon exploited the advantage of bipedalism; and in association with increased brain size and the development of a manipulative hand, he was making rapid cultural progress.

After the Middle Pleistocene period, with its comparative wealth of fossil remains, there follows in the fossil record a critical gap which is slowly being closed as field-workers yearly produce more finds. The gap is crucial to the human evolutionary story since it is about here that the beginnings of our own species *Homo sapiens* will be found.

The known possible linking forms come from both Europe and East Africa. In 1965 an occipital bone, the back part of a skull, was recovered from Vértesszöllös, near Budapest, in deposits dated at about 350 000 years B.P. It is large, suggesting a large brain size, but thick and primitive in some respects (Fig. 16). This mixture of features indicates mosaic evolution. This fossil has been classified as an *erectus/sapiens* intermediate from Southern Europe. From Britain, there is the well-known Swanscombe skull found in the 100-ft Great Inter-glacial gravels of the Thames estuary; it has been dated at about 250 000 years B.P. (Fig. 17). This skull is remarkably sapient in many ways and still forms an important part of the earliest evidence of our own species from anywhere in the world. Not unlike the Swanscombe skull is a skull from Stein-heim, in Germany, also from interglacial river

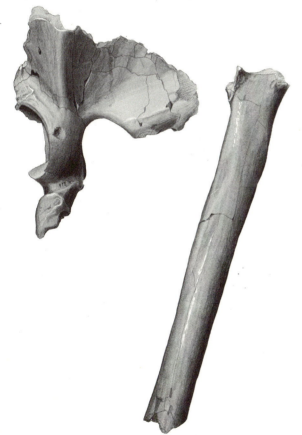

FIG. 15. The pelvis and femur of *Homo erectus* from Olduvai Gorge. The femur is similar to that known from Choukoutien. The pelvis is a unique specimen and bears a remarkably stout iliac pillar.

FIG. 16. The occipital bone from Vértesszöllös, Hungary, shows both *erectus* and *sapiens* features. It is regarded as an example of mosaic evolution that may be an *erectus/sapiens* intergrade. (Courtesy of A. Thoma.)

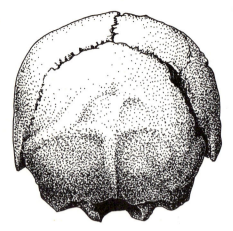

FIG. 17. The early *Homo sapiens* skull from Swanscombe, Kent. The skull is well rounded and lacks heavy muscle attachments.

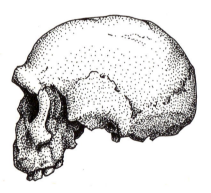

FIG. 18. The early *Homo sapiens* skull from Steinheim, near Stuttgart, West Germany. It is badly crushed but the brow-ridges are modest and the teeth small.

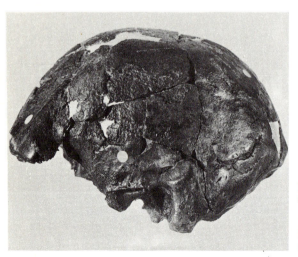

gravels (Fig. 18). The principal differences between the earlier *Homo erectus* skulls and these two sapient skulls are their more rounded shape, larger skull volume, and in the case of the Steinheim skull the relatively reduced brow-ridges and advanced form of the teeth. Teeth are useful and important indicators of human evolutionary stages; those of the Steinheim skull are small, modern in shape, and show a reduction in the size of the third molar, or wisdom tooth. The East African remains that deserve consideration are two newly found skulls from the Omo region of Southern Ethiopia (Fig. 19). These and some postcranial remains have been provisionally assigned to the Upper Middle Pleistocene of East Africa, perhaps 100–130 000 years B.P. One of the skulls shows a number of primitive features while having a large cranial capacity, the other is much more modern in its general form. These skulls are still being analysed; they are important early sapient finds at a crucial point in human evolution.

To attempt to trace a firm connection between *Homo erectus* and *Homo sapiens* by means of the Swanscombe, Steinheim, and Omo remains is probably not only premature but unwise. Only general evolutionary trends, such as overall cranial enlargement and rounding and continuing facial and dental reduction can be traced at this stage. In taxonomic terms, however, there seems little doubt that Swanscombe, Steinheim, and Omo men must all be allocated to *Homo sapiens*.

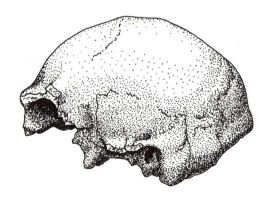

FIG. 19. The Omo II skull from southern Ethiopia. An example of early *Homo sapiens* from East Africa.

FIG. 20 (left). One of the Upper Pleistocene skulls from Solo, Java. Its features are mixed, but it is usually regarded as a representative of *Homo sapiens* from the Far East.

The modern men of the Upper Pleistocene

The amount of fossil material available increases markedly as the Upper Pleistocene is reached, and it is more widely spread. Remains of fossil *Homo sapiens* are known from the Far East, Asia, Africa, the Near East, and Europe, but it was not until about 20–25 000 years B.P. that modern man penetrated the Americas and Australasia.

The fossils of this period from the Far East are from Solo (Java), Niah (Borneo), and once again from Choukoutien (Peking), though from a much later layer than the previous *Homo erectus* skulls.

The Solo remains (Fig. 20) were found in river deposits; with the eleven skulls were stone implements and some mammalian bones. The skulls are all thick-walled and of the same general shape having sloping foreheads, divided supraorbital ridges, strong occipital ridges, and pronounced mastoid processes. In some respects these resemble the earlier *Homo erectus* skulls but in others, such as cranial capacity, they are more advanced.

The Niah skull was found 2·4 m beneath the floor of a very large cave in North Borneo; with it were stone tools. Radio-carbon dating suggests that it is about 39 000 years old. There is no doubt that this is a sapient skull, with small brow-ridges, a depressed nasal root, and a rather straight forehead. It resembles the skulls of modern aboriginal Australians.

The Upper Cave at Choukoutien yielded later sapient remains of eight individuals with some stone tools and ornamental objects. The skulls were damaged, but some authorities have seen in them the early features of the Mongoloid races. They were all lost during the Second World War.

The Upper Pleistocene fossil remains of man from Africa come from many sites but by far the best preserved and most important are the Broken Hill skull and limb bones from Rhodesia (Fig. 21). The cranium is very heavily built, with massive brow-ridges, a flattened vault, and a prominent occipital ridge. The face is large and there is a large palate bearing worn and carious teeth. In some respects this skull resembles the Neanderthal skeletons from Europe, but the limb bones are quite different and essentially modern in most of their features.

The European material from this period is particularly interesting since it includes a group formerly widely known as a separate species, *Homo neanderthalensis* or Neanderthal man. These men lived in Europe and the Middle East during the Last Gla-

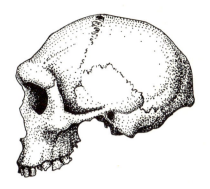

FIG. 21. The Broken Hill skull from Zambia (Northern Rhodesia). An Upper Pleistocene representative of the African segment of evolving *Homo sapiens*.

ciation, perhaps 50 000 years B.P., and they can be distinguished in their most marked form by a number of morphological features of the skull and postcranial bones. The classic Neanderthaler was a short thick-set individual with powerful limbs, a somewhat large braincase, a chinless jaw, and a large face. Several of these features, in particular the short, thick body, have been described as adaptations to cold, related to the harsh environment in which he lived.

The first Neanderthal find was in 1856, just prior to the publication of Darwin's *Origin of species* and well in advance of the application of his theory to human evolution. It came from the Neander valley near Dusseldorf in Germany and provoked much argument. Some regarded the remains as diseased, others thought them the bones of an imbecile, or even the remains of an 'ancient savage and barbarous race'. It was not until 1908 that another skeleton of the same general character was found, this time in France at La Chapelle-aux-Saints (Fig. 22), and this specimen led to the view that the Neanderthalers were a separate species of man, stooping in posture with a bent-kneed gait and a forward-jutting head and neck. This postural analysis was shown to be incorrect many years later when more Neanderthal skeletons were examined; the earlier misconceptions were based unknowingly on pathological bones—the poor 'Old Man' from La Chapelle-aux-Saints suffered from deforming osteoarthritis. Later finds, such as the La Ferrassie skeletons, have shown that Neanderthal man was fully upright, bipedal, and capable of leading an active life as a hunter under the difficult climatic conditions of a periglacial region.

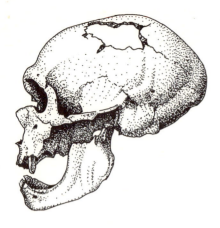

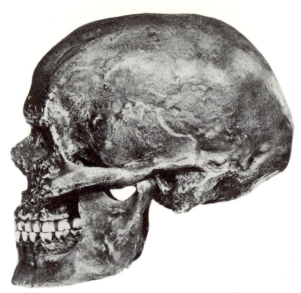

FIG. 22. The Neanderthal skull from La Chapelle-aux-Saints. It has a large cranial capacity and a large face.

FIG. 23 (right). The skull of Cro-Magnon man from the Dordogne region of southern France. It is modern in form but has low orbits and a compressed nasal root.

FIG. 24. The 'Willendorf Venus', an early sculpture in the round which may be a fertility or cult figure.

The Neanderthalers were widely spread in Europe and the Near East, from Gibraltar to Czechoslovakia and from Germany to Israel. Many Neanderthal sites are known from the limestone caves of the Dordogne region of France. Culturally Neanderthal man was capable of stone-tool manufacture and fire-making, as was *Homo erectus*, but also he buried his dead and even performed some mystical rites shown by the deliberate arrangement of goats' horns or bears' skulls at some burial sites.

Curiously Neanderthal man disappears very abruptly from the fossil record, to be replaced by a distinctively more modern type of man. The reasons for his disappearance have been the subject of argument. He may have been overcome by some natural catastrophe, or driven out by more advanced peoples, or overrun and absorbed into the new sapient populations invading from the Near East. The last explanation is the most likely on present evidence. At Mount Carmel in Israel, two sites known as Tabūn and Skhūl were found side by side in two adjacent caves. The Tabūn skeleton is very Neanderthal in appearance; the Skhūl remains are more modern but with mixed features. At first it was suggested that the two skeletons were sapient/Neanderthal hybrids, or perhaps the extremes of the range of a single variable population; but it has now been shown that the more modern Skhūl remains postdate the Tabūn skeleton by as much as 10 000 years and we may conclude that following the recession of the ice of the Last Glaciation more modern

14

forms of man replaced the classic Neanderthalers.

This new group is named after its 'type site' in the Dordogne, Cro-Magnon. The earliest known remains are dated at about 32 000 years B.P., but the first representatives probably entered Europe 5–10 000 years before this. The Cro-Magnon people were quite different from the Neanderthalers in that they were tall, slender, and modern in skeletal form, with rather long 'five-sided' skulls. In most other respects they were very similar to the Indo-European peoples of today (Fig. 23).

They appear to have hunted and gathered food and they made a wide range of tools and weapons from both stone and bone. They also produced the first known decorative art and sculpture. Small engravings, bas-reliefs, and sculptures of animals and fertility figures have been found (Fig. 24). They created the magnificent painted caves of Lascaux and many other sites. Presumably these works of art were more than purely decorative; they may have been thought by Cro-Magnon man to have had mystical or magical functions such as influencing the outcome of the hunt or even an amorous adventure (Fig. 25).

In any event there is no doubt that the Cro-Magnon peoples possessed such a high degree of technical skill and artistic ability that they must be regarded as fully modern men, both in physical appearance and the sophistication of their cultural attainments.

Following the Cro-Magnon peoples came Mesolithic man and then Neolithic or 'new stone age' man. These latter people are known from many sites but anatomically they are not particularly distinctive. Culturally they must have been a fascinating group, for it was they who began co-operative food production, they who domesticated useful animals, and probably they who originated the division of labour into socially useful groups. Civilization had begun, with the environment being utilized rather than fought.

The fossil history of man is of necessity incomplete, for although every year we learn more of our fossil past, every year we find new questions to answer. Only further research, in both field and laboratory, will improve our understanding of how we came to be what we are: man.

FIG. 25. An example of cave art that shows the skill of Cro-Magnon man of the Upper Pleistocene period.

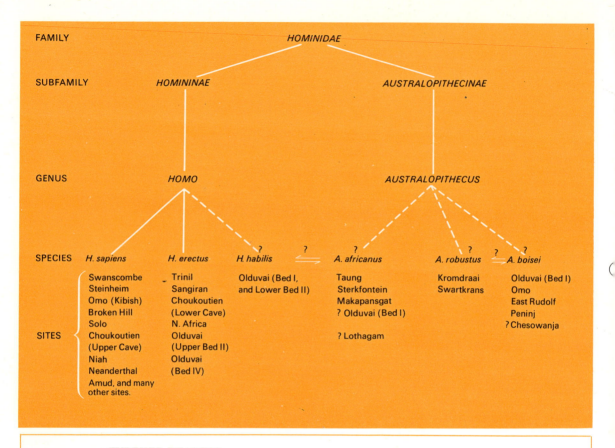

	HOMININAE			AUSTRALOPITHECINAE		

FAMILY — HOMINIDAE

SUBFAMILY — HOMININAE / AUSTRALOPITHECINAE

GENUS — HOMO / AUSTRALOPITHECUS

SPECIES — H. sapiens | H. erectus | H. habilis | ? | A. africanus | A. robustus | ? | A. boisei

SITES						
H. sapiens	H. erectus	H. habilis	A. africanus	A. robustus	A. boisei
Swanscombe	Trinil	Olduvai (Bed I, and Lower Bed II)	Taung	Kromdraai	Olduvai (Bed I)
Steinheim	Sangiran		Sterkfontein	Swartkrans	Omo
Omo (Kibish)	Choukoutien		Makapansgat		East Rudolf
Broken Hill	(Lower Cave)		? Olduvai (Bed I)		Peninj
Solo	N. Africa				? Chesowanja
Choukoutien	Olduvai		? Lothagam		
(Upper Cave)	(Upper Bed II)				
Niah	Olduvai				
Neanderthal	(Bed IV)				
Amud, and many other sites.					

FURTHER READING

General

CLARK, W. E. LE GROS (1964). *Fossil evidence for human evolution*. Chicago University Press.
——(1967). *Man-apes and ape-men*. Holt, Rinehart, and Winston, London and New York.

DAY, M. H. (1967). *Guide to fossil man*. Cassell, London.

HOWELL, F. CLARK (1966). *Early man*. Time-Life International.

HOWELLS, W. W. (1960). *Mankind in the making*. Secker and Warburg, London.

NAPIER, J. R. and NAPIER, P. H. (1967). *Handbook of living primates*. Academic Press, London and New York.

PILBEAM, D. R. (1970). *The evolution of man*. Thames and Hudson, London.

See these titles in the Oxford Biology Readers series:

1. *Some general biological principles illustrated by the evolution of man*. Sir Gavin de Beer.
10. *Studying the past by pollen analysis*. R. G. West.
11. *Homology: an unsolved problem*. Sir Gavin de Beer.
18. *The origin of Chordates*. Q. Bone.
22. *Adaptation*. Sir Gavin de Beer.
28. *Primates and their adaptations*. J. R. Napier.

32

3

Oxford Biology Readers
Edited by J.J. Head and O.E. Lowenstein

The Mysterious Origin of Flowering Plants

Kenneth R. Sporne

Oxford University Press, Ely House, London W.1

GLASGOW   NEW YORK   TORONTO   MELBOURNE   WELLINGTON

CAPE TOWN   SALISBURY   IBADAN   NAIROBI   DAR ES SALAAM   LUSAKA   ADDIS ABABA

BOMBAY   CALCUTTA   MADRAS   KARACHI   LAHORE   DACCA

KUALA LUMPUR   SINGAPORE   HONG KONG   TOKYO

Kenneth Sporne is a Fellow of Downing College, Cambridge, and University Lecturer in Botany. He is author of *The morphology of pteridophytes* (Hutchinson, third edition 1970) and *The morphology of gymnosperms* (Hutchinson 1965).

PHOTOSET AND PRINTED IN GREAT BRITAIN BY

BAS PRINTERS LIMITED, WALLOP, HAMPSHIRE

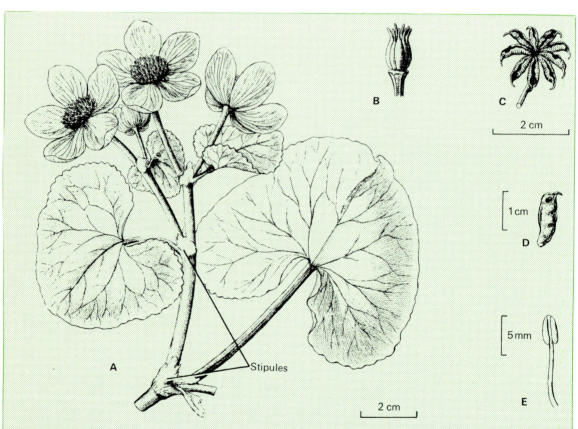

FIG. 1. *Caltha palustris* (marsh marigold). A, part of the plant, showing the net-veined leaves, each with stipules at the base of the petiole, and the flowers whose perianth segments (varying in number from 5 to 8) are all petaloid and whose stamens vary in number from 50 to 100.
B, flower after the perianth and stamens have fallen, leaving follicles (varying in number from 5 to 13) attached to the receptacle.
C, ripe follicles splitting open to release seeds.
D, single ripe follicle. E, stamen.
(From Reichenbach)

Flowering plants are the dominant land plants of the world today. More than 250 000 species have been identified, ranging in size from *Wolffia* (a relative of duckweed, only 1 mm across) to species of *Eucalyptus* (Australian gums, some of which are among the tallest trees in the world). Yet there is scarcely any evidence that they were in existence before Cretaceous times (only 135 million years ago), since when their rapid spread has been one of the most important events in the history of life upon the earth. As Takhtajan remarks, it had a decisive influence on the whole future of the terrestrial animal world and, in the final reckoning, made possible the appearance of man.

The earliest writings about the structure and fundamental nature of flowering plants were those of Theophrastus, in the third century B.C., but, of course, a great deal must have been known prior to this, if only to enable primitive man to distinguish edible from poisonous species. Indeed, because of man's dependence upon flowering plants for food, the study of botany was, until comparatively recently, devoted entirely to them, to the virtual exclusion of other kinds of plants. In spite of this, however, our knowledge of the group is still lamentably incomplete. Theories without number have been put forward concerning the origin and subsequent evolution of flowering plants, but none has received universal approval. Darwin, in a letter to Hooker, written in 1879, made the following comment: 'The rapid development, as far as we can judge, of all the higher plants within recent geological times is an abominable mystery', and the situation has scarcely changed since then, in spite of the remarkable advances that have been made in the twentieth century.

What are flowering plants?
Angiosperms (flowering plants) and gymnosperms (conifers, cycads, etc.) together constitute the Spermatophyta (seed-plants), the difference between them, supposedly, being the extent to which the ovules (immature seeds) are exposed at the time of pollination. In gymnosperms the pollen grains have direct access to the ovule, whereas in most angiosperms the ovules are enclosed within carpels and the pollen lodges on a stigmatic surface. Here it germinates to form a pollen tube which carries the male nuclei to the egg nucleus within the embryo sac of the ovule. However, there are several angiosperms in which the carpels are

not completely closed at pollination (e.g. *Drimys*, related to *Magnolia*) and there are some, even, in which the carpels are wide open.

To many people, the term 'flower' brings to mind a structure containing sepals, petals, stamens, and carpels. However, there are many flowers with a single whorl of perianth members, as in *Caltha*—marsh marigold (Figs. 1 and 2), and there are some completely without a perianth, e.g. each individual flower in the 'male' catkin of *Corylus*—hazel (Fig. 3). The extreme of simplicity in flower construction is reached, among dicotyledons, in *Euphorbia*—spurge, where the 'male' flower consists of a single stamen and, among monocotyledons, in *Lemna*—duckweed, where the 'female' flower consists of a single unilocular ovary.

It is, in fact, almost impossible to devise a satisfactory formal definition of the angiosperms,

FIG. 2. Flowers of *Caltha palustris*, illustrating protandry, in which the stamens (lower flower) mature before the carpels (upper flower). This difference in timing favours cross pollination.

3

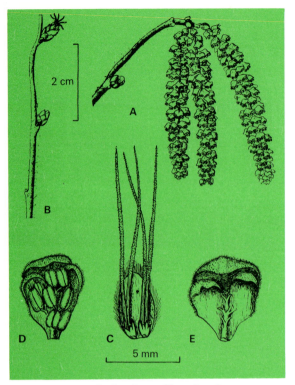

Fig. 3. *Corylus avellana* (hazel)
A, group of 'male' catkins. B, twig, bearing a single
'female' catkin. C, bract detached from 'female' catkin,
showing the two subtended flowers.
D and E, 'male' flowers removed from the catkin, showing
bract, bracteoles, and four bifid stamens. (In E, the
anthers have been removed.) (A and B from Reichenbach; C, D, and E after Prantl.)

for most of the criteria that might be used are
subject to exceptions. Thus, while the vast majority
of angiosperms possess vessels in their xylem,
there are several genera without them and, furthermore, vessels are not peculiar to this group of
plants, for they occur among gymnosperms in the
Gnetales and among pteridophytes in some ferns,
horsetails, and club-mosses. One of the most
peculiar and widespread features shown by
angiosperms is the arrangement within the embryo sac of eight nuclei (egg, two synergidae, two
polar nuclei, and three antipodals), but there are
many exceptions to this. Another characteristic is
the process known as *double fertilization* leading
to the formation of endosperm round the developing embryo, but it must be confessed that only a
small proportion of the total number of angiosperm species has, so far, been checked for the

occurrence of this process. At least it can be said,
however, that the phenomenon is not known to
occur in any other group of plants. The same is
true of the formation, in the phloem, of sieve tubes
and companion cells from a common mother cell.

Origins: monophyletic or polyphyletic?
The first question to be asked in any discussion of
the evolution of a group of organisms is whether it
is monophyletic or polyphyletic, that is, whether
it had a single origin or multiple origins. Concerning the angiosperms there is wide disagreement.
Thus there are some who believe that monocotyledons and dicotyledons had separate origins,
while others believe that catkin-bearing plants
arose independently from the rest of the angiosperms. However, the majority favour a single
origin, arguing that angiosperms have so many
peculiar attributes (the eight-nucleate embryo sac,
double fertilization, sieve tubes and companion
cells, stamens, carpels, etc.) that the statistical
probability of their being brought together more
than once in evolution is extremely small.

The fossil record gives little direct help in this
matter for, among the earliest remains of angiosperms, from Middle Cretaceous deposits, forty-
nine modern families are represented, including
four families of catkin-bearing plants and five
monocotyledon families. Although the reliability
of many of these identifications is questionable,
since many of the remains are of leaves alone, the
conclusion is nevertheless inescapable that a wide
range of flowering plants was already in existence
when the group first appeared. So far (with the
possible exception of some recently described
palm-like stems from the Jurassic) not a single
bona fide fossil flowering plant has been identified
from earlier deposits. The search for fossils of 'the
original angiosperms', of course continues, but the
thoughts of botanists have also turned to the
possibility of recognizing precursors of flowering
plants among gymnosperms.

Possible precursors
1. The Gnetales. The Gnetales have been favoured
because they alone among gymnosperms possess
vessels. It is a fascinating group consisting of three
genera, *Gnetum*, *Ephedra*, and *Welwitschia*, so
different from each other in growth habit and
general morphology, that at first sight it seems
incredible that they should be classed together.
Most of the forty species of *Gnetum* (Fig. 4) are

4

lianes, although a few are trees (e.g. *Gnetum gnemon*, cultivated in Malaysia for its edible seeds). One of the most striking features is the appearance of their leaves, which are almost indistinguishable from those of an angiosperm in having a broad lamina with reticulate venation. *Ephedra* (Fig. 5) is represented by some forty species, all of which are 'switch-plants' with minute leaves. Several are important medicinally as a source of the drug Ephedrin. *Welwitschia* is represented by a single species, restricted to a narrow coastal desert region of south-west Africa.It has been described as the most bizarre of all plants, for its stem resembles a gigantic turnip, reaching a diameter of more than 1 metre, with two opposite leaves that persist throughout the life of the plant, growing continuously from a basal meristem.

Of the three genera, it is *Ephedra* which has been taken to represent the forerunner of the angiosperms, for Wettstein and his followers claimed

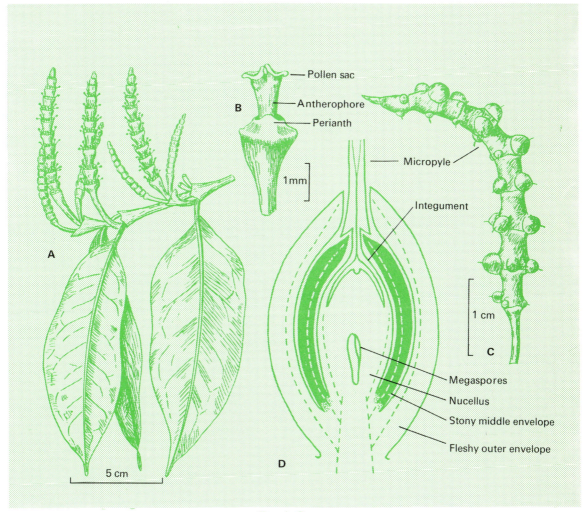

FIG. 4. *Gnetum*
A, twig of *G. gnemon* with pollen-bearing organs. The broad leaves are very similar to those of flowering plants.
B, 'male flower' of *G. africanum*, showing the perianth round the base of the antherophore and the two terminal pollen sacs, from which the pollen has been shed.
C, ovuliferous spike of *G. africanum*, showing the whorled arrangement of the ovules and their long micropyles.
D, longitudinal section (diagrammatic) of ovule of *G. gnemon*, showing the three envelopes—integument, stony middle envelope, and fleshy outer envelope—surrounding the nucellus (= megasporangium) and megaspores. (A after Firbas, modified from Karsten and Liebisch; B and C after Waterkeyn; D based on Pearson.)

5

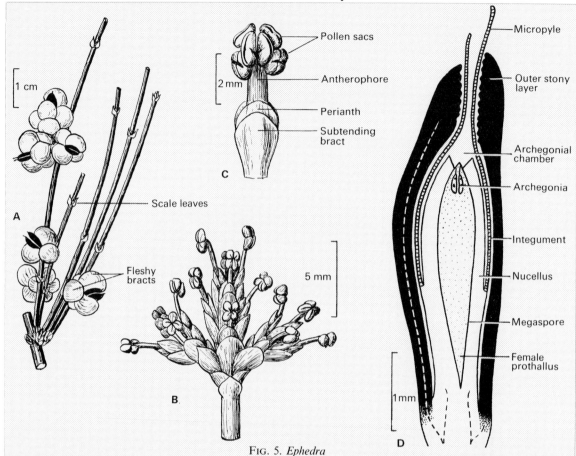

FIG. 5. *Ephedra*

A, portion of *E. americanum*, showing the opposite pairs of scale leaves and the ovules (black) partially enclosed by fleshy bracts.

B, 'male' reproductive organs of *E. alata*, showing opposite and decussate bracts with 'male flowers' in their axils.

C, 'male flower' of *E. fragilis* with several pollen sacs on an antherophore at the base of which is a thin perianth and a subtending bract.

D, Ovule (diagrammatic) of *E. foliata*, showing the outer stony layer, the integument prolonged into a micropyle, the nucellus, the single megaspore containing the female prothallus within which are archegonia. Pollen is drawn into the archegonial chamber, where it germinates to form a tube, which conveys male nuclei to the egg nuclei to the archegonium.

(A after Le Maout and Decaisne; B and C, after Eichler; D based on Maheshwari.)

that similarities can be found with *Casuarina* (Australian she-oak) and hence with catkin-bearing plants generally. The 'male' reproductive organs consist each of a single microsporangiophore (often called a stamen, or an antherophore) surrounded at its base by a papery cup-shaped envelope (usually called a perianth) and subtended by a bract. At its apex is a group of septate microsporangia, from which pollen grains are released at maturity. The 'female' reproductive organs consist each of a single ovule with two envelopes outside the nucellus. By analogy with the 'male' organs it is sometimes suggested that the innermost of the two envelopes surrounding the nucellus represents a single integument, while the outer one represents a perianth. Commonly, there are two such ovules, subtended by two pairs of bracts which, after fertilization, swell up to form a juicy fruit-like body, attractive to birds, within which the seeds are almost hidden.

At the time of pollination, the integument is prolonged into a narrow tubular micropyle at the mouth of which a mucilaginous drop of fluid is secreted. Pollen grains, blowing in the wind, become trapped in the drop, which is then re-absorbed so that they are drawn down inside the ovule. Within a few hours, the pollen grains germinate to produce pollen tubes which penetrate the archegonium, where fusion takes place between one of the male nuclei and the egg nucleus. The other male nucleus may sometimes fuse with the ventral canal nucleus, a process taken by some to be homologous with the fusion in flowering plants between one of the male nuclei and the polar nuclei; but no endosperm results from this so-called double fertilization in *Ephedra*. The whole process is, therefore, on a level no higher than that in *Pinus*.

Gnetum and *Welwitschia* are both at a supposedly higher level of evolution in that no archegonia are formed. Instead, it seems that any nucleus near the apex of the female prothallus can behave as an egg nucleus, waiting passively for the arrival of the pollen tube in *Gnetum*, but in *Welwitschia* being carried actively by *prothallial tubes* to meet the pollen tubes somewhere in the nucellar cap. So-called *secondary endosperm* is formed in *Welwitschia*, but no male nuclei are involved in its formation; it is merely a further development of female prothallial tissue subsequent to fertilization.

The theory which derives catkin-bearing plants from *Ephedra*-like plants is known as the pseudanthium theory. It involves the further derivation of the typical buttercup-like flower by condensation of a catkin, which is itself an inflorescence of unisexual flowers. Some morphologists find acceptance of this evolutionary process so difficult that they have favoured a double origin for flowering plants, deriving catkin-bearing plants from *Ephedra*-like ancestors, while buttercup-like flowers are derived from completely separate origins.

However, it is necessary to look in more detail at the vessels of the Gnetales, for it was those which first turned people's attention to the group as possible forerunners of flowering plants. There seems to be little doubt that they evolved from pitted tracheids, for intermediates occur between vessels and tracheids (Fig. 6), whereas it seems to be equally clear that angiosperm vessels (Fig. 7) evolved from scalariform tracheids. Further-

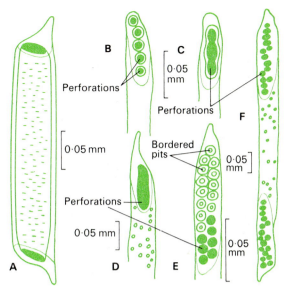

FIG. 6. Vessels in Gnetales
A, vessel of *Welwitschia*, showing simple perforation plates.
B – D, vessels of *Gnetum africanum*, showing varying numbers of perforations (illustrating possible stages in evolution).
E, vessel end-plate of *Ephedra major*, showing bordered pits as well as perforations.
F, vessel of *E. californica* with multiperforate end-plates.
(B – D, after Duthie; E, after Thompson; F, after Esau.)

more, as already mentioned, there are several angiosperm genera with homoxylous wood (i.e. lacking vessels altogether). The conclusion, therefore, is inescapable that angiosperm vessels evolved separately from gnetalean vessels and that when searching for possible forerunners of the flowering plants, one should look not for a group *with* vessels but for one *without* them.

The Gnetales are virtually without any fossil record, for the only known remains are of *Ephedra* pollen from Eocene deposits and some pollen rather like that of *Ephedra* and of *Welwitschia* from the Permian. By contrast, there are several groups of gymnosperms which have been claimed by various authors to be forerunners of angiosperms and which are known only from their fossil remains, viz. Bennettitales, Pentoxylales, and various families of pteridosperms (e.g. Glossopteridaceae and Caytoniaceae). However, before discussing them, a word of caution is necessary about the dangers of basing phylogenetic speculations on fossil remains. Our knowledge of fossil plants is necessarily limited to those parts of the plant which, at a particular stage in

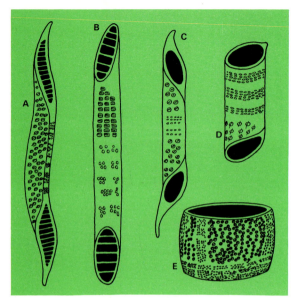

FIG. 7. Vessel evolution in flowering plants
Vessel elements are illustrated as follows:
A, from *Betula*, elongated and narrow, with an oblique
   scalariform end-plate (primitive).
B, from *Liriodendron*, with fewer bars in the end-plate.
C, from *Quercus*, with a simple perforation in an oblique
   end-plate.
D, from *Acer*, in which the end-plate is less oblique.
E, from *Quercus*, short and wide, with a transverse end-
   plate (advanced). (Based on Eames and MacDaniels.)

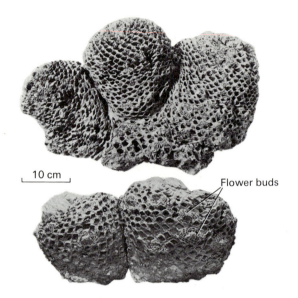

FIG. 8. *Cycadeoidea*
Photographs of branching trunks of two Cretaceous
species, showing persistent leaf bases. A, *C. marshiana*.
B, *C. superba*. (Note the 'flower buds' buried between
leaf bases.)
(From Wieland.)

its life history, happened to become fossilized.
Parts which are not in organic contact when discovered must be placed in separate genera (e.g. one
for the stem, one for the leaf, one for the root).
Gradually, as more specimens are found, it may
become possible to build up a picture of the whole
plant, but there are very few instances where this
has been achieved. The dangers of reading too
much into such fragmentary evidence cannot be
over-emphasized.

2. The Bennettitales. The Bennettitales (Fig. 8) first
appeared in the Triassic, more than 180 million
years ago, and survived for about 80 million years,
becoming extinct in the Cretaceous. Interest was first
aroused because the best known genus, *Cycadeoidea* (Fig. 9), had reproductive organs that were
superficially like the flower of an angiosperm, with
an elongated receptacle bearing perianth-like
bracts and a whorl of pollen-bearing microsporophylls below an ovuliferous region. Comparisons
were drawn with the flower of *Magnolia* (Figs.
10 and 11), imagination ran riot, and hypothetical

intermediates were invented which did little to
further the cause of morphology. Instead, they
tended only to bring the subject into disrepute
among more critical botanists.

Cycadeoidea must have looked like a rather
stumpy palm tree, with a barrel-shaped trunk,
covered with persistent leaf bases and bearing a
crown of leathery pinnate leaves. Early accounts
of the pollen-bearing organs suggested that they
consisted of a whorl of pinnate 'sporophylls' fused
into a cup near the base, but with free distal portions, infolded while in bud. It was supposed that,
at maturity, they unfolded, allowing the pollen to
be shed from the many-chambered capsules
arranged along their pinnae. However, recent work
has shown that these ideas were mistaken, having
been based on specimens from which the outer
regions had been eroded away. The present view is
that the so-called sporophylls were fleshy and
formed a cylinder completely overtopping and
enclosing the ovuliferous region. There are signs of
of an abscission zone at the base, which would have
allowed the whole 'male' structure to be shed at
maturity, thereby exposing the ovuliferous region,
in which large numbers of stalked ovules and
interseminal scales were packed tightly together

8

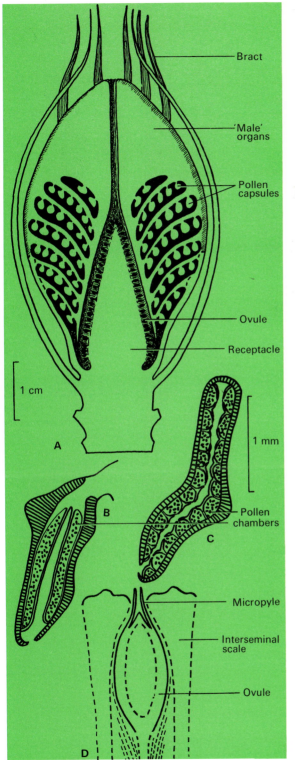

Bract

'Male' organs

Pollen capsules

Ovule

Receptacle

1 cm

1 mm

A

B

C

Pollen chambers

Micropyle

Interseminal scale

Ovule

D

5 cm

FIG. 10. *Magnolia*
Photograph of a flower of *M. soulangeana* (× ⅓), showing the perianth members and the elongated receptacle bearing stamens below and carpels above.

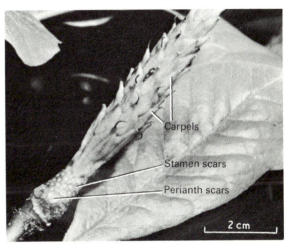

Carpels

Stamen scars

Perianth scars

2 cm

FIG. 11. *Magnolia*
Photograph of the flower-receptacle (× ½) of *M. soulangeana* after the perianth and stamens have fallen. Above the perianth scars there are spirally arranged stamen scars and spirally arranged carpels.

FIG. 9. *Cycadeoidea*

A, diagrammatic reconstruction of bisexual 'flower', as if cut in half longitudinally, showing central conical receptacle covered with ovules and interseminal scales, surrounded by fleshy 'male' organs (fused at the base but free higher up) with cavities containing pollen capsules. The whole 'flower' was enveloped by numerous bracts.

B and C, pollen capsules cut longitudinally and transversely to show the elongated pollen chambers within.

D, ovule with micropyle between interseminal scales (believed to represent sterile ovules).

(A, based on Delevoryas; B and C, after Wieland; D, after Harris.)

9

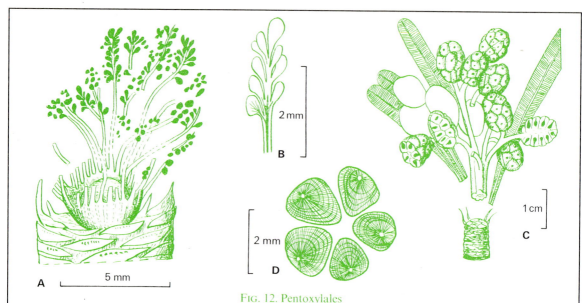

Fig. 12. Pentoxylales

A, reconstruction of the 'male flower' *Sahnia nipaniensis*, showing the branched microsporophylls fused at the base into a cup.
B, detached microsporophyll with numerous microsporangia.
C, reconstruction of branch with leaves and seed-bearing structures, said to resemble those of the monocotyledon family Pandanaceae.
D, transverse section of stele of the stem *Pentoxylon sahnii*, made up of several (often five) separate strands, each with its own cambium and secondary wood.
(A and B after Vishnu-Mittre; C and D after Sahni.)

over the central receptacle.

It is hard to understand how such a structure ever came to be compared to a *Magnolia* flower. The pollen-bearing organs were not in the least like stamens and there was nothing in the ovuliferous region corresponding to carpels. Presumably, the chief reasons were the occurrence (rare in gymnosperms) of a hermaphrodite structure and the arrangement (as in angiosperms) of the pollen-bearing organs beneath the ovules. But such similarities are far too superficial to support ideas of close relationship.

3. *The Pentoxylales.* Whereas the Bennettitales were almost world-wide in distribution, the Pentoxylales (Fig. 12) are known only from Jurassic deposits in India and New Zealand. Whether they were shrubs or small trees is not known, but they were certainly woody plants and their stems (*Pentoxylon*) had several—often five—separate conducting strands, each with its own cambium. This feature has suggested a possible link with monocotyledons which, although lacking normal secondary thickening, are characterized by having many vascular bundles in their stems.

The leaves associated with *Pentoxylon*, known as *Nipaniophyllum*, were strap-shaped and had a prominent midrib, from which lateral veins, parallel to each other, departed at an obtuse angle. Their pollen-bearing organs, *Sahnia*, were similar to those of the Bennettitales in that they were arranged in a single whorl and were fused into a cup near the base, but the distal parts bore spirally arranged stalked sporangia. The seed-bearing structures must have looked like mulberries, for they consisted of about twenty sessile seeds attached to a central receptacle, each having a fleshy envelope (sarcotesta) round a central 'pip' (sclerotesta).

Meeuse claims that there are close affinities between the Pentoxylales and the monocotyledon family Pandanaceae ('screw-pines'), particularly in their seed-bearing structures. It is, however, only possible to agree with this if, at the same time, one is prepared to abandon the carpel as a valid concept having universal application among angiosperms. According to him, the envelope round the ovule in *Pandanus*, *Urtica* (nettle), and *Piper* (pepper), which is normally called the carpel, is homologous with the sarcotesta in the Pentoxylales and with the cupule in other gymnosperms. The typical carpel of most other flowering plants

10

is regarded by him as a composite structure, made up of an ovule-bearing branch fused to a supporting bract.

4. The Glossopteridaceae. Somewhat similar interpretations of the classical carpel have recently been developed by Melville, for whom the Glossopteridaceae (Fig. 13) have a special significance. While stating that he does not regard them as direct ancestors of flowering plants, he nevertheless believes them to be the closest early relatives that have so far been discovered. They formed an important element in the flora of Gondwanaland (the large continent embracing the present-day regions of South America, South Africa, India, Australasia, and Antarctica, which lay to the south of the Tethys Sea in late Palaeozoic and early Mesozoic times). They had tongue-shaped leaves which ranged in length from a few centimetres to several decimetres and they had reticulate venation.

During the past twenty years, some nineteen species belonging to six genera of reproductive organs associated with such leaves have been described. However, details of their internal structure are not known because no petrified specimens have yet been discovered. According to Plumstead, most of the reproductive bodies had a double structure, consisting of two appressed scale-like valves. These were borne at the tip of a pedicel, either in the axil of a leaf or adnate to its midrib. One valve had on its inner surface a group of sacs containing hard tissue, suggesting that they were seeds. The other may have borne microsporangia or it may have had a protective function.

While it is widely accepted that the Glossopteridaceae were seed-plants, this is not fully proven. Nevertheless, Melville, having rejected the classical carpel theory, develops his gonophyll theory around the basic concept of a branched ovuliferous pedicel in the axil of a leaf. From such a prototype, closely resembling the reproductive organs of the Glossopteridaceae, it is possible by mental processes to derive all manner of ovuliferous structures. Likewise, he derives a wide range of staminate structures from a branched microsporangiate pedicel in the axil of a leaf.

Much of Melville's reasoning is based on a study of the path taken by vascular bundles and their branching pattern within the receptacle of the flower. A branching vascular bundle is seen by him as the 'ghost' of a branching structure which, during the course of evolution, has long since lost

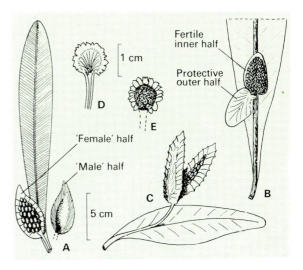

Fig. 13. Glossopteridaceae
A, leaf of *Glossopteris indica* and its reproductive organ *Hirsutum dutoitides*, with 'female' half bearing seeds and (perhaps) 'male' half bearing pollen sacs.
B, leaf of *G. stricta* and its reproductive organ *Plumsteadia* (*Cistella*) *stricta*, with its sessile inner half bearing seeds and its outer half protective.
C, leaf of *G. tortuosa*, bearing two halves of the reproductive organ *Scutum rubidgeum*.
D and E, two halves of the reproductive organ *Ottokaria buriadica*.
(All after Plumstead.)

its identity as a separate organ. To this extent, his thought processes resemble those of an earlier morphologist, Saunders, who believed that the number of separate vascular bundles leaving the central vascular cylinder is equal to the number of individual organs that they supply. A simple example is provided by the vascular supply to sepals, where it frequently happens that ten bundles supply a whorl of five. Five bundles are opposite the sepals and become their midribs, while the other five alternate with them and fork into two bundles, each providing a marginal vein for the two adjacent sepals. If Saunders's logic is pushed to the extreme, then one must conclude that such a whorl of sepals is really made up of ten units instead of five.

Similar thought processes led Saunders to interpret ovaries as made up, not of n classical carpels, but of $2n$ units of varying kinds, (solid carpels, semi-solid, and valve carpels). By this means, she was able to provide an ingenious interpretation of the so-called 'false septum' in the ovaries of the Cruciferae but, despite this, her

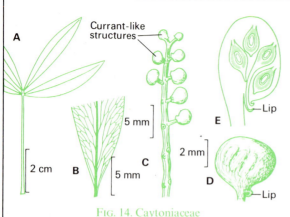

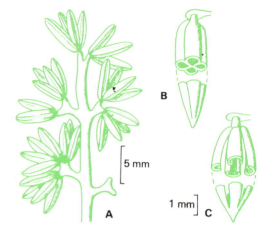

FIG. 14. Caytoniaceae
A, leaf (*Sagenopteris phillipsii*) with two opposite pairs of leaflets.
B, portion of leaflet of *S. colpodes*, showing reticulate venation.
C, 'fruit'-bearing rachis of *Caytonia nathorstii* with currant-like structures containing seeds.
D, sketch of a compressed 'fruit', showing the lip, which at one time was thought to act as a stigma.
E, diagrammatic longitudinal section of 'fruit' to show channels from the lip to the seeds.
(A, B, C, and E after Harris; D after Hamshaw Thomas.)

FIG. 15. Pollen-bearing organs of Caytoniales
A, 'male' sporophyll of *Caytonanthus arberi*.
B and C, restorations of synangia, cut across to show the four pollen sacs and the mode of dehiscence, which is quite unlike that of the anthers of flowering plants.
(A, after Hamshaw Thomas; B and C after Harris.)

ideas did not receive general acceptance. Anyone who has looked at a bathroom loofah must surely agree that one should not pay too much attention to the exact path taken by each and every vascular bundle. Indeed, there are some morphologists who have taken the extreme view that vascular patterns should be ignored altogether.

5. *The Caytoniaceae*. Hamshaw Thomas did not go quite as far as some other critics of the classical carpel concept, but he did suggest that the follicle might not necessarily be the most primitive type of carpel. He was led to this by his study of a most interesting group of fossil plants, the Caytoniaceae (Fig. 14). Their reproductive organs were first described in 1925 from Jurassic rocks of Yorkshire, where they were associated with leaves that had been known for a long time under the name *Sagenopteris*. Then, a few years later, they were found in places as far apart as Greenland and Sardinia. Scarcely anything is known of their growth habit, except for some hints that the leaves were borne on twigs rather than on massive trunks. They were compound leaves, with a slender petiole, and frequently had two pairs of leaflets, each with a midrib and a network of veins, rather like that of *Glossopteris* (and of many dicotyledons, too).

The pollen-bearing organs, *Caytonanthus* (Fig. 15), consisted of a dorsiventral rachis with opposite, or sub-opposite, branching pinnae, whose terminal branchlets each bore a single pendulous synangium made up of four fused sporangia. Comparisons were drawn with the four pollen sacs of a typical angiosperm stamen, but the mode of dehiscence was quite different.

The seed-bearing organs, *Caytonia*, consisted of a slender dorsiventral rachis, some 5 centimetres long, bearing two rows of currant-like bodies which, for convenience, may be called 'fruits', since they probably had a juicy pulp surrounding the seeds. Hamshaw Thomas was unable to find any pollen grains in the micropyles of any of his Yorkshire seeds and concluded, therefore, that the 'fruits' must have been closed at the time of pollination. Near the base of each 'fruit' there was a flange, which he supposed might have acted as a kind of stigma on which the pollen grains lodged. Accordingly, he supposed that the primitive angiosperm carpel might also have had a basal stigma. From such a carpel, it would have been only a short evolutionary step to an achene, such as the strawberry 'pip', with its gynobasic style. The typical follicle would then represent a more advanced kind of carpel (Fig. 16). However,

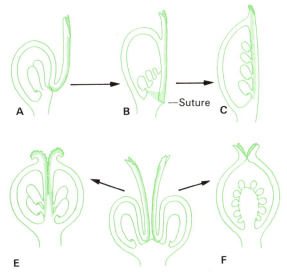

FIG. 16. Evolution of angiosperm ovaries, according to Hamshaw Thomas.

A, carpel of *Chrysobalanus* (a tropical genus in the Rosaceae) showing the gynobasic style, which Hamshaw Thomas suggested had evolved from the lip of a *Caytonia* 'fruit'. Reduction of the ovule number to one would result in achene, such as the pip of a strawberry (*Fragaria*).

B, intermediate stage, showing fusion of style, with suture at the base, leading to C, which represents a typical follicle, with an apical stigma.

D, a group of carpels with gynobasic styles.

E and F, two ways in which fusion of several carpels could occur to give a gynoecium with axile placentation (E) or one with free-central placentation (F).
(All after Hamshaw Thomas.)

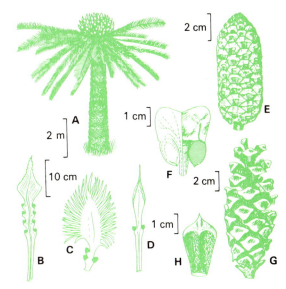

FIG. 17. Cycadales

A, 'female' plant of *Cycas media*, showing terminal group of ovuliferous sporophylls and crown of pinnate leaves, borne on a stout trunk.
(Note the alternating bands of leaf scars and sporophyll scars on the trunk.)

B, 'female' sporophyll of *C. circinalis* with many ovules.

C, 'female' sporophyll of *C. revoluta* with fewer ovules.

D, 'female' sporophyll of *C. circinalis* var. *papuana* with two ovules.

E, 'female' cone of *Zamia floridana*, showing corky tips of spirally arranged sporophylls.

F, single sporophyll of *Zamia skinneri*, showing the two reflexed ovules. (the left half dissected to show vascular supply).

G, 'male' cone of *Stangeria paradoxa*, showing spirally arranged sporophylls, each with microsporangia on the lower surface.

H, 'male' sporophyll of *Cycas rumphii*, detached from cone axis.
(B, C, D, and E, after Schuster; E, after Wieland.)

when Harris examined material from Greenland, he found abundant evidence that the pollen grains had direct access to the micropyle of the ovule, thereby establishing that *Caytonia* was still only at the level of a gymnosperm. It was certainly not an angiosperm, nor is there much support, nowadays, for the suggestion that it might even have been a forerunner of the angiosperms. Most morphologists have been unable to accept the suggestion that the achene is more primitive than the follicle; and this may well be due to the continuing influence that Goethe has had on botanical thought.

Goethe's essay, entitled 'The Metamorphosis of Plants' and written in 1790, developed the idea that all the lateral appendages of flowering plants (i.e. leaves, bracts, sepals, petals, stamens, and carpels) have the same fundamental nature. In subsequent years, this came to be interpreted as meaning that they all had the nature of leaves. It

was then only a short step to the idea that carpels have evolved from leaves. So it came about that the follicle was accepted, almost without question, as the most primitive kind of carpel. It was seen as an inrolled leaf with marginal ovules. One result of this was to direct attention to yet another group of gymnosperms, the Cycadales, as the possible forerunners of the flowering plants.

6. The Cycadales. At the present day, the Cycadales (Fig. 17) are represented by about sixty-five species, belonging to nine genera, which are restricted to Central America, South Africa, eastern Asia, and Australia. Most of them resemble palm trees, in

13

having a stout trunk with a crown of pinnate fronds, although some have subterranean tuberous stems. All are dioecious and, in one species at least, the 'sex' of the individual plant is determined by X and Y chromosomes.

The microsporangia are thick-walled and are crowded on the under side of sporophylls that are arranged spirally around a central axis to form a cone. In most genera, the ovules are also borne in compact cones, in which each sporophyll bears two reflexed ovules. In *Cycas*, however, there is an alternation of fronds and loosely arranged sporophylls throughout the life of the 'female' plant. Furthermore, the sporophylls are much more leaf-like than in the other genera and they may bear as many as eight marginal ovules. It is commonly accepted that this is the primitive kind of sporo-phyll, from which the others evolved by a reduction in the number of ovules to two and the suppression of the distal lamina.

Very recent work gives support to this idea, for in 1969 Mamay published an account of two new Permian fossils (not yet named), one from Kansas and the other from Texas. The former is interpreted as the rachis of a laminate frond bearing two opposite rows of ovules, as in some species of *Cycas*. The latter is of particular interest in relation to theories of angiosperm evolution, for the lamina of the sporophyll appears to have been partially inrolled over the ovules, which were attached, not to the margin, but to the surface of the lamina (Fig. 18). These new finds provide a hint that the most primitive kind of carpel may bear its ovules superficially, instead of marginally, a view

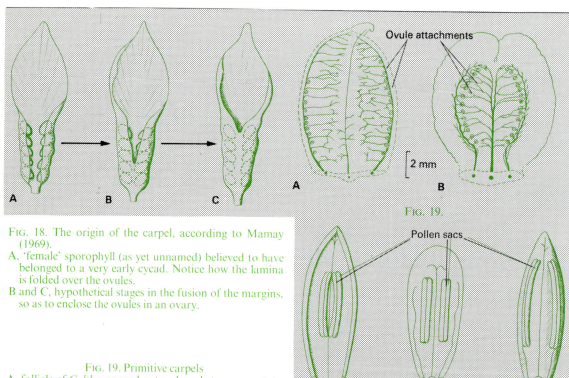

FIG. 18. The origin of the carpel, according to Mamay (1969).
A, 'female' sporophyll (as yet unnamed) believed to have belonged to a very early cycad. Notice how the lamina is folded over the ovules.
B and C, hypothetical stages in the fusion of the margins. so as to enclose the ovules in an ovary.

FIG. 19. Primitive carpels
A, follicle of *Caltha*, opened out and made transparent, to show the venation and the marginal attachment of the ovules. Note that the venation pattern is very different from that of a foliage leaf.
B, conduplicate carpel of *Drimys*, opened out to show the superficial position of the ovules. The margins of the carpel in this species are merely appressed and not fused as in follicles.
(Broken lines represent cut surfaces.) (B, after Bailey and Swamy.)

FIG. 20. Primitive stamens
A, *Austrobaileya* (upper side). B, *Degeneria* (under side). C, *Magnolia* (upper side).
Note how, in all three, the stamen is dorsiventral and the pollen sacs are superficial (i.e. some distance from the margin).
(All based on Canright.)

which has already been expressed by some other American workers, who claim that the carpels of *Drimys* represent a very primitive type.

The so-called 'conduplicate' carpel of *Drimys piperita* (Fig. 19) can be interpreted as a leaf-like structure with its lamina folded along the midrib, so as to bring the hairy stigmatic margins into loose contact (it will be remembered that in this genus the carpel is never completely closed, even at pollination). When opened out flat, it is seen that the ovules are attached superficially, some distance from the margin. It has also been suggested that the most primitive stamens have a flattened lamina and bear their pollen sacs in a superficial position, as in *Magnolia* and some related genera (Fig. 20). An attractive feature of these ideas is that they suggest basic similarities between the 'male' and the 'female' parts of the flower. It is of particular interest that *Drimys* should be among those with the most primitive carpels, for it is also one of the few genera whose wood lacks vessels.

A statistical assessment of primitiveness

My own contributions to the problem of the evolution of flowering plants confirm that *Drimys* and other relatives of *Magnolia* are very primitive. They involve a statistical analysis of the occurrence of floral and vegetative characters among dicotyledons. It has been known for a long time that many such characters are correlated (i.e. they occur together more often than would be expected on a random basis). Thus, it is well known that bilateral symmetry is much commoner in flowers with fused petals than in those with free

Character correlations in dicotyledons	1. Woody habit	2. Scalariform end-plates	3. Scalariform side-walls	4. Apotracheal parenchyma	5. Unstoreyed wood	6. Leaves alternate	7. Stipules present	8. Secretory cells	9. Leuco-anthocyanins	10. Flowers unisexual	11. Flowers actinomorphic	12. Petals free	13. Stamens pleiomerous	14. Carpels pleiomerous	15. Carpels free	16. Axile placentation	17. Pollen binucleate	18. Seeds arillate	19. Two integuments	20. Integument bundles	21. Ovules crassinucellate	22. Endosperm nuclear
1. Woody habit		+		■		+		+	+	+	+	+	+	+			+	+		+		+	+
2. Scalariform end-plates	+		+	+	+				+		+						+					
3. Scalariform side-walls		+			+				+		+					+						
4. Apotracheal parenchyma	■	+			+	+			+		+		+				+				+	−
5. Unstoreyed wood		+	+	+							+											
6. Leaves alternate	+			+									+					+				+
7. Stipules present								+	+			+	+	+		+		+	+		+	+
8. Secretory cells	+						+		+		+	+						+	+	+	+	+
9. Leuco-anthocyanins	+	+	+	+			+	+			+	+	+				+		+	+	+	+
10. Flowers unisexual	+		+		+				+		+								+		+	
11. Flowers actinomorphic	+	+	+	+	+			+	+	+			+	+				+	+		+	
12. Petals free	+		+				+	+	+		+			+		+		+	+		+	+
13. Stamens pleiomerous	+			+		+	+	+	+		+	+			+	+		+	+	+	+	+
14. Carpels pleiomerous	+						+	+			+	+	+		+	+		+	+		+	
15. Carpels free													+	+		■			+			
16. Axile placentation	+		+				+				+			+	■							
17. Pollen binucleate	+	+		+					+													
18. Seed arillate						+	+					+	+	+					+	+	+	+
19. Two integuments	+						+	+	+		+	+	+	+				+			+	+
20. Integument bundles			+					+	+				+					+			+	+
21. Ovules crassinucellate	+						+	+	+	+	+	+	+					+	+	+		+
22. Endosperm nuclear	+			−		+	+	+	+			+	+					+	+	+	+	

FIG. 21.
This table shows the extent to which twenty-two characters (floral, vegetative, and biochemical) are correlated. Each + represents a statistical correlation of significance greater than 50:1 between a pair of characters, among the families of dicotyledons. It is suggested that these characters are all indicators of primitiveness.

petals. It may well be true that these two particular characters may have a functional connection, associated with insect pollination, but it is hard to imagine how there could be any such functional connection between the woody habit and the presence of two integuments in the ovule, for example. So far, I have found twenty-two such characters (Fig. 21) that are involved in a total of ninety-eight correlations, all of which are shown by a χ^2 test to have a probability of significance greater than 50:1.

Now, by definition, primitive families possess more than the average number of primitive characters. In other words, primitive characters are not randomly distributed; and it can be shown that this fact alone would lead one to expect primitive characters to be statistically correlated. Equally, of course, the argument applies to advanced characters, so the problem is, therefore, to decide whether the characters listed in Fig. 21 are primitive rather than advanced. Here the fossil record can be of some assistance. Many lists of fossil angiosperms have been published and they show a higher proportion of these characters than a random distribution would lead one to expect. Accordingly, one concludes that they are more likely to be primitive than advanced; and the inclusion of the woody habit among them gives further support to this idea. From what is known of rates of mutation in relation to generation time, one would expect a much slower rate of evolution in trees than in herbs.

If our conclusions are correct, then we can use the twenty-two characters to make a rough assessment of the relative primitiveness of the various families of dicotyledons, using a mark scheme ranging from zero to 100. The Magnoliaceae, along with the recently discovered Austrobailey-aceae, come out with the lowest score of 20. The Rhizophoraceae (a family of mangrove plants) are also very low, at 21. Then come the Winteraceae (including *Drimys*) at 26. The Malvaceae and Rosaceae are at 27 and it is particularly interesting to find the catkin-bearing family Fagaceae as low as 29, for it suggests a very early divergence in evolution, even if not a diphyletic origin of dicotyledons.

A similar analysis of monocotyledons was carried out by a colleague, Lowe, whose calculations suggest that each of Hutchinson's three main subdivisions of the group has at least one very primitive family. Thus, the Calyciflorae include

Butomaceae (e.g. *Butomus*—flowering rush) with a score of 25, the Corolliferae include Amaryllidaceae (e.g. *Narcissus*) at 15, while the Glumiflorae include Juncaceae (rushes) at 33. The low score achieved by the Butomaceae is interesting, in view of Hutchinson's belief that the monocotyledons branched off from the dicotyledons at the level of the Ranales and that the closest affinities are between the Butomaceae and the Ranunculaceae.

Conclusions

Summing up the present position regarding the origin of the flowering plants, one can say that several different lines of evidence confirm the view that the Magnoliaceae, or their allies, are the most primitive and that, of all the gymnosperms that have been hailed as the forerunners of the flowering plants, cycads are the most promising candidates.

FURTHER READING

General

EAMES, A. J. (1961). *Morphology of the angiosperms*. McGraw-Hill, New York.

SPORNE, K. R. (1969a). *The morphology of gymnosperms*. Hutchinson, London.

TAKHTAJAN, A. (1969). *Flowering plants, origin and dispersal*. Oliver and Boyd, Edinburgh.

For reference

AXELROD, D. I. (1952). A theory of angiosperm evolution. *Evolution, Lancaster, Pa.* **6,** 29–60.

HUTCHINSON, J. (1959). *The families of flowering plants*, 2 vols. Clarendon Press, Oxford.

LOWE, J. (1961). The phylogeny of monocotyledons. *New Phytol.* **60,** 355–87.

MAMAY, S. H. (1969). Cycads: fossil evidence of Palaeozoic origin. *Science, N.Y.* **164,** 295–6.

MELVILLE, R. (1962 and 1963). A new theory of the angiosperm flower. *Kew Bull.* **16,** 1 and **17,** 1.

PLUMSTEAD, E. P. (1958). Further fructifications of the Glossopteridae and a provisional classification based on them. *Trans. Proc. geol. Soc. S. Afr.* **61,** 51–96.

SPORNE, K. R. (1969b). The ovule as an indicator of evolutionary status in angiosperms. *New Phytol.* **68,** 555–66.

THOMAS, H. HAMSHAW (1934). The nature and origin of the stigma. *New Phytol.* **33,** 173.

3

Oxford Biology Readers
Edited by J.J.Head and O.E.Lowenstein

Studying the Past by Pollen Analysis

R.G. West

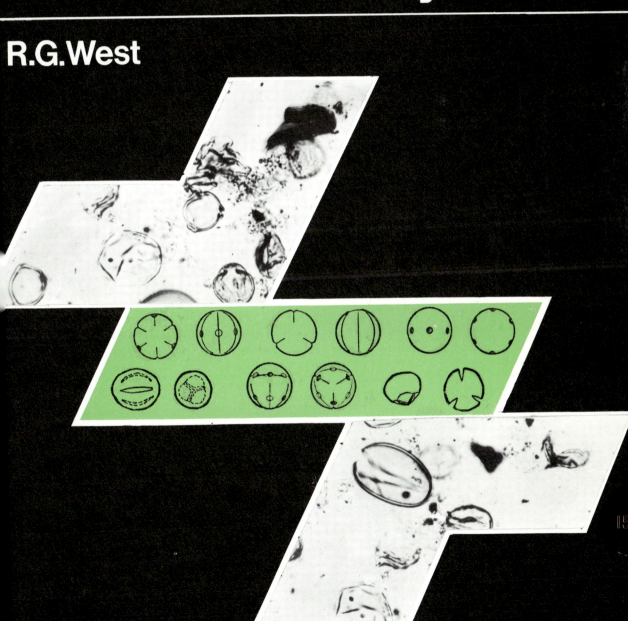

Oxford University Press, Ely House, London W.1

GLASGOW   NEW YORK   TORONTO   MELBOURNE   WELLINGTON

CAPE TOWN   SALISBURY   IBADAN   NAIROBI   DAR ES SALAAM   LUSAKA   ADDIS ABABA

BOMBAY   CALCUTTA   MADRAS   KARACHI   LAHORE   DACCA

KUALA LUMPUR   SINGAPORE   HONG KONG   TOKYO

R. G. West, F.R.S., is Director of the Subdepartment of Quaternary Research in the University of Cambridge, and author of *Pleistocene geology and biology* (Longmans 1968).

PHOTOSET AND PRINTED IN GREAT BRITAIN BY

BAS PRINTERS LIMITED, WALLOP, HAMPSHIRE

During the course of the last century the search for fossils in peat and lake sediments led to the discovery of pollen grains and spores (of sizes commonly 15–50 μm), well preserved as microfossils (Fig. 1). They showed the great morphological diversity of pollen grains and spores characteristic of living plants, which made it possible to refer them to genera and species of living plants.

Pollen analysis is concerned with the study of fossil assemblages of pollen grains and spores that have been isolated from sediments deposited in the recent past or as far back as the Palaeozoic era. It includes the study of assemblages of pollen grains and spores found in other circumstances, such as in honey, but primarily its application is in the study of vegetational history. In this Reader pollen analysis is taken to include the study of pollen of the seed-bearing higher plants and of spores of lower plants.

The fossil occurrence of pollen grains results from the extremely resistant nature of the outer part of the pollen grain wall, the exine. The substance composing it is known as sporopollenin, and its chemical nature is only now becoming understood. Fine structure and chemical studies of the developing pollen grain within the anther indicate that sporopollenin is deposited as the result of polymerization of carotenoids and carotenoid esters in such a way that very intricate patterns of exine structure and surface sculpturing are formed.

Pollen morphology

Variation in exine structure and sculpture, together with variation in the number, shape, and symmetry of the apertures in the pollen grain wall, gives the morphological diversity of pollen grains. In the lower plants the spore wall shows a similar diversity in structure and surface pattern. The symmetry of apertures and scars is related to the arrangement of the pollen grains and spores in the tetrad formed by meiosis from the mother cell

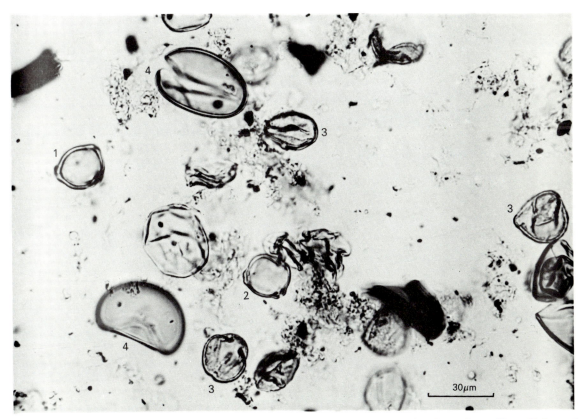

FIG. 1. Pollen preparation from a late Flandrian peat in Scotland. (1) 3-pored grain of *Corylus avellana*, (2) 3-pored (septate pores) grain of *Betula*, (3) 3-furrowed grain of *Quercus*, (4) monolete Filicales spores.

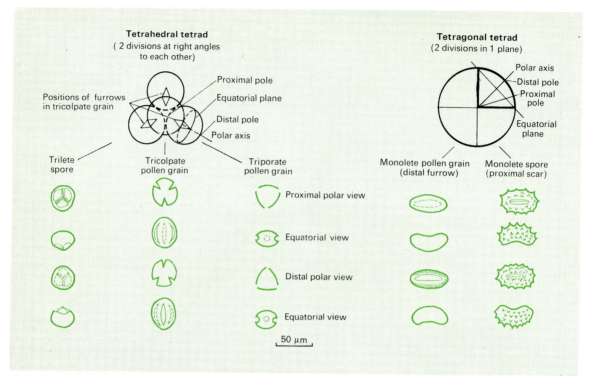

FIG. 2. Tetrad types formed during meiosis and the consequent symmetry of furrows and pores in pollen grains and dehiscence scars in spores. Dashed features are on the reverse side. (Partly after Erdtman.)

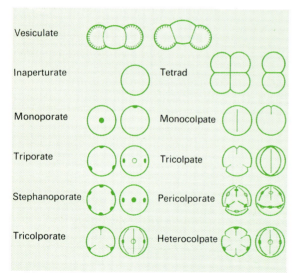

FIG. 3. Examples of the morphological classes of pollen grains, based on number and arrangement of furrows and pores. In the pairs of drawings of the monad pollen grains, the left drawing shows the polar view and the right drawing the equatorial view. Black pores and thicker furrows on the obverse side, outline pores and thinner furrows on the reverse. (After Faegri and Iversen.)

(Fig. 2). In the pollen grain the apertures may be furrows, pores (which are believed to be evolved from furrows), or a combination of both. Examples of the morphological classes of pollen grains on the basis of aperture type and symmetry are shown in Fig. 3. This type of classification is the basis for the description and identification of fossil grains.

Added to this is the whole range of characters dependent on exine structure and sculpture. This may be simple, or complex and highly differentiated. The divisions of the exine are shown in Fig. 4. The ektexine, with its stratification into foot layer (sometimes absent, as in Fig. 4), columellae, and tectum is the main vehicle for structural and sculptural variation. Examples of this variation are shown in Fig. 5. A plethora of terms have been applied by botanists in attempts to differentiate the exine structure, both in descriptive and developmental terms. This has led to unnecessary confusion; in part it is a result of lack of knowledge of the details of the development of the exine, but this is under active investigation with the aid of the electron microscope.

Isolation and identification of pollen grains

Since sporopollenin is chemically resistant, chemical methods can be used to separate pollen grains and spores from other components of sediment. The preparation processes are complicated and involve removal of lignin by oxidation, of cellulose by hydrolysis, and of mineral matter by hydrofluoric acid (or by flotation). Starting with a sediment sample of perhaps 0·5 cm³, a preparation sequence will end with the isolation of the pollen assemblage under a cover slip on a microscope slide (Fig. 1). Examination under a high-power microscope (× 400 to × 1000), using both transmitted and phase-contrast microscopy, reveals enough detail of the morphology of the grains for identification to be attempted. Identification is by comparison with prepared slides of pollen grains of living taxa whose specific identity is known with certainty. A reference collection is indispensable for identification, though keys and pollen atlases are a substantial preliminary aid. In some genera (e.g. *Plantago*) it is possible to identify species since each species has its own particular morphological characteristics. In other cases only generic (e.g. *Ulmus*) or family (e.g. Chenopodiaceae) identification is possible. With pre-Quaternary pollen assemblages, where relationships of fossils to living plants are uncertain, names founded on morpho-

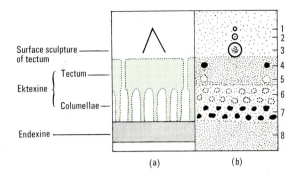

(a)   (b)

FIG. 4. Structure of the exine of a tectate pollen grain. (a) Section through an exine, a few μm thick. The two principal divisions of the exine are shown: the lightly staining endexine, and the darker staining ektexine, divisible in this example into columellae and tectum, which is perforate. Exines of other species may be much simpler (see Fig. 5). (b) The optical expression of the exine structure in (a), as the microscope is focused down from the topmost (outer) level at 1 to the endexine level at 8. Each component of the structure is seen in optical section. Perforations appear dark at higher levels, then light. Solid structures first appear light, then dark. The structural complexity of the exine can thus be analysed by the light microscope. (After Erdtman.)

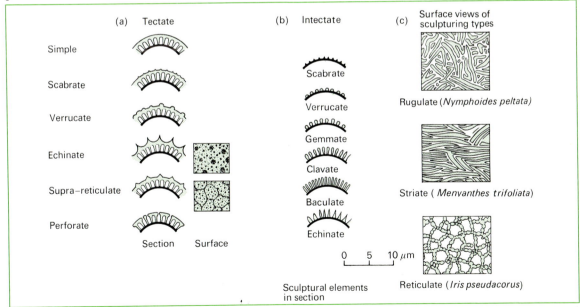

FIG. 5. Structure and sculpture of exines. Ektexine tinted, endexine black. (a) Structure of various tectate exines. (b) Sections through intectate exines. (c) Surface views of some sculpturing types. (After Faegri, Iversen, and Troels-Smith.)

5

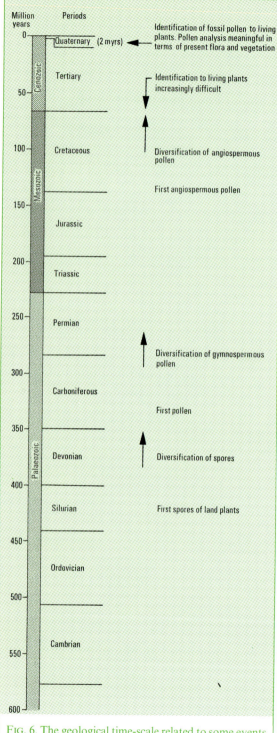

logy are used for fossil taxa.

Preservation of pollen grains

Since the pollen grain exine is resistant it may have a long geological life once it is incorporated into a sediment, but only if the grains avoid mechanical attrition and chemical changes such as oxidation. Environments which favour preservation are those where sediment is being deposited near or below the water table, as on the growing surface of a wet peat mire, in lakes, or in the sea. Here oxidation can be minimal. With the passage of geological time, the grains will remain preserved as long as chemical or mechanical alteration of the rock is not too great. Spores are known dating back to the Silurian period, more than 400 million years ago (Fig. 6). Studies of the morphological changes of spores during subsequent ages make possible some understanding of their evolutionary development, and of that of the plants associated with them. Because of the wide dispersal of spores in sediments, this approach to evolutionary analysis promises to be rewarding.

The pollen spectrum

The principal freshwater environments where pollen grains are now being deposited and preserved are on wet peat mire surfaces and in lakes. If we consider a rather shallow and small lake, first formed by the melting out of a block of ice in a moraine of the last glaciation some 12 000 years ago, the sediment within the lake, perhaps some 12 m thick, will have been deposited over the life of the lake. By making a borehole in this series of sediments (Fig. 7), we can obtain a vertical sequence of samples of sediment, each of which will have been deposited at a particular time. Each sample can then be analysed for pollen and spore content, with each grain or spore being identified as the prepared slide is traversed on a mechanical stage under the high-power microscope. The number of grains counted varies between 150 and 1000, depending on the number available and on sampling error considerations. A *pollen diagram* can then be constructed showing how the *percentage* representation of each taxon identified varies with the passage of time. This is a graphical expression of pollen analyses at successive levels. An example is shown in Fig. 8.

Analyses of this kind were first produced by the Swedish geologist von Post in about 1916. It was quickly realized that here was a huge potential for

Fig. 7. Hiller peat borer, raised after sampling Flandrian peat at Meare Poole, Somerset. The sampler chamber is open and the contents under field examination. (Godwin.)

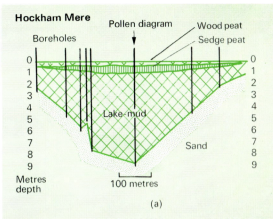

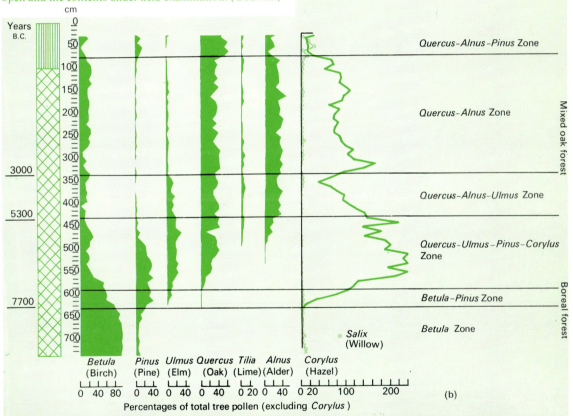

Fig. 8. A lake-filling at Hockham Mere, Norfolk, and pollen diagram from its centre (after Godwin). (a) Section across the lake-filling, proved by boreholes. The change in sediment near the top indicates that open water conditions with mud deposition were followed by the growth of sedge fen, forming sedge peat, then fen woods, forming wood peat. This is a normal hydrosere succession. (b) Pollen diagram from the centre of the lake-filling. Results of pollen analyses for six trees and two shrubs at successive levels are expressed as percentages of total tree pollen. Pollen assemblage zones and equivalent modern forest types are shown on the right. Age of certain pollen zone boundaries shown on the left. The diagram indicates the sequence of forest assemblages in the area during the last 10 000 years.

7

environmental analysis, and the study spread rapidly. The discovered sequence of changing pollen frequencies could be divided into pollen zones, each characterized by a particular assemblage. The sequence of assemblages was then compared with forest formations as seen in northwest Europe today. The sequence of forest changes (see Fig. 8) was considered a reflection of changing climate since the last glaciation, showing first the development, after a tundra phase, of forest with

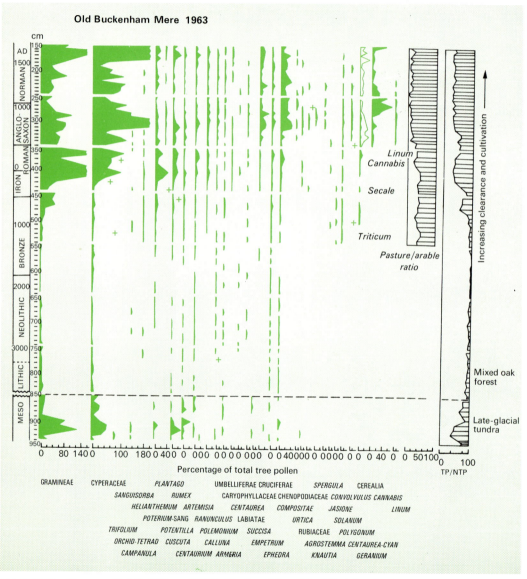

FIG. 9. Pollen diagram of a lake filling at Old Buckenham Mere, Norfolk (after Godwin), showing the relative frequencies of pollen of herbaceous plants throughout the sequence. The basal sediments are of late-glacial age, formed before the mixed oak forest of the climatic optimum of the Flandrian (postglacial) stage had spread to East Anglia. In this time high frequencies of herbaceous pollen are reduced, as seen in the tree pollen (TP) to non-tree pollen (NTP) ratio on the right of the diagram. High frequencies of herb pollen are also seen towards the top of the diagram. These relate to the spread of agricultural activities following the arrival of man. Pollen grains of many cultivated species are recorded. Certain taxa (to the left) are associated with pasture, others (on the right) with arable cultivation, so giving a ratio of pasture/arable pollen.

Betula and *Pinus*, then the establishment of mixed oak forest, associated with a climatic optimum (about 2°C warmer in July than today), and finally a period of forest interference and reduction associated both with man's agricultural activities (starting with the Neolithic), and climatic deterioration since the climatic optimum (Fig. 9).

The steps in the process of studying a sample of pollen are therefore (1) pollen identification, (2) the production of a pollen spectrum, from which is deduced (3) the vegetational and floristic significance and hence (4) the physical environment from which the sample was drawn and (5) the changes in the total environment over a period of time.

Pollen analysis is an extremely powerful tool for the investigation of vegetational, floristic, and climatic changes which took place in the past. A pollen diagram shows fine changes in pollen input in sediments. These can be interpreted in terms of floristic and climatic change, not only in the period of some 12 000 years since the last glaciation, but, given the right sediments, back through the series of ice ages and associated temperate interglacial periods (which occupied the last 2 million years) into the Tertiary, Mesozoic, and Palaeozoic (Fig. 6). Once we get back into the Tertiary and older periods, identification of pollen grains with living taxa becomes increasingly difficult; consequently the pollen taxa found are only referable to successively larger taxonomic groups. We can summarize the insight into past environmental change revealed by pollen analysis under the headings biological, geological and environmental, though these overlap.

Knowledge of biological changes. Pollen analysis can elucidate the history of vegetation, particularly the sequence of gross vegetational changes since the last glaciation (Figs. 8 and 9), during the last and previous glaciations, during the temperate interglacials, and back into the Tertiary. It can also show the evolution of plant communities and the rearrangement of genera which have accompanied climatic change and which have led to the communities we see today; and the history of particular species known by their pollen, with changes in distribution controlled by climatic and other changes.

The history of agricultural practices can be deduced from replacement of forest pollen by pollen of weeds and cultivated plants. This is known in considerable detail from the Neolithic (about 3000 B.C.) onwards from pollen analysis of

closely spaced samples in sediments formed near sites of archaeological occupation (see Fig. 9).

Knowledge of geological changes. The sequence of pollen zones has been used as a scale to characterize stratigraphical sequences. Because pollen is widely dispersed this characterization is regional, and provides a good basis for correlation. Before the advent of radiocarbon dating in 1949 pollen analysis was the most important basis for the correlation of biogenic deposits of the last 10 000 years. This method is still important for deposits beyond the range of radiocarbon dating (about 50 000 years), e.g. in interglacial and glacial periods and in the Tertiary and earlier. Other geological results can be correlated with those given by pollen analysis. For example, pollen analysis through a series of sediments reveals changes of salinity associated with changes in sea level; these can be correlated with other geological evidence of changes in sea level.

Knowledge of environmental changes. The pollen diagram has become an important source of much information about environmental history (changes in vegetation, climate, and sea levels). This provides the background for faunal change and for the evolution and development of man's civilization.

Modern developments in pollen analysis
In the last decade, and very actively at present, the methods of pollen analysis have come under much closer scrutiny. A great deal has been gained from perhaps rather crude methods and the appetite is whetted for even greater gains to be made by their refinement. The development of other techniques, notably radiocarbon dating, has also made certain refinements possible.

The major question is: how can pollen input (that is, the quantity of pollen settling into deposits) be more closely related to the distribution pattern and frequency of the species which give rise to it? There are three approaches to the finer analysis of this relationship:

1. Investigating the relationship between present-day pollen sedimentation and present-day vegetation in an area.
2. Making measurement of pollen input and pollen concentration in sediments more quantitative by replacing percentage frequency calculations with counts of numbers of grains/cm^2 sediment surface/year of deposition.

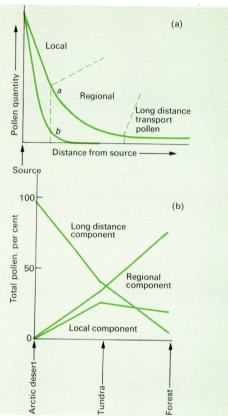

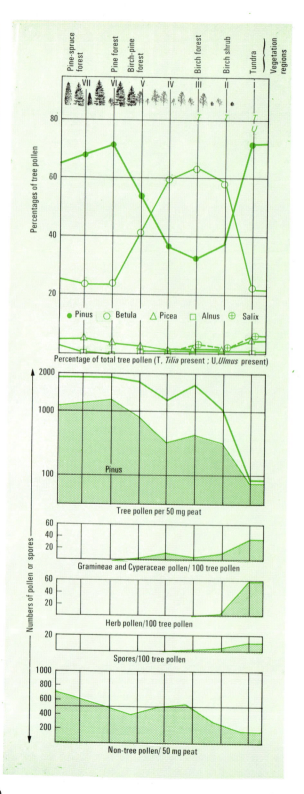

FIG. 10. Components of the atmospheric pollen rain and deposition. (a) Curves showing quantities of pollen dispersed at increasing distance from pollen source. Local, regional, and long distance components are indicated; curve *a*, a widely dispersed grain such as that of *Pinus sylvestris*; curve *b*, locally dispersed grains, such as the tetrads of *Empetrum*. (b) Components of atmospheric pollen deposition in relation to increasing regional pollen productivity. Arctic desert (or oceanic area), no vegetation, long-distance component predominates. Tundra, a degree of local and regional pollen production, but not high; still a large long-distance component. Forest, regional pollen production high.

FIG. 11. Results of pollen analysis of surface peat samples from different vegetational regions in north-east Finland, along a S (left) to N (right) transect (Aario). The diagram shows the percentage (of total tree pollen) frequencies of the important tree genera, and the tree and non-tree pollen content measured in 50 mg of peat. The low pollen deposition in the tundra is composed of a long-distance transport element of genera highly productive of pollen (e.g. *Pinus*) and a smaller locally produced herbaceous element. The diagram reveals the difficulties of interpreting percentage tree pollen spectra. The tree pollen spectra from the pine forest are similar to those from the tundra, but the herbaceous spectra from these two regions are very different. Any interpretation should be based on total pollen spectra, with the probabilities of long distance transport carefully considered.

10

3. Investigating such factors determining the relation between vegetation and pollen input as (i) pollen production in different species; (ii) pollen dispersal mechanisms of different species; (iii) pollen dispersal processes, including air transport, terrestrial transport, water transport, recirculation in lakes, and downwash in peats; and (iv) the preservation of pollen and its differential destruction in the sediment.

Let us now consider what can be learnt from these approaches and how they can refine the kinds of interpretation already described.

Present pollen sedimentation. Present pollen rain can be determined by putting out pollen traps in the form of jars or slides, or by the collection of moss cushions or surface sediments of lakes, and analysing the material caught or collected. The spectrum obtained can then be compared with the frequency and pattern of distribution of species in the surrounding vegetation. Studies such as these have been made in various vegetation formations, for example tundra, forest, and steppe. Ideally, it would be best to determine both annual pollen deposition and the abundance and distribution of each species in the pollen catchment area, but this is a complex and difficult task.

The results obtained so far can be summarized as follows. In each area the atmospheric pollen rain can be divided into three components (Fig. 10); a local component, a regional component, derived from the vegetation within, say, 50–100 km,

and a component of long-distance transport, perhaps from several hundred km off. The relative size of these components depends on the pollen production and dispersal characteristics of the species involved. In tundra, with plants which have a measured low (annual) rate of pollen production, the long-distance component may much exceed local and regional production. This is because boreal forest genera such as *Pinus* and *Betula* produce much pollen, which is well dispersed (Figs. 10 and 11).

	Point Barrow (mean of 2 samples)	Wretton
	(Percentages of total pollen)	
*Picea**	+	−
*Pinus**	−	6
*Alnus**	7	−
Betula	6	3
Salix	3	−
Gramineae	45	35
Cyperaceae	30	34
Compositae	1	4
Cruciferae	1	4
Epilobium	+	−
Ericaceae	1	1
Rumex	+	−
Others (herbs)	6	13

* long-distance origin

FIG. 13. Pollen analyses of surface samples in coastal tundra at Point Barrow, Alaska (Livingstone) and from last-glacial organic sediments at Wretton, Norfolk.

					R values				
Pohl's relative pollen production (*Fagus* : 1)		Iversen's pollen production groups		At 4 sites in Minnesota forest (Janssen)				In forest of Eastern Canada and northeast U.S.A. (Livingstone). Mean of 22 sites
Alnus	17·7	A. High	*Pinus*	*Pinus*	—	—	1·26	9·3	6·8
Pinus sylvestris	15·8		*Betula*	*Betula*	7·6	3·8	2·1	1·9	4·8
Tilia cordata	13·7		*Alnus*	*Quercus*	1·5	2·0	2·1	5·7	146
Corylus	13·7		*Corylus*	*Ulmus*	0·51	1·8	1·7	4·0	26
Betula	13·6			*Tilia*	0·75	0·35	0·73	0·31	—
Carpinus	7·7	B. Moderate	*Picea*	*Picea*	—	—	0·3	0·28	0·74
Quercus	1·6		*Quercus*	*Larix*	0·015	0·04	0·01	0·03	—
Fagus	1·0		*Fagus* *Tilia*	*Fagus*	—	—	—	—	0·93
		C. Low	*Ilex* *Viscum* *Lonicera*						

FIG. 12. Relative pollen production and R values.

Analyses from traps or surface samples reveal the transport capabilities of grains of particular species. It is known, for example, that some genera in shrub-tundra, e.g. *Empetrum*, contribute only locally to the pollen rain, while other pollen taxa of the region, such as Gramineae, Cyperaceae, and *Rumex acetosella*, are dispersed widely, producing a regional component of the pollen rain. Dispersal ability is related to factors to be discussed later.

Margaret Davis has introduced the term R value for forested areas.

$$R = \frac{\text{Pollen percentage of the species}}{\text{Vegetational percentage of the species}}$$

Pollen diagrams can be corrected by multiplying pollen frequencies by the R value (Fig. 12). This gives a correction to the pollen curves which is based on the present-day relation between pollen rain and frequency of the species in question in the surrounding vegetation. This assumes that the observed R values can be extrapolated back in time even though the frequency and distribution of the genera concerned may have changed very substantially. R values for the same genus differ in different forest regions (see Fig. 12), but they are nevertheless useful when interpreting the import-ance of particular genera in the vegetation at particular times. Similar corrections can be applied using factors based on relative pollen production.

Surface sample pollen analyses help in the interpretation of fossil pollen assemblages. By drawing parallels, past vegetation may be reconstructed. For example, fossil pollen spectra in southern England beyond the area covered by the ice of the last glaciation are like spectra from coastal tundra from Point Barrow, Alaska (Fig. 13), but not like spectra from other types of tundra in northern Europe.

Detailed studies such as these show that past assemblages are not necessarily reproducible in modern pollen rains. This is hardly surprising and indicates that the present assemblages have evolved from those in the past, to the accompaniment of migration and climatic change and the adjustment of species to each other and to a changing environment.

Measurement of pollen content of sediments. Careful preparation methods allow the total number of pollen grains from each identified taxon to be counted in a given volume of sediment (Fig. 14). A known quantity of pollen grains exotic to the sediment being analysed is added to a known volume of

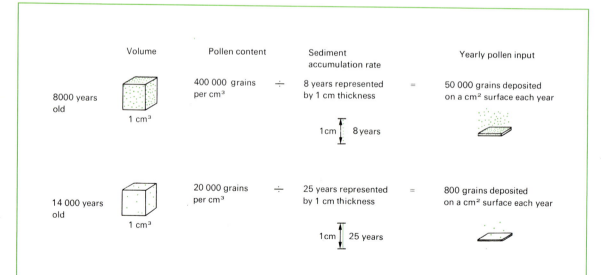

FIG. 14. Comparison of pollen input rates at two levels in a lake sediment core (after Davis). The lower relates to a time of low pollen input during the tundra period after the retreat of the ice of the last glaciation in southern New England; the upper to a forest period of high pollen input. Measurements by Ritchie of pollen deposition in the sedgemoss tundra of the Canadian arctic are lower than the fossil tundra spectra by a factor of ten. Possibly the higher pollen productivity of the ancient tundra is related to its lower latitude.

the sample. The observed ratio of exotic to fossil pollen assists in the computation of the original fossil pollen content.

There are many inherent statistical errors, but their magnitude can be estimated, and it is possible to express within certain limits of error the pollen frequencies of each taxon per volume of sediment. If we then carry out radiocarbon dating of a series of vertical samples of the sediment from the same core, it is possible to work out the average rate of sedimentation over the time involved, and so express the pollen analyses as grains/cm² sediment surface/year. The passage of time can also be estimated by counting the annual deposition layers found in some sediments. The pollen content spectrum can be used to construct a diagram (Fig. 15) which has an advantage over the *relative* (%) pollen diagram (Fig. 8) in that changes of a taxon are a result of change of input. In the relative diagram changes are affected by changes in fre-

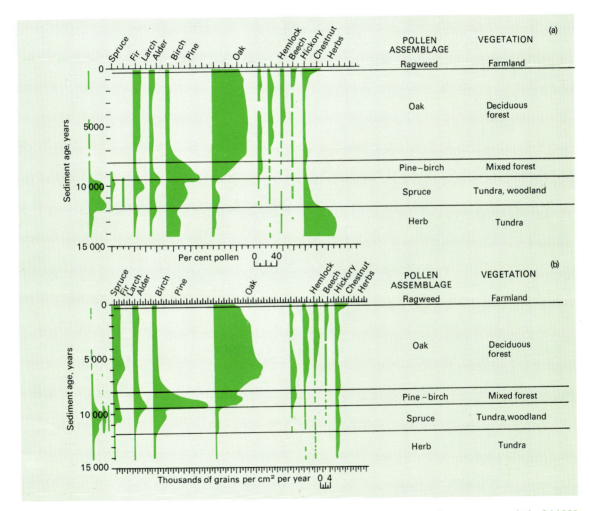

FIG. 15. Pollen diagrams from a lake infilling from southern New England; sedimentation covers a period of 14 000 years since the retreat of the ice of the last glaciation in that area (after Davis). (a) Pollen content expressed as percentage total pollen counted. (b) Pollen content expressed as thousands of grains deposited per cm² sediment surface per year. The difference between the diagrams is greatest in the first 6000 years covering the period of rapid changes in pollen productivity. Note, for example, the low pollen productivity of the tundra period, with a low herb deposition, compared with the high percentage representation of herbs in (a). The pine curves are also very different; in (b) the measured deposition is low, compared with higher percentages in (a). Diagram (b) is thus far more informative than (a) regarding vegetational history in the first 6000 years.

quencies of other taxa. The pollen content diagram is particularly informative when there are considerable changes in pollen content such as occur on changes from tundra to forest. Here, *percentage* changes may mask the very significant changes in content involved. The difference between the content and relative diagrams at such a transitional period is shown in Fig. 15. Another example of a significant change in pollen content is found at the time of deforestation associated with the Neolithic.

Pollen content analysis is a difficult procedure. Circumstances sometimes demand the extra information it provides, but relative (%) pollen analysis still has its place.

So far pollen content studies have mostly been made on lake sediments, partly because the methods were developed principally by Margaret Davis and others for use in the temperate parts of the United States. Here lake sediments have been studied rather than mires such as bogs and fens. Lake environments may subject pollen grains to differential sorting in currents; and recirculation and redeposition of the surface sediments may take place, as discussed later. This may interfere with the simple relation between pollen content of the sediment and the pollen rain produced by the

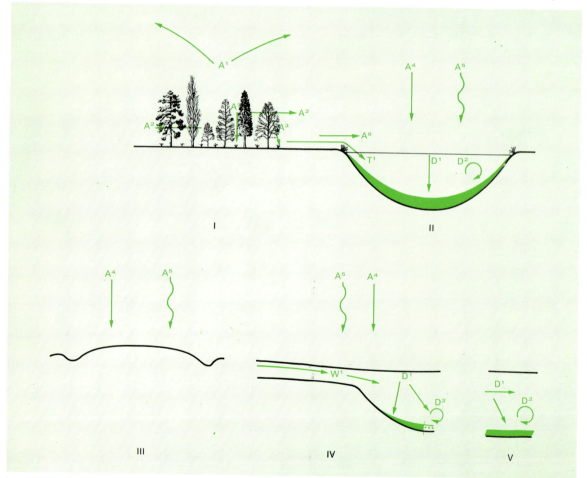

FIG. 16. Paths of pollen dispersal and deposition in different environments. I, Pollen dispersal from vegetation. II, Pollen deposition in a lake. III, Pollen deposition on a growing raised bog surface. IV, Pollen deposition in an estuary. V, Pollen deposition in the sea distant from land; low atmospheric contribution. A^1, atmospheric dispersal. A^2, dispersal in the trunk space. A^3, dispersal to the ground. A^4, atmospheric fall-out. A^5, atmospheric rain-out. A^6, filtering effect in lateral dispersal. T^1, terrestrial dispersal of component A^3 (and components A^4 and A^5). W^1, transport by water in rivers. D^1, deposition in water, accompanied by sorting by current action and differential flotation. D^2, recirculation of pollen, already once deposited, by reworking of sediments.

14

vegetation. In fact a raised bog surface would be a far more suitable environment for pollen content analysis; here there is a direct contact between the receptive sediment and the pollen rain. However, it is difficult to extract pollen from the more fibrous *Sphagnum* peats, and the sequence of sediments is usually far more complicated than in the centre of a deep lake basin. Both these difficulties militate against the application of pollen content analysis to peats; but if they were overcome, the results would be of great interest both intrinsically and in relation to results already obtained from lake environments.

In assessing results from pollen content analysis, we must remember the errors present both in the preparation technique and in the determination of rates of sedimentation. If the rate changes by a factor of two for a time between two radiocarbon dates, the increase in thickness will be averaged over the period between the dates. Even when pollen content has been determined, we still have the problem of relating it to the surrounding vegetation from which it came.

Factors controlling the production, dispersal, sedimentation, and preservation of pollen grains. The production of pollen by different species is very variable, and is related to the mode of pollination. Insect-pollinated plants are usually low pollen producers, wind-pollinated plants high pollen producers. Relative pollen production by different tree genera is known (Fig. 12), and these figures should be borne in mind when interpreting pollen diagrams. The *total* pollen production by a particular species in a particular area depends on the number of individuals, the frequency of flowering, the position of the plants in relation to dispersal possibilities, and the size and other dispersal characteristics of the pollen grain itself.

Pollen productivity of a particular species in relation to the vegetation as a whole is extremely complex and difficult to analyse, particularly because the special local factors at each site make generalizations from individual results difficult. Similarly, each pollen catchment area is unique, and analysis of dispersal and sedimentation is not easy. Results are strictly relevant to the study area only, but some generalizations from them may be possible. These studies are important and have hardly been begun.

The pollen in a catchment area may have arrived by air or water. Various possibilities are shown in

Fig. 16. The relative contributions of air and water vary from site to site, and at one site from season to season. Closed lakes contain pollen from the air and from terrestrial run-off. Lakes which receive and discharge water contain also a water-transported component whose importance will depend on hydrological factors and topography. A raised bog surface receives most pollen by air transport, though it may be redistributed by water pool networks which the bogs contain. Marine and estuarine sediments receive mostly water-dispersed pollen, especially if the shore is distant.

Each particular site has its own properties in regard to pollen deposition. Since air, land, and water transport disperse different taxa differently, it is important to understand these properties. For example, the heavy tetrads of pollen grains of the Ericales are not dispersed far in the air, but they are resistant to corrosion and may accumulate in soil, where run-off and water transport may concentrate them in a lake basin. Again, marine deposits often show high frequencies of winged conifer grains, which float and are easily transported by water.

Once pollen has been deposited in water or on the mire surface there are other processes which may affect the final pollen spectrum in the sediment. In lakes, differential dispersion depending on grain density/surface area properties may occur, and there may be recirculation of pollen by water movements within a lake basin and by animal disturbance of surface sediments. The extent of these changes is not known. In raised bogs, there may be downwashing through peat for a short distance, but this effect is believed to be small or nil. Downwash and differential destruction of pollen makes interpretation of the pollen analyses of soil profiles difficult.

Measurements on air dispersal of pollen have been made by H. Tauber of the Danish National Museum. He set out traps in particular parts of a forest and on a neighbouring lake. He assessed the amount of pollen dispersed at the level of the tree trunks (trunk space), and compared it with the total pollen catch on the lake surface (Fig. 16). During a particular experiment the trunk space component was 42% of the catch on the lake surface. Tauber has also demonstrated other effects significant in the process of dispersal, such as the filtering effect of lakeside vegetation on laterally dispersing pollen and a reflotation effect of grains impacted on twigs during the summer and released

15

in the autumn by the disturbance of the twigs.

These considerations of pollen transport and deposition have a bearing on the choice of a site for pollen analytical work. For regional vegetational studies, lakes that are not too large and are situated in closed basins are usually considered most suitable. Here the local vegetation may not contribute so much pollen as on a raised bog site, which enables a more satisfactory separation of the local and regional components to be made. The absence of a through flow of water reduces the amount of water-transported pollen. On the other hand, a raised bog site has the advantage that it is possible to identify the local pollen component more closely from the preserved macroscopic plant remains it contains.

Outlook for future research

In these attempts to refine the process of pollen analysis and the interpretation of its results, the detailed studies mentioned under the three categories above have all introduced new possibilities for the investigation of vegetational history and the history of particular plant associations. They indicate clearly where the difficulties lie, and how some of them might be solved. Generalizations will come in due course; but since every site is unique with regard to its vegetation and the features about it which affect pollen production and sedimentation, extrapolations into the past are always open to question. For example, the present-day situation regarding sedimentation, vegetation, and environment at a particular site is very different from that obtaining some 10 000 years ago, when tundra may have been present at latitudes now occupied by forest.

Improvement of interpretation lies not only with the pollen analyst, but also on improved knowledge of the autecology of identified species and the sociology of the communities concerned. It also lies with related activities which give a more complete view of vegetational and environmental history. These include, for example, studies of macroscopic plant remains from pollen-analysed sediments (the frequencies of which can be treated quantitatively as with pollen); chemical and physical studies of fresh water sediments; and analyses of faunal changes. Pollen analysis is one tool, perhaps the most powerful one, in the reconstruction of past ecological conditions, but it should not be considered in isolation.

Recently attempts have been made to apply data-recording and computer techniques to pollen analysis. Pollen diagrams from different sites have frequently to be compared, and this is difficult when they are constructed and zoned in different ways. Standardization of procedure should make it possible to record the analytical information on tape so that comparisons can be made by suitable programmes with automatic graph-plotters.

In order to reconstruct the changes in vegetation in a given region (which are related to climatic and other environmental changes), comparisons between one region and another are essential. The first of these comparisons are now being made in the Flandrian (the postglacial) period in Britain and elsewhere. The patterns which emerge will be of fundamental importance for the understanding of the present British flora and its regional content. They should reveal the effects of changing climate on the vegetation and flora of different regions, an important prerequisite for understanding the relation between environment and plants. Eventually it should be possible to carry the same type of analysis back into the past glacial and interglacial periods of the Pleistocene, and perhaps even into the Tertiary, to the time when our present genera are first found in the fossil record.

FURTHER READING

General

GODWIN, H. (1956). *History of the British Flora.* Cambridge University Press.

WALKER, D. and WEST, R. G. (eds.) (1970). *Studies in the vegetational history of the British Isles.* Cambridge University Press.

WEST, R. G. (1968). *Pleistocene geology and biology.* Longmans, London.

For reference

CHALONER, W. G. (1967). Spores and land plant evolution. *Rev. Palaeobotan. Palynol.* **1**, 83–93.

——— and MUIR, M. (1968). Spores and floras. In *Coal and coal-bearing strata* (eds. D. Murchison and T. S. Westoll). Oliver & Boyd, Edinburgh.

DAVIS, M. B. (1969). Palynology and environmental history during the Quaternary Period. *Am. Scient.* **57**, 317–32.

FAEGRI, K. and IVERSEN, J. (1964). *Textbook of pollen analysis.* Munksgaard, Copenhagen.

LIVINGSTONE, D. A. (1968). Some interstadial and postglacial diagrams from eastern Canada. *Ecol. Monogr.* **38**, 87–125.

10